IAN MORTIMER

Dr Ian Mortimer is the *Sunday Times* bestselling author of *The Time Traveller's Guide to Medieval England* and *The Time Traveller's Guide to Elizabethan England*, as well as four critically acclaimed medieval biographies, and numerous scholarly articles on subjects ranging in date from the twelfth to the twentieth centuries.

He was elected a Fellow of the Royal Historical Society in 1998. His work on the social history of medicine won the Alexander Prize (2004) and was published by the Royal Historical Society in 2009. He lives with his wife and three children in Moretonhampstead, on the edge of Dartmoor.

D0308134

ALSO BY IAN MORTIMER
AND AVAILABLE FROM VINTAGE

The Greatest Traitor: The Life of Roger Mortimer,
1st Earl of March, Ruler of England, 1327–1330

The Perfect King:
The Life of Edward III, Father of the English
Nation

The Fears of Henry IV: The Life of England's
Self-made King

1415: Henry V's Year of Glory

In the Time Traveller's Guide Series

The Time Traveller's Guide to Medieval England:
A Handbook for Visitors to the Fourteenth
Century

The Time Traveller's Guide to Elizabethan
England

IAN MORTIMER

Human Race

10 Centuries of Change on Earth

VINTAGE BOOKS

London

1 3 5 7 9 10 8 6 4 2

Vintage
20 Vauxhall Bridge Road,
London SW1V 2SA

Vintage is part of the Penguin Random House group of companies
whose addresses can be found at global.penguinrandomhouse.com.

Penguin
Random House
UK

Copyright © Ian Mortimer 2014

Ian Mortimer has asserted his right to be identified as the author
of this Work in accordance with the Copyright, Designs and Patents
Act 1988

First published in Vintage in 2015
First published in hardback by The Bodley Head in 2014 with the
title *Centuries of Change: Which Century Saw the Most Change and
Why it Matters to Us*

www.vintage-books.co.uk

A CIP catalogue record for this book is available
from the British Library

ISBN 9780099593386

Typeset by Palimpsest Book Production Ltd, Falkirk, Stirlingshire
Printed and bound by CPI Group (UK) Ltd, Croyon CR0 4YY

MIX
Paper from
responsible sources
FSC® C018179

Penguin Random House is committed to a sustainable future
for our business, our readers and our planet. This book is made
from Forest Stewardship Council® certified paper.

To my children
and all my descendants.
This is the book I feel I was born to write.
That doesn't mean it is the book you were born to read –
but it might help.

Contents

Acknowledgements

First, I would like to thank my editor, Jörg Hensgen, for his sterling work on this book. And for his patience. And for his advice on theology. I'd also like to thank my agent, Jim Gill, and my commissioning editor, Stuart Williams, for their encouragement and support.

A special thank you is due to John and Anne Casson. They commissioned my talk to celebrate the 1100th anniversary of the diocese of Exeter in 2009, which allowed me to formulate my thinking on the conclusion of this book. They also sold us the old house that appears in the conclusion of the thirteenth-century chapter and occasionally thereafter. Not only has it been a useful point of reference; it is thought-provoking to live in a building for which some physical or documentary evidence remains for eight of the ten centuries described in these pages.

Several people gave advice in the course of putting this study together and bringing it to completion. I would particularly like to thank Professor Jonathan Barry for his comments on the book as a whole, Dr Paul Dryburgh for his suggestions for the medieval chapters, Dr Andrew Hinde for his advice on my approach to population levels, and Jonathan Camp for his feedback on my views of various philosophers. I am also grateful to Martin Amis for suggesting that I address Stephen Pinker's arguments on the decline of violence in the sixteenth century.

I would like to thank my mother, Judy Mortimer, for her rundown of how the introduction of electrical appliances affected our family's standard of living in the twentieth century. I am also very grateful to my cousins, Charles and Sarah Read, for letting me stay with my family in their holiday cottage in Suffolk, during which time there were several power cuts, each of which increasingly informed the section dealing with our reliance on electricity.

I would also like to thank all the people – too many to name – who, over the years, have patiently discussed the subject of change with me and expressed their views. This book would not have come into existence if they had not been so engaged by the idea. I'd particularly like to thank a near neighbour in Moretonhampstead, Maya Holmes, who first mentioned the significance of the photograph *Earthrise* to me. I also owe a thank you to James Kidner for telling me about the Gallup poll on the importance of religion around the world; to Canon Bill Girard for introducing me to some of Pope Benedict's writings; to Dr Marc Morris for his words of wisdom on the Norman invasion and the process of civilising the English; and to Nick Hasell for the story of the retired farmers choosing wellington boots as the most important change in their lifetimes. I would also like to thank the BBC newsreader who presented the résumé of the twentieth century that evening in late December 1999, and uttered the words that inspired this book.

Finally, and most of all, thank you, Sophie.

Introduction

> Printing, gunpowder and the compass – these three
> have changed the whole face and state of things
> throughout the world.
>
> Francis Bacon, *Novum Organum* (1620)

One evening towards the end of 1999, I was at home watching the news on television. After the presenter had delivered the main stories of the day, she started to introduce a résumé of what I thought would be the events of the past twelve months, as usual on such late-December evenings. That year, however, she began a review of the whole twentieth century. 'As we draw to the end of the century that has seen *more change than any other* . . .' she began. I caught those words in my mind, held them there, and started thinking about them. What do we really know about change? I wondered. What makes this presenter so confident that the twentieth century saw more change than, say, the nineteenth, when railways transformed the world? Or the sixteenth, when Copernicus suggested that the Earth rotates around the Sun, and Luther broke the Christian Church in two? Soon black-and-white movies, a mushroom cloud, space rockets, cars and computers began to fill my television screen. The presenter's statement that the twentieth century had seen more change than any other was clearly based on the assumption that 'change' is synonymous with technological development – and that the twentieth century's innovations were without parallel.

In the years that have passed since that day, I have talked about 'change' with a great many people. When asked the question 'Which century saw the most change?', almost everyone agrees with the newsreader: surely it is the twentieth. Some people laugh at the very idea that I could even consider it to be any other. When pressed to explain, they usually respond by pointing to one or more of five twentieth-century inventions: flight, the atomic bomb, the Moon landing, the Internet and the mobile phone. They seem to believe that these modern achievements make everything else that went before inferior, and that change in previous centuries was barely noticeable by comparison. This seems to me to be an illusion – in respect of the assumptions that modern achievements represent the most significant changes and that pre-modernity was relatively static. Just because a certain development reached its apogee in the twentieth century does not mean that it was then that it changed at the fastest rate. The illusion is further reinforced by the instinct to prioritise events that we have seen with our own eyes, either in the flesh or on TV, over events that do not have a living witness.

Only a small minority of people immediately see the potential candidature of a century other than the twentieth. This is normally because they have a specialism that makes them acutely aware of the consequences of an earlier technological development – be it the stirrup, the horse-drawn plough, the printing press or the telegraph. I have not kept count, but it would be a fair approximation to say that when I have posed the question 'Which century saw the most change?', 95 per cent of people have answered 'the twentieth century' for technological reasons; most of the remainder have suggested an earlier century on the strength of a different technological reason; and just a handful of individuals have mentioned a non-technological event prior to 1900, such as the Renaissance or the campaign for the rights of women. As far as I remember, no one has ever suggested a century before 1000, even though one could make a good case for the fifth, which saw the collapse of the western half of the Roman Empire.

Some people reply by asking a question of their own: 'What do you mean by *change*?' On the face of it, it is an obvious response. But it is also a curious one. Everyone knows what change is – an alteration of state. Yet when asked to identify the century that saw the greatest change, people seem to lose their grasp of the meaning of the word. The collective human experience over a long period of time is just too huge in scope for us to think of the myriad changes it incorporates – taken together, all the different factors are unmeasurable. We can calculate certain specific changes across the centuries – life expectancy at birth, reproductive rates, longevity, height, per capita calorific intake, labourers' average wages – and for a large proportion of the last thousand years we can measure such things as church attendance, levels of violence, relative wealth and literacy; but to measure any one of these things accurately we have to isolate it from all the other aspects of our lives. We cannot measure differences in ways of living. It is like measuring love.

Actually, it is considerably more difficult than measuring love. At least love can be related to a scale – say, from considering sending a Valentine's Day card to launching a thousand ships to win back your loved one. Lifestyles cannot be related to a scale. Any quantifiable change that might be considered the most significant can be countered by another quantifiable change. For example, the twentieth century certainly saw the greatest growth in expectancy of life at birth: it increased by more than 60 per cent in most European countries. But against this it could be held that individual men and women had much the same potential lifespan in previous centuries. Even in the Middle Ages, some men and women lived to 90 years of age or more. St Gilbert of Sempringham died in 1189 at the age of 106; Sir John de Sully died in 1387 at 105. Very few people today live any longer than that. True, there were comparatively few octogenarians in the Middle Ages – 50 per cent of babies did not even reach adulthood – but in terms of the maximum lifespan *possible*, there was little change across the whole millennium. As soon as people try to find a measurable fact with which to answer the 'greatest change' question, other measurable facts get in the way. Why select one rather than another? As the example of life expectancy as opposed to life potential shows, it is purely a matter of personal preference.

This might suggest that the question is nothing more than a parlour game: a matter of curiosity and amusing debate, along the lines of 'Who was the greatest king of England?' But actually it is a serious matter. As I have tried to show in my *Time Traveller's Guides*, understanding human society in different periods of time gives us a more profound view of the nature of mankind than the relatively superficial impressions we get by looking at the way we live today. History helps us to see the full range of our capabilities and inabilities as a species; it is not just a nostalgic look back on the way things were. You cannot get the present in perspective without looking at the past. It is only through looking back to the fourteenth century, for example, that we can see how resilient we are in the face of adversities as cataclysmic as the Black Death. It is only through looking back to events such as the Second World War that we can see how innovative, highly organised and productive we can be when faced with a massive crisis. Similarly, looking at the history of Western governments over the last hundred years teaches us how myopic and short-termist we are in today's Western democracies, in which politicians pander to the whims of society and seek instant solutions to our problems. Only dictators plan for a thousand years. It is history that teaches us how violent, sexist and cruel our own societies have been – and could be again. While historical study has many purposes, from understanding how our modern world has evolved to learning how we entertain ourselves, the most profound purpose of all is to reveal something of the nature of humanity, in all its extremes.

This book is my somewhat belated response to the question implied by that newsreader in December 1999. However, I ought to say that in attempting to determine the century that has seen more change than any other, I have set certain parameters. The first is that I deliberately retain the ambiguous and vague definition of 'change' so as to encompass the greatest range of potential developments that might be considered within each century. Only in the Conclusion do I attempt to disentangle and grade them. The second is that I consider just ten centuries: the millennium that constituted the run-up to the year 2000. This is not to deny the importance of earlier periods but rather to keep Western culture in focus. I did not want this book to become yet another list of 'turning points' in world history. The third is that the book is about change within Western culture, which is largely a product of the countries that constituted Christendom in the Middle Ages. I only expand the study to a wider context in those centuries in which the inheritors of that Latin-writing world themselves reached beyond the oceans. Thus in this book 'the West' is not a geographical unit but an expanding cultural network originally centred on the Christian kingdoms of medieval Europe. Obviously I do not mean to belittle medieval cultures outside Europe – this book is about change, not pre-eminence. If I had considered my question since the birth of *Homo sapiens* as a species, Africa would have featured heavily. If I had considered it from the end of the last Ice Age, the Middle East would have figured more prominently. If I had attempted to chart all the significant ups and downs of human civilisation, factors such as the use of tools, the control of fire, the inventions of the wheel and the boat, and the development of language and religion would all have been taken into consideration. But these are other histories, and beyond the parameters of this book.

While this book is not a history of the whole world, it is also not a comprehensive history of a set of countries or a region. Many of the greatest events in national histories do not feature here, or are mentioned only in passing. Although certain invasions marked significant national changes – the Norman conquest of England, for example, or the arrival of US Commodore Perry in Tokyo Harbour in 1853 – these were relatively local events. Geographically specific elements *can* be part of the main story (for instance, the Italian Renaissance and the French Revolution) but most of them are peripheral to my central question. German unification was of little importance to, say, the Portuguese, and the Norman invasion of England was of no great interest to Sicilians, who had a Norman invasion of their own with which to contend. Similarly, the rise of slavery in America and the Caribbean appears only in a subsection of the chapter on the seventeenth century. This is because the resurgence of slavery took place on the periphery of what was then the West. Seventeenth-century Europeans were more directly affected by the less substantial white slave trade, which saw hundreds of thousands of people from Western Europe stolen by Barbary pirates and sold into slavery in North Africa. But even that did not affect Western culture as much as the five major changes selected for that chapter. The resurgence of slavery, like many national battles, should certainly feature in any world history but this book is no such thing. It is a synthesis of thinking about the development of the West in order to answer a specific question.

This focus on the question means that certain individuals and themes are also given less prominence than they usually receive in general history books. Friends and colleagues have asked 'How can you ignore Leonardo da Vinci?' and 'How can you leave out music?' Although Leonardo was an astonishingly talented man, his technological speculations had almost no impact on anyone in his lifetime. Very few people read his notebooks, nor did they build his inventions. His only important legacy was his painting, but, frankly, I don't see that my way of life would be very different today if one or two Renaissance painters had not been born. Had no one painted portraits that would be a different matter, but the influence of one individual artist is relatively small compared to the impact of, say, Luther or Copernicus. As for music, it is common in every country and has been so for more than a thousand years. Instruments, tunes and harmonies might have altered in form, and there is a case to be made for the ability to record music being a profound change, but the production of music is one of the great constants in human life, and interesting more for its ubiquity than its ability to alter the way we live.

It seems self-evident that the most important changes are those that go beyond national boundaries, entertainment and spiritual values. The most significant ones have an impact far outside their own fields. A scientist who *only* affects other scientists is, in the context of this book, comparatively inconsequential; likewise a historian who only influences our ideas about the past, or a great philosopher whose ideas only affect other thinkers. A friend of mine who knows far more about philosophy than I do thought it strange to read a book that pays so much attention to Voltaire and Rousseau but hardly mentions Hume and Kant, whom he considers far more important. But as he readily acknowledged, this is not a history of philosophy. It just so happens that the messages that Voltaire and Rousseau circulated had a direct impact on the political thinking of the eighteenth century. Kant is barely mentioned for much the same reason that Mozart scarcely appears: his legacy did not directly touch on one of the key changes of the last three centuries. The Parisian revolutionaries in 1789 did not storm the Bastille demanding that the nobility obey Kant's 'categorical imperative'; their leaders were inspired by Rousseau's social contract.

In the course of writing this book, I have repeatedly encountered one particular problem. Many of the most important developments in Western culture do not fit neatly within the borders of a single century. So should we consider the development in question when it started or when it had its greatest impact? Do we locate an invention when it was invented or when it became ubiquitous? There is no easy answer to this. On the one hand, it seems obvious that an invention does not change the world until it becomes widely used. Thus the internal combustion engine is described in relation to the twentieth century, not the nineteenth. On the other hand, however, if you only describe a development when its use became common, you ignore its early impact. Most people in the West were unable to read before the nineteenth century, but it would be a grave mistake to ignore the earlier developments in education, particularly those of the thirteenth and sixteenth centuries. Also, if we delay describing some developments until they become ubiquitous, they tend to bunch up, creating the false sense of a sudden surge of change in a later century and an equally artificial sense of stasis in the previous one. To describe the Industrial Revolution wholly as a nineteenth-century phenomenon, for example, would be to diminish the sense of industrial change in the eighteenth. It would also ignore people's awareness of techno-logical change happening around them, which considerably pre-dated the point at which they themselves started wearing machine-made clothes. Thus a degree of flexibility has been employed. In response to that newsreader's assertion in 1999, I find it more important that readers understand the wide range of changes that took place over many centuries, rather than setting some arbitrary rules that result in a misrepresentation of the past.

In 2009, I was commissioned to deliver a lecture to celebrate the 1,100th anniversary of the foundation of the diocese of Exeter, in south-west England. I took as my theme the question at the heart of this book: which of the last eleven centuries saw the most change? For that occasion I felt I needed not just to illustrate the various changes that had taken place since AD 909 but also to come to some conclusion. In the course of preparing the talk, a pattern emerged from my study that left me thinking that a certain threshold was passed within the time frame under consideration, and that this would continue to affect humanity for ever. The conclusion of this book has developed that original insight. I believe that if humanity survives another thousand years, the change I have selected as the most profound will be regarded as an archetypal moment in human history – as important as the ancient inventions that formed our culture: language, writing, fire, the boat, the wheel and religion.

Rethinking the question in the years since 2009, and walking through high-shelved corridors and library halls to research this book more thoroughly, I have felt overwhelmed by the scholarship of our society, especially the output of the last 60 years. In one library I was struck motionless by the feeling of never being able to know enough to write a book like this properly. Several centuries have threatened to overwhelm me, towering above me like huge shadows. I faced a wall of books about the Crusades and felt as nameless and insignificant as the people hacked to death in the streets of Jerusalem in 1099. I walked into a room full of books about eighteenth-century France and almost despaired. Any historian who does not retain a degree of humility in confronting so much evidence is deceiving himself, and anyone who does not admit to his or her inadequacy in writing authoritatively about the human past on this scale is a fraud. Of course I would very much like to know *everything* in order to supply the most thorough and well-informed answer possible to the question I have raised, but there is only so much information the human mind can retain. In my case I have had the advantage of working in the field of English history since my teenage years, first as an amateur, then as a student, an archivist, and finally as a professional historian and writer. As it is *English* history that I have researched for thirty years, there is an inevitable imbalance in this book in that most of the statistics I quote relate to England, but my choice of changes has not been limited to those that have affected this country. Rather I have selected subjects that affected a major part or all of the West and I have used English facts and figures where they illustrate the practical aspects of a change or to convey a sense of proportion. That seemed better than ignoring my field of expertise in order to even out the geographical imbalance.

It may well be that you do not agree with my choice of the century that saw the most change. It may well be that you remain defiantly convinced that none of the wars, famines, plagues and social revolutions of the past are as significant as being able to use a mobile phone or buy your weekly groceries via the Internet. It does not matter. The aim of this book is to provoke discussion about what we are and what we have done over the course of a thousand years, as well as what we are capable of doing and what is beyond our capabilities, and to estimate what our extraordinary experiences over the last ten centuries mean for the human race. If a few more people discuss such questions, and thereby realise something about human nature over the long term and consider how that insight can be applied to the future, it will have succeeded.

Ian Mortimer
Moretonhampstead, Devon
July 2014

The Eleventh Century

I am writing these words on the top floor of a three-storey house in a small town called Moretonhampstead – or 'Moreton' as most people call it here – which is situated on the eastern edge of Dartmoor in Devon, south-west England. It bore the same name, Moreton – 'the place in the moor' – in the eleventh century. However, the name and the granite bedrock on which it is built are about the only things that have not changed over the intervening period. One thousand years ago there were no three-storey houses here. There weren't even any two-storey ones. The dozen or so families resident in the area lived in small rectangular huts of stone and earth. The one room was heated by a central hearth, from which smoke billowed up to the blackened rafters. These houses were cut low into the hills to avoid the weather coming off the moor, and were roofed with thatch of bracken or straw. The inhabitants lived a tough life, eating mainly vegetables, cheese and the hardy grains that they could grow in the acid soil, such as rye, oats and vetches. No one could read or write; there were no priests here, no parish church. There may have been a crudely carved granite font in the house belonging to the king's bailiff, and a cross where an itinerant preacher would tell stories from the New Testament, but that was all. Although some twenty religious communities are known to have existed in Devon at that time, the nearest two were the bishop's modest cathedral at Crediton, some 13 miles to the north, and a small monastery in Exeter, 13 miles to the east. Neither of these amounted to much more than a small oratory attended by a handful of priests. The visit of a holy man to Moreton would have been a rare event. So would a feast.

The differences between ways of life then and now are all the more profound when you start to examine the things we take for granted. For example, virtually everything I own was purchased at some point, whether by me, my friends or my family. My predecessor living in Moreton in the year 1001, on the other hand, might never have handled money in his life. It *did* exist, in the form of silver pennies – King Ethelred the Unready minted considerable numbers of them to pay the invading Danes – but for a householder living in Moreton in 1001 there was little to buy: he had to make most things himself. If he wanted a bowl, he had to carve one out of wood. If he wanted a cloak, he had to obtain wool from local sheep, twist it by hand into thread, weave it into cloth, and finally tailor it. If he wanted to dye his new cloak, he had to prepare the colours from natural plant dyes, such as woad (blue) or madder roots (red). If he had to pay for any of these things, it would be an exchange in kind: he would probably offer animals, skins, meat or eggs – or that bowl he had so laboriously carved. There was simply very little need for cash: most householders only needed it to pay rent to their lord or to acquire something like a cauldron, a knife or an axe that could not be made locally. As a result of this scarcity of coin, hardly any silver hoards from this period have been found in the West Country. Coin production in Europe as a whole was very small, but in Devon it was almost unknown.[1]

The one place where you would have needed silver pennies was a market town. In the early eleventh century, however, there were only four such places in the whole of Devon: Exeter (13 miles), Totnes (22 miles), Lydford (on the other side of a trackless and boggy moor), and Barnstaple (38 miles). Even travelling the relatively short distance to Exeter, the nearest of these, would have been difficult. It was dangerous for a man to be alone on the forest paths, for he risked being attacked by thieves or even by wolves, which were still roaming wild in England. The trackways were rough and you would have had to ford the River Teign, which in winter had the force to sweep people off their feet. It was also risky to leave your property and family unattended back home, as they might be set upon by outlaws. As a consequence, ordinary people in 1001 did not travel far. The social structures that would require their descendants to journey considerable distances – the courts, Parliament, fairs and networks of religious orders – barely existed. People living in this far-flung corner of Christendom stayed among their own kind, where they felt safe: neighbours and kinsmen were the only people on whom they could depend to protect them and their families, to trade fairly with them, and to help them in times of famine.

In this way we begin to touch upon the real differences between my way of life and that of my predecessors in Moretonhampstead. The human race in 1001 was not just illiterate, superstitious, ignorant of the outside world and devoid of spiritual supervision; it faced continual hardships and dangers. Hunger and deprivation were widespread. Society was violent, and to protect yourself you had to meet force with force. In addition to the home-grown thieves and murderers, Vikings had attacked England intermittently over the last two centuries. In 997 they burnt the small market town of Lydford, on the north-western side of Dartmoor, and destroyed the abbey of Tavistock to the south-west. In 1001 they returned to Devon and attacked and burnt Exeter before turning east (fortunately for Moreton) and destroying the villages of Broadclyst and Pinhoe. But there was no guarantee that they would not return next year, sail up the River Exe to Exeter, and then try their luck to the west. King Ethelred could not have dragged his army along the remnants of Roman roads as far as Devon and through the forest paths to Moreton quickly enough to save the villagers from such attacks, even if he had wanted to do so. If the Vikings were to return, all the villagers could do was gather up their children and run and hide on the barren moor or in the woods.

How representative is this description of other parts of Christendom? As you would expect, there were significant variations even within England. If you travelled the 13 miles over the hills from Moreton to Crediton, you would find a more heavily populated manor, where the bishop of Devon was the lord. In his house you would even discover a couple of manuscript books: one about the early Christian martyrs and the other an encyclopedia compiled by the ninth-century French scholar Hrabanus Maurus. If you left Crediton and travelled into Exeter, you would find merchants and priests living within the old Roman walls. There was a market at the centre of town but still you would have been struck by the agricultural appearance of the place, which was home to fewer than a thousand souls. Winchester, then the capital of England, had a population of about 6,000. London, the largest urban settlement in the kingdom, had more than 10,000, many of whom resided in Lundenwic, or Aldwych, the port to the west of the city. In the south-eastern counties there were more people, more churches, and thus more priests than in Devon. There, coins were more regularly used and markets more common. Kent, for example, had 10 boroughs or places with a market (3.5 per 500 square miles, compared to Devon's 0.8), with a commensurately greater level of local travel. Even some long-distance journeys were undertaken: London toll regulations refer to traders arriving from Normandy. But although the Viking attacks had not entirely extinguished international trade, their threat was universal. And so was the fear of violence.

Further afield, you would have found even greater variations. Differences in economic prosperity and urban sophistication were to be seen all across Europe. With regard to religion, in 1001 Christendom was on the brink of achieving its familiar pan-European form. Wales, Scotland and Ireland were all independent Christian countries but with violent internal divisions even more marked than those of England. Scandinavia was only partially converted to Christianity, with areas of Norway resisting conversion. In eastern Europe, the kingdom of Poland had become Christian in 966. The kingdom of Lithuania remained pagan, as did the Slavs, but the kingdom of Kiev – lorded over by the Rus, the Vikings who gave their name to Russia – had started to turn to Christianity in 988. The Magyars lived in what is now Hungary. A century earlier, they had pushed into western Europe, fighting their way through the Holy Roman Empire into Burgundy and France, where they continued to raid until 955. In 1001 they too were in the process of being converted to Christianity by the recently crowned King Stephen I, who had defeated his pagan uncle. In the north of Spain, the Christian kingdoms of León (including Castile) and Navarre (including Aragon), and the independent county of Barcelona had started the Reconquista: the fight to recover what is now Spain and Portugal from the Muslim caliphate of Córdoba that would last until the end of the fifteenth century. Thus Christendom was expanding rapidly from its central core into northern, eastern and southern Europe – but not without a daily breaking of the commandment 'Thou shalt not kill.'

The central core of the Christian world was dominated by the Holy Roman Empire, which stretched from the north coast of Germany all the way south to Rome and included Austria, northern Italy and Lotharingia (comprising the Low Countries, eastern France and the Rhineland). It was governed by the Holy Roman Emperor, who was often the ruler of one of its many constituent duchies, magravates, counties or kingdoms. In his capacity as emperor, however, he was an elected spiritual monarch, chosen by a college of archbishops and secular lords. The empire's neighbour to the west, the Christian kingdom of France, was ruled by the recently established dynasty of Hugh Capet, but it was only about half the size of modern France. To the south-east there was the independent Christian kingdom of Burgundy, which reached from Auxerre to Switzerland and down to the Mediterranean coast of Provence.

It was in the Mediterranean kingdoms that daily life was most markedly different from England. Córdoba was one of the richest and most sophisticated cities in the world, with levels of trade and learning that outstripped anything to be found in Christendom; perhaps as many as half a million people lived there. The architecture was on a truly splendid scale – as shown by the Great Mosque, which still stands today. It was said that the caliph's library housed more than 400,000 volumes. In Italy, people were living in towns more or less as they had done in the days of the Roman Empire. It was home to the largest trading settlements in western Christendom: Pavia, Milan and Amalfi each had perhaps 12–15,000 inhabitants, with the maritime states of Venice, Pisa and Genoa following close behind.

The only part of Christendom as wealthy and sophisticated as the caliphate of Córdoba was the Byzantine Empire, in particular its capital city, Constantinople, which was at the height of its prosperity in the early eleventh century. Estimates vary wildly, but its population was probably around 400,000. It also had a highly developed judicial system, economic links across the Middle East, and jaw-dropping wealth. From the Great Palace, the emperor in 1001, Basil II, commanded an area that covered the whole of the north-eastern Mediterranean coast, including southern Italy, most of the Balkans, Greece and Anatolia (modern Turkey) as far as the border with Palestine. He also ruled the Greek islands, Cyprus, Crete and part of the northern coast of the Black Sea. Near the Great Palace was the church of St Sophia, with a massive dome 182 feet in height – it was by far the largest building in Christendom. Works of art had been collected from all over the ancient world by the fourth-century Roman emperor Constantine and placed here to adorn the city he had made his capital. Ancient Greek bronze sculptures stood near Ancient Egyptian obelisks. In 1001, Rome, the original capital of the empire, was insignificant by comparison: its walls surrounded an area only half the size of Constantinople, its artworks had fallen or been stolen, and sheep and cattle now grazed among the ruins on the city's famous hills. As for the rest of Christendom, the sophisticated Byzantines regarded them as mere barbarians.

Given these extremes – from a handful of self-sufficient farmers struggling to get by in their earth-walled houses on the wet hills of Moreton to the gilded brilliance of Muslim Córdoba and the huge wealth of Christian Constantinople – it might seem impossible to identify developments that changed the whole of the nascent Western world. And yet despite everything that divided them, they had more in common than contemporaries would have realised. When the bishop of Barcelona wished to buy two rare books from a Jew in 1043, he did not pay in silver but with a house and a piece of land, showing that non-monetary purchases might be undertaken even by the educated and prosperous citizens of the Mediterranean region.[2] If a famine gripped Europe, everyone suffered – including the Byzantines, who saw high prices and reduced trade. If a disease spread anywhere in Christendom, it killed rich and poor alike. And no one, anywhere, was ever free from the violence of the time. England was conquered by Duke William of Normandy in 1066, and it was another Norman, Robert Guiscard, who occupied the southern Italian possessions of the Byzantine Empire in 1060–8. True to the saying that 'he who has the most has the most to lose', the Byzantine emperor, Romanos Diogenes, was captured at the Battle of Manzikert in 1071, and as a result of this defeat, Anatolia was lost to the Seljuk Turks. While he was still a captive, a *coup d'état* toppled him from power in Constantinople. Later he was blinded, and died of his injuries in a monastery. Frankly, he would have been safer in Moreton.

The growth of the Western Church

There is no doubt that most scholars would identify the rise of the Roman Catholic Church as the single greatest change of the eleventh century. It was a consequence, in part at least, of the states on the periphery of Christendom turning to the Church of Rome. This geographical expansion underpinned the rise of the papacy as a pan-European power, with wide-ranging political and moral authority. It also led to an increase in the power of the Church generally, and thus brought about a series of changes that affected the whole of society.

Without this growth, the Middle Ages would not have unfolded in the way they did.

Between 955 and 1100, Western Christendom doubled in size. It was not an instantaneous transformation: many places resisted the Christian faith for decades, but over the period almost the whole of western Europe came to live and worship under the cross. The reasons for this are complex; no doubt missionary zeal played its part, but a more important factor was the desire among rulers either to stabilise their realm against violent neighbours, or to extend their authority by conquering new lands. To do either of these things they needed alliances, and the Catholic Church provided a moral framework within which to establish bonds of trust. And as more princes adopted the Catholic faith, the Church became ever more powerful and attractive – a snowball effect – making the localised pagan religions insignificant. On top of this, rulers saw advantages in adopting a religion that was essentially a dictatorship. The Catholic Church reinforced a monarch's own authority and, through its hierarchical philosophy, helped him stabilise and control his kingdom.

In return, such a rapidly growing patrimony naturally increased the political power of the pope, though it intensified his rivalry with the patriarch of Constantinople. Nominally the pope, as the successor of Rome's first bishop, St Peter, took precedence, but this primacy was rarely stated overtly and thus was open to question. In an attempt to clarify matters, Pope Leo IX sent a delegation to Constantinople in 1054, charging its members with requiring the patriarch, Cerularius, to acknowledge the supremacy of the Roman pontiff. The fine political balance of the past was upset – and so was Cerularius. He flatly denied that the Roman Church had authority over the Byzantine Empire. The Roman delegates duly excommunicated him; Cerularius replied by excommunicating them. From that moment on, the Roman Catholic and Greek Orthodox churches went their separate ways. Hence the year 1054 is seen as a momentous date in the history of the Church. In reality it was just the formal recognition of a division that had been growing for centuries. Significantly for the pope, however, it was the patriarch who was soon on the back foot, following the collapse of Byzantine authority in Italy in the 1060s and the loss of Anatolia in 1071.

The growing power of the pope also put him at loggerheads with the Holy Roman Emperor. In 1001, there was as yet no official mechanism for the appointment of a new pope. Sometimes the noble families of Rome would choose one of their own; sometimes they would accept the nomination of the Holy Roman Emperor. The emperor maintained the right to appoint the man he thought best suited for the position, whether that meant a direct appointment or a rigged election. As a consequence, there was often conflict, with the pope occasionally being deposed by the emperor and replaced with an imperial sycophant. In 1046, when Henry III succeeded as king of the Germans and came to Rome to be anointed as emperor, he found three popes simultaneously claiming the title of pontiff. These were Benedict IX, who had sold the papal title but then refused to give it up; Gregory VI, the man who had purchased it from him; and Sylvester III, the local choice. Not wanting his imperial title to be sullied by doubt, Henry III summoned the Council of Sutri, at which all three popes were deposed; the emperor then appointed his own confessor as the next pope, Clement II. But the problem of appointment soon arose again. In 1058, two rival popes, Benedict X and Nicholas II, went to war with each other. The following year, Nicholas II prevailed and issued the papal bull *In nomine Domini*, which laid down the rule that all future popes were to be elected by the College of Cardinals, in secret, without the Holy Roman Emperor's intervention.

In nomine Domini was just the first of a series of reforms promulgated by the Church's cardinals, the most prominent of whom was Hildebrand, who would later become Pope Gregory VII. These reforms tended to set priests apart from all other men. Catholic clergy – from parish priests to bishops – were now prohibited from marrying. They were required to look like Latin clergymen – tonsured and clean-shaven – and to speak like them too, using only Latin in their religious services. They were prohibited from buying and selling ecclesiastical offices, a practice known as simony. They were removed from the jurisdiction of secular courts and tried by their bishop's ecclesiastical court, in which they could not be sentenced to death. Most important of all, the reforms proscribed lay investiture. This meant that, in theory, no secular lord could appoint an ecclesiastical official; the senior clergy, including all the bishops and archdeacons in Christendom, had to be appointed by the pope. Gregory VII even extended his authority over the Holy Roman Emperor: he twice excommunicated Henry IV and on one occasion forced him to walk barefoot across the Alps in a hair shirt to beg forgiveness from him at Canossa. Although the reforms took time to come to fruition – many resisted the ban on clerical marriage and some rulers refused to give up the right to appoint senior clergymen – their impact was huge. By 1100, the Church had become an independent political and religious body that incorporated all the kingdoms from Norway to Sicily and from Iceland to Poland, and exercised varying degrees of influence over what people were beginning to call 'the Latin World'.

The growth of papal power was accompanied by an intensification of Church authority at grass-roots level. Priests were established as permanent ecclesiastical agents within communities. Religious focal points – which, as we have seen in Moreton, in 1001 had been little more than crosses for public preaching and baptismal fonts in manor houses – were increasingly turned into churches in the proper sense of the word, paid for by the communities that wished to worship in them. We have already seen that at the opening of the eleventh century, the bishop of Devon, who nominally administered an area of 2,590 square miles, exercised authority over barely two dozen established priestly communities. The bishop of Paderborn in Lotharingia similarly administered only 29 churches within his diocese of 1,158 square miles.[3] By 1100, however, their successor bishops each supervised hundreds of parishes. In some parts of England the process of parochialisation was almost complete by the end of the century. No fewer than 147 churches are mentioned in Domesday Book (1086) for the county of Kent; however, the contemporary collection of documents, Domesday Monachorum, suggests that there were twice that number, showing the parish system was already fully formed there.[4] Much the same can be said for Sussex, where at least 183 of its eventual total of 250 medieval parish churches had been built by 1100.[5] The populous and rich areas of Norfolk and Suffolk had even more churches.

In addition to the priests who served these congregations, a whole hierarchy of senior clergy was appointed. Archdeacons served under bishops as spiritual administrators of a given region. Deans were chosen to supervise collegiate churches. Northern European bishops were no longer situated in quiet places somewhere in the countryside; they moved into towns, in emulation of their southern European counterparts. In England, the bishop of Crediton moved to Exeter in 1050; the bishop of Dorchester relocated to Lincoln in 1072; the bishop of Selsey moved to Chichester in 1075; the bishop of Sherborne went off to Old Sarum (later Salisbury) in 1078; and the bishop of Elmham moved first to Thetford in 1072 and then to Norwich about 1095. By 1100, the English bishops were all established in towns – places with better infrastructure and faster communications. You could say that, whereas in 1001 it was rare to see any priests, in 1100 it was hard to get away from them.

In addition to the parish priests, archdeacons, bishops and archbishops, the pope presided over the monks whose numbers were growing with extraordinary rapidity. In the early tenth century, the duke of Aquitaine had founded a monastery at Cluny along revolutionary lines. Like other monasteries of the time, it was to follow the Rule of St Benedict; unlike other monasteries, however, it was to do so very strictly. Cluniac monks were forbidden from sexual intercourse and all forms of corruption, including simony and nepotism, and fell under the direct jurisdiction of the pope. Yet the most important element of the Cluniac way of life was a return to prayer as the monks' key occupation. Labourers were employed to work in the monastery's fields, which freed the monks to perform the liturgy. The new monastic model appealed to members of noble families, who considered manual labour to be beneath them. Cluny thus soon attracted both followers and wealth. It also drew very capable men to be abbots. More monasteries linked to the mother house of Cluny were established, and in the eleventh century this became the first proper monastic order in Christendom. At its height, the Clunic order comprised nearly a thousand monasteries throughout Europe. The whole project demonstrated the power of an organisation of religious houses under a single leadership. Soon even stricter monastic orders were established, such as the Carthusians in 1084 and the Cistercians in 1098.

If it seems that an army of monks and priests was on the move, headed by their papal commander-in-chief, then that is not far from the truth. In 1095, at Clermont in France, Pope Urban II held a council that marked a watershed in the growing influence of the Latin Church. The Byzantine emperor, Alexis Comnenus, buckling under the force of the Seljuk Turks in Anatolia, had asked Urban II to exert pressure on the nobles in the west to send military aid to their Christian brothers in the east. What a turnaround! At the beginning of the eleventh century, Constantinople had thought of Rome as uncivilised; in 1054, its patriarch had excommunicated the pope's legates; but in 1095, the

Byzantines saw the pope as their potential saviour. Urban II, keen to heal the divide of the Greek Orthodox and Roman Catholic Churches, and hoping finally to assert papal supremacy over the whole of Christendom, was eager to assist. On 27 November, he preached a sermon to a large crowd in which he urged Christian men to desist from fighting each other and to go to Jerusalem to recover the holy seat of their lord, Jesus Christ, from the rule of the Fatimid caliph. His appeal received an ecstatic response and led directly to several waves of armed pilgrimage to the Holy Land, most notably the First Crusade. This expedition of Frankish and Norman nobles swept through Anatolia and Syria, conquering the great city of Antioch on the way to Jerusalem, which fell to their army on 15 July 1099. This in itself was extraordinary. Just imagine setting out from France on foot today for Jerusalem. Now imagine doing it without any guide-books, phrasebooks or money, facing incredible heat and large numbers of heavily armed enemies. And imagine doing it without having ever travelled more than a few miles from your native village. The very drama of setting out on a Crusade is impossible for us fully to appreciate. But urging people to fight their way to the ends of the Earth is a measure of how far the Church had come.

This had been an astonishing century for the Catholic Church. At the start of the period the pope had been hired and fired by the emperor. He could rely very little on the kings and dukes of Christendom who were embroiled in war. He could not exert his authority because necessary administrative and communication structures were patchy or non-existent. Where there were priests, they often defied religious expectations by using their own languages and customs, buying and selling offices, marrying, and behaving altogether like secular men. However, by the end of the century, the Catholic Church was united, centralised, organised, powerful and expanding. It was able to bring the emperor barefoot across the Alps and even to instigate the conquest of the Holy Land. It promoted literacy, book composition and the stimulation of intellectual activity across the whole continent. But its real triumph was in exerting its authority at grass-roots level. The eleventh century was when the Catholic Church changed from being simply a faith into which people were baptised to being a vast, organised system that governed how they lived and died.

Peace

There is a measure of irony in describing the First Crusade and then claiming that one of the greatest changes of the eleventh century was the growth of peace. The irony is all the greater when you consider that one of the reasons for the peace – or, rather, the lack of conflict – was the agency of the Church, which in 1095 had urged its members to wage war. Nevertheless, if you compare the Europe of 1001 with that of 1100, the latter is so much less violent that there can be no doubt that peace was one of the biggest changes of the eleventh century.

To understand this apparent contradiction it is necessary to consider the violence in context. Yes, the eleventh century saw a whole succession of wars, but the very word 'war' is significant here. When the Vikings attacked Devon in 997 and 1001, they were not fighting a war; they were carrying out the daily acts of aggression celebrated by a society whose culture endorsed violence as a way of life. Likewise, when the Magyars invaded the Holy Roman Empire and the Muslim general Almanzor attacked the kingdom of León, these hostilities were part of ongoing cultural conflicts. Each side understood itself naturally to be the mortal enemy of the other. When the pagan kingdoms on the perimeters of Christendom were converted, they swapped outright enmity for cultural fraternity and cautious coexistence. They might still go to war, and, as we have seen, even popes might fight each other. However, these were political wars to resolve a specific dispute, limited in their duration; conflict was no longer a daily occurrence. The Norman conquests of England (1066–71) and southern Italy (over the course of the eleventh century) were among the last Viking-style invasions of established Christian kingdoms. A permanent state of regular killing, which once had existed throughout Christendom, had by 1100 been pushed to the periphery and beyond, into the lands of pagan adversaries: the Córdoban Muslims, the Seljuk Turks and the heathen Lithuanians and Slavs.

The process by which European society shrugged off the old culture of universal violence was not just down to old pagan enemies becoming new Christian friends. It was also due to socio-economic changes, most notably the emergence of the feudal system. At the time of the Viking and Magyar invasions, European armies had fought on foot; they were no match for fast-moving maritime attackers or marauders on horseback. In order to preserve their lands and people, the rulers of Europe thus created specialist forces of mounted soldiers, whose armour, manoeuvrability and rapid redeployment could hold invaders in check. But warhorses were exceedingly expensive, and so was the chain mail that such knights were increasingly required to wear. Moreover, years of training, starting from boyhood, were required to fight on horseback. Thus the great nobles of Europe endowed these knights and their families with considerable estates: a new privileged class was created to sustain this military force. The feudal system was a reciprocal arrangement by which local communities fed and equipped their armed lords and were protected by them. This process, which was well under way by 1001, made it increasingly difficult for invaders to exploit the hitherto poorly defended European peasantry. The word 'feudal' might have negative connotations in the modern world, but in 1100 it meant that Christian Europe was better defended than it had been before.

It has to be said that the establishment of a class of feudal lords to defend Christendom against external enemies encouraged violence of a different kind, namely between the lords themselves. Indeed, the Norman chronicler William of Poitiers contrasted the frequent bloodshed in feudal Normandy, where neighbouring lords fought each other, with the relative peace of Anglo-Saxon England before 1066. However, there were a number of pressures on lords that encouraged peace. Several of these were pioneered by the Church, which actively pursued innovative ways of dampening violence. For example, a lord could be forced to do penance when he committed an act of unspeakable cruelty. One man who has gone down in history as committing more than his fair share of such acts is Fulk Nerra, Count of Anjou. On discovering that his wife had committed adultery with a farmer, he burnt her at the stake in her wedding dress. When he instigated a massacre so horrifying that the Church could not let it pass without recrimination, he was ordered to undertake a pilgrimage to Jerusalem. After another atrocity, he was made to build a new monastery where priests would pray for his soul. By the time he died in 1040, Fulk had completed two pilgrimages to Jerusalem and founded two monasteries. While the Church clearly failed to curb all of his violent tendencies, it did manage to make an example of his black reputation, forcing him to perform these penances. Others no doubt took note. And one wonders how many more violent acts Fulk would have committed if there had been no restraint at all upon his temper. Ironically, on the strength of his pilgrimages and monastic foundations, some historians have referred to him as pious.

A second strategy devised by the Church to curtail violence was the movement known as the Peace of God, which had started in France in the late tenth century. This entailed the forceful use of religious propaganda – particularly the carrying of saints' relics in processions – to urge lords to make peace, and to protect women, priests, pilgrims, merchants and farmers from the destruction of war. We may take a cynical view of such measures but in the eleventh century superstitious belief in the power of relics to inflict mortal damage on a person who broke his religious oath was not just widely held, it was intensely believed. It took on a special significance around the year 1033, which was understood to mark the millennium of Christ's death. Another movement was that of the Truce of God. This initially prohibited all warfare from Friday night until first thing on Monday morning, as well as on the feast days of major saints. In the 1040s it was extended to prohibit men over the age of twelve from fighting between Wednesday evening and Monday morning and throughout Lent and Advent. It has to be said that it was not wholly successful: the Battle of Hastings was fought on a Saturday, so both Saxons and Normans broke the Truce of God. Nevertheless, it was constantly reaffirmed by ecclesiastical councils. Consequently lords were regularly reminded that the Church did not condone fighting among Christians. As Pope Urban made clear in his sermon at Clermont at the end of the century, if Christians really had to fight then they should direct their energies against the enemies of Christendom.

It is easy for us to mock churchmen who thought they could ban warfare from teatime on Wednesday until nine o'clock on Monday morning but these movements had more force than the Church alone could exert, for they advocated a morality that kings also endorsed. It was hugely destructive to a ruler's interests if his vassals directed their resources and energies against each other. For this reason, both William the Conqueror and the Holy Roman Emperor, Henry IV, championed the Truce of God. In addition, secular rulers had methods of their own to preserve peace. In England, William purposely distributed his lords' manors in small, separate parcels across the country, reducing their ability to control any stretch of land as a sub-kingdom. In addition, every manor was held personally from the king, so the bonds of fealty kept his vassals at peace. These bonds also meant that a king's expectations of propriety had to be obeyed. Whether it was a Monday morning or a Saturday afternoon, the rules of war applied, and they often included the Church's proscriptions. The slow process of containing and controlling Christian violence had begun.

The discontinuation of slavery

The French historian Marc Bloch declared that the disappearance of slavery constituted 'one of the most profound [transformations] . . . that Mankind has ever known'.[6] Without a doubt, slavery's demise was a considerable change in European society over the period 900–1200, but it was a complex process. For a start, as that time range indicates, it was not a sudden and complete 'abolition': there were still slaves in the West in the thirteenth century, and several centuries after that in eastern Europe. Also, not all slaves experienced the same conditions of slavery: different countries had different laws restricting what a man might do with his slaves. On top of these factors, it is not always clear how slavery relates to other forms of servitude, especially the position of the serf or unfree peasant. Nevertheless, significant steps to limit slavery were taken in the eleventh century that led to its gradual disappearance in the West – hence the introduction of the subject here.

Slavery is an ancient institution, and medieval slavery had its origins in the Roman Empire. The Roman legal principle of *dominium* held that ownership of things went far further than simply owning them; it extended to doing what you liked with them – and 'things' included people. After the Western Roman Empire broke up in the fifth century, each of the new kingdoms that emerged applied its own limits to this principle, and both slaves and slave owners fell under the kingdom's own laws. Different rules developed about whether a free woman who married a slave became a slave herself and vice versa; or whether a man who married a slave not knowing her status was at liberty to leave her. In some places a man was entitled to sell his wife into slavery, thereby annulling the marriage. If a man sold himself into slavery, it did not necessarily follow that his wife and children also became slaves, for they had been born free; but nor did they necessarily retain their freedom. In some kingdoms, a man who killed his slave had to perform penance, its severity depending on whether the slave was guilty of transgression or the master was just indulging himself. Some laws required a man who had two sons by one slave girl to set her free. Certain regions accepted that slaves could keep any money they earned, so that they might buy their freedom. The laws of King Ine of Wessex, adopted by Alfred the Great, laid down that if a master forced his slave to work on a Sunday, the slave was automatically set free and the master fined thirty shillings.

Among all these variations, there is one fundamental difference that distinguishes the slave from the unfree villein or serf of the feudal system. The lord of a manor could impose restrictions on what his serfs did, whom thcy might marry, where they could go and what land they worked, but this was by virtue of their attachment to the manor. The serf was bound to the land, and his services and duties were inherited, transferred or sold with it. It was therefore an indirect form of servitude, and that implied several other important differences. The powers of the lord were limited by custom, and the serfs on a manor thus enjoyed certain rights. A slave, on the other hand, was property, pure and simple. He or she could be bought and sold independently of his or her spouse, or they might be transferred as a pair. He or she could be beaten, maimed, castrated, raped, forced to work all hours (except, as mentioned, in some kingdoms, on Sundays) and even killed, without any measure of recrimination against the owner. It wasn't simply the case that slaves were second-class citizens. The peasants were the second-class citizens. Slaves weren't citizens at all.

You would have thought that the Christian Church might have suppressed slavery. It was torn, however. On the one hand there was the view expressed by Pope Gregory the Great in the late sixth century: that mankind was created free by nature and therefore it was morally just to restore men and women to the freedom to which they had been born. On the other, there were men like St Gerald of Aurillac three centuries later, who freed many slaves at the time of his death but while alive regarded them as his property – as is shown by his very unsaintly threats to mutilate a number of them for not being sufficiently obedient.[7] Part of the problem was that, as we have seen, the Church had limited influence in the early eleventh century, and lacked the power and means to hold unscrupulous lords to moral account. But the fundamental issue was that slaves were property. If the Church was reluctant to give away its own property, how could it urge wealthy individuals to give up theirs? In towns such as Cambrai, Verdun and Magdeburg, the bishop was even paid a tax on the sale of every slave. In order to grow and exert its authority, the Church needed the help of wealthy individuals whose prosperity depended on the labour of slaves. Such men were unlikely to support a Church that denied them their wealth. The Church was thus trapped between its moral mission and its need for money and influence.

So what changed over the course of the eleventh century? To answer this question we must examine why people were enslaved in the first place. To begin with, men and women captured in war were frequently sold as slaves. This was standard practice both within Christendom and beyond the periphery. Christian English slaves were sold to Denmark in the reign of Cnut. English slaves, captured by pirates, were sold in Ireland. Irish and Welsh slaves were sold in England. Our word 'slave' derives from the Slavs, who had not yet been converted to Christianity and were thus very vulnerable to the raids of Christian slavers. But not all enslavement was due to war: some people sold themselves into slavery. That slavery could be a self-inflicted condition may come as a shock to us today but sometimes people had no choice: they sold themselves or members of their family to avoid starvation. For others still, slavery was a form of punishment. A robber caught stealing might be made his victim's slave rather than being put to death. In some kingdoms, traitors were punished with a sentence of slavery. Clergymen who attempted to justify slavery argued that it was more merciful to enslave a criminal or defeated soldier than to hang him. Hrabanus Maurus, the author of one of the books to be found in the bishop's house in Crediton in 1001, stressed this very fact.

A number of social developments brought this situation to an end. First there was the Church's promotion of peace. With a diminished level of conflict, there were fewer opportunities to enslave one's enemies. There was also a long period of economic growth: wasteland was cleared, marshland drained and new manors founded, and there was an altogether greater volume of trade. It stands to reason that if the two principal causes of enslavement were cultural conflict and extreme poverty, and Europe experienced less of both, then slavery was likely to decrease. Increased prosperity also led to the growth of urban life in Germany, France and Italy in the late eleventh century; slaves could now run away to a large town and sell their labour. In addition, lords were less keen to take responsibility for feeding slaves who were not productive; the feudal peasant, bound to the land, who worked for the lord for free but fed himself, was a more economic unit. On top of these factors, the growing wealth and power of the Church gradually allowed it to strengthen its moral stand. The Peace of God movement included the provision that slaves who ran away at its gatherings could be permanently free. The practice of punishing criminals with enslavement also began to disappear. Finally there was the impact of individual rulers' policies. Several contemporary writers state that William the Conqueror firmly believed that slavery was barbaric and that he took measures to stop the trade in slaves.[8] At the end of his reign, 6 out of 28 men in the manor of Moreton were still described as slaves (*servi*), but in the country as a whole, slaves represented about 10 per cent of the population. The Church reinforced the Conqueror's anti-slavery message after his death. In 1102, the Synod of London declared that 'never again should anyone engage in the infamous business, prevalent in England, of selling men like animals'. By that time slavery had all but disappeared from France, central Italy and Catalonia.[9] Although it persisted in the Celtic countries for another century or so, and in eastern Europe for much longer, the practice of selling human beings in the marketplace, which had been normal in the West since prehistory, was rapidly coming to an end.

Structural engineering

The fourth great change of the eleventh century still characterises the towns we live in today. Broadly speaking, in 1001, buildings in Western Europe were small and architecturally unambitious, conforming to the style and scale of examples of Roman construction that had survived from the ancient world. Most cathedrals were barely the size of a large parish church today, with wooden roofs no more than 40 feet above the ground. By 1100, however, architects and structural engineers had transcended the limits of their Roman forebears, developing the style we know today as Romanesque. Hundreds of huge buildings, some with vaulted ceilings over 70 feet high and towers of over 160 feet, had been built across Europe; hundreds more were under construction. Similarly, in 1001, there were very few defensive buildings that we would recognise as a castle; by the end of the century, there were tens of thousands. In the eleventh century, people in Europe learnt to build strong walls and lofty towers – and they did so in every corner of Christendom.

By now it will hardly come as a surprise to learn that the growing ambitions of the Church were a major influence on these developments. The rebuilding of Cluny, the Burgundian church at the heart of the rapidly spreading Cluniac order, had started in 955. When it was dedicated in 981, it was enormous – breathtakingly so for the time – with seven bays in its nave and side aisles. In the early eleventh century, it carried on growing and acquiring new features, such as a narthex and a tunnel-vaulted ceiling (which was good for carrying plainsong). Another huge tunnel-vaulted church was simultaneously under construction at the abbey of St Philibert in Tournus, 20 miles from Cluny, and in 1001, work began on the church of St Bénigne in Dijon, 80 miles to the north. The reason why these magnificent early Romanesque churches first appeared in Burgundy may have been a desire to build fireproof stone buildings after the attacks of the Magyars in the middle of the tenth century. On the whole, however, the motivation for new construction normally comes down to money, and Cluny certainly had plenty of it. The routes from Italy to northern France passed through the region, bringing merchants and pilgrims – and their cash. But whatever the reason, these three churches, completed in the first two decades of the eleventh century, proved hugely influential. In the case of Cluny, this was because priors from daughter houses regularly returned to the order's mother house. They now wanted a church of their own like that of Cluny, and from here word spread beyond the limits of the Cluniac family.

Another religious inspiration for the new architecture was the Reconquista, the recapture of Spain from the Muslims. Santiago de Compostela, in the kingdom of León, had been a major pilgrimage destination since the ninth century, but in 997, it had been sacked by Almanzor, the military ruler of Muslim Spain. Almanzor had died in 1002, and the caliphate of Córdoba never fully recovered from the infighting that followed. The Christian kingdoms of Spain saw their chance and went on the offensive, pushing the frontier deep into Spain and securing the land for the Christian faith by building castles and churches. They encouraged knights to come to fight in their religious war. Pilgrims could once more make the journey to Santiago de Compostela in relative safety. A number of impressive Romanesque churches were constructed along the main routes that these travellers took through France on their way to Spain, at Tours, Limoges, Conques and Toulouse, as well as at the final destination itself. These towns gathered money from those who visited and spent it on enlarging their churches, so that future knights and pilgrims should stand in awe at the wonder of God. As the century went on, the successes of the Reconquista encouraged ever more visitors, whose donations further fuelled the church-building boom.

The Romanesque style spread with extraordinary rapidity from central France, as patrons and masons elsewhere saw what they might now build in stone. In Normandy, Duke William oversaw the consecration of Jumièges Abbey in 1067, and together with his wife he founded two large abbey churches at Caen. The Holy Roman Empire also embraced the new fashion eagerly, shifting away from Carolingian-style churches towards the construction of massive Romanesque cathedrals like that at Speyer, begun about 1030, where the emperors were now buried. The huge wealth generated in Italy by the emerging merchant states of Pisa, Florence, Milan and Genoa ensured that southern Europe did not miss out. Pisa Cathedral was begun in 1063, and the 160-foot (48.8-metre) nine-stage tower of Pomposa Abbey dates from the same year. Venice, which had always looked to Constantinople and the East rather than to France for inspiration, also began to build on a massive scale. St Mark's Basilica was begun in 1063, and although it was designed along Byzantine lines, in its scale it was clearly influenced by the new churches of France and Germany. By the end of the eleventh century, even England had been caught up in the fever of building cathedrals and abbey churches. Nothing of any significance had been constructed before the Norman Conquest, but that momentous event was the start of many important rebuildings and foundations. The Cluniacs began to build their first monastery in England at Lewes in about 1079. Apart from Edward the Confessor's church at Westminster, every single cathedral and abbey church in the kingdom was rebuilt within fifty years of the Normans' arrival.[10] The lasting marks of this transformation include parts of St Albans Abbey (now Cathedral, begun c.1077), Gloucester Abbey (now Cathedral, begun 1087), Winchester Cathedral (begun 1079), Durham Cathedral (begun 1093) and Norwich Cathedral (begun 1096).

So what, you might ask: does it matter? After all, exchanging one building style for another is hardly a huge change in the way people live their lives. But the importance here lies not in the symbolism of building loftier churches, but in the technology that made them possible – innovations in structural engineering. Building high churches of stone, with vaults capable of withstanding the attacks of savage Magyars intent on burning everything in their path, had obvious military uses. Thus it should come as no surprise that the development of large-scale Romanesque architecture runs hand in hand with that of the castle.

The castle became the physical embodiment of feudalism. When a king endowed a lord with a manor, he thereby made the lord responsible for the safe keeping of the people who lived there. And to protect their land, its people and their produce, from the late tenth century lords started building fortified residences in stone and wood. The earliest castle we know of is Doué-la-Fontaine, which was fortified about 950, probably as a result of the rivalry between the counts of Blois and the counts of Anjou. At the beginning of the eleventh century, Fulk Nerra built Langeais and more than a dozen other castles in his county of Anjou. These were mostly square stone donjons with thick walls and entrances at first-floor level, to withstand enemy attacks. An impregnable castle meant that it was very unlikely that a lord would lose control of his territory, even if it was overrun by his enemies. All he had to do was wait until they ran short of food or were off guard, so they would abandon the siege or could be beaten in a surprise attack. Castles thus quickly became the nails by which kings and lords could fasten control on a region and ensure its long-term security and stability. Throughout the century, as improvements in structural engineering led to higher and stronger towers, so a lord's feudal ties to his land grew stronger.

Just how important the castle was in Europe can be seen by what happened in regions that had to do without them. In 711, the Visigothic kingdom of Spain had simply melted under the heat of the Muslim invasion, having no castles in which the population could take shelter. As we have seen, the Viking and Magyar attacks of the ninth and tenth centuries left local communities helpless. And Norman chroniclers ascribe the failure of the English resistance in 1066 to their lack of castles. The only defences that William the Conqueror had to negotiate were the old walled burghs – fortified towns – but these were few and far between. Although the gates were barred against William at Exeter in 1068, the citizens were not strong enough to man such a long wall and withstand his forces. Soon after their surrender,

William built a castle to control the city. In London, he constructed three castles to exert his authority, the Tower being the outstanding survival, and at York he built two castles to guard the city. In all, by 1100, more than 500 castles had been built in England. The country had changed from being an almost defenceless kingdom to one bristling with towers. The same change took place throughout Europe. In every city in Italy, for example, the tall towers of the most powerful families reached up like outstretched arms proudly punching the sky. It became increasingly difficult for one king simply to conquer another's land by force alone. The conquest of Normandy by the French in 1204 and that of English-controlled Gascony in 1453 prove that it was not impossible, but in most places the land was so heavily defended with castles that success depended on more than mere military prowess; it required the local lords to change sides. In this way feudalism's physical manifestation contributed to the security of Europe, and further strengthened the peace that was beginning to spread through Christendom.

Conclusion

We have seen how some of the key features that we associate with the medieval period – papal supremacy, parish organisation, monastic orders, castles and great cathedrals – were barely present in 1001 but fully formed by 1100. But the old world came to an end in other ways too. The eleventh century saw a profound change in the nature and extent of war and violence, and the beginning of the end of slavery. What is perhaps most extraordinary is the degree to which the Church took a role in all these things. Even the end of the Viking invasions can ultimately be associated with the influence of the Church, as Christianity spread into Scandinavia and beyond.

What did all that mean for my predecessors in Moreton? Over the course of the century, priests would have visited more regularly, building the first church here around the year 1100. It was a small structure, dark inside, with a crudely carved granite frieze around the outside showing the tree of life, abstract swirls and mythical monsters. Primitive though it may have appeared to a traveller from Byzantium, it connected Moreton permanently to the rest of Christendom. As elsewhere in Christian Europe, here parishioners heard sermons on morality and godliness as part of their way of life. After the seat of diocesan administration had shifted to Exeter in 1050, the building of a new cathedral there brought learning to the region on an unprecedented scale. From the time of its foundation, there were at least 55 books in its library, donated by Bishop Leofric. The building of a royal castle not only established Norman control, through the watchful eye of the king's sheriff, it also impressed the king's authority on the city. All the major Norman lords with lands in Devon had houses in Exeter, and in 1087 it attracted a new Benedictine monastery. The city's market grew accordingly to serve the expanding population, and this in turn encouraged the clearing of woods and moorland in order to grow more produce. Those leaving Moreton no longer had to fear the Vikings, and the prosperity of Exeter meant they had many reasons to make the 13-mile journey more regularly. Doing so, with newly minted silver pennies in their purses, would have made them aware that they were no longer eking out a living at the very fringe of Christendom; they were part of a much larger whole.

The principal agent of change

Major changes in society are rarely the work of one mind, and never one hand. Most of the great developments of the past were not the product of a single genius but of a great number of people who thought along the same lines and saw similar opportunities. It is therefore almost impossible to align social change with individual decision-making. Like the nature of change itself, which is easy to define on a small scale but impossible when there are so many factors, it is difficult to identify the true impact of an individual on a continent over a whole century. Nevertheless, it is a salutary exercise to consider individual contributions, if only to see how limited they were and how change was due to many thousands of decision-makers.

In 1978, a popular American writer, Michael Hart, selected the hundred people whom he considered were the most influential of all time.[11] It was a fairly arbitrary list and was sorely lacking in intellectual rigour (as indicated by his inclusion in the second edition of the earl of Oxford as the author of Shakespeare's plays). However, as a boy, I found it quite stimulating, and that was surely the author's intention.

His list contained two individuals from the eleventh century: William the Conqueror and Pope Urban II. While William's decision to invade England in 1066 undoubtedly marks him out as the principal agent of change for anyone living in this country, his actions were of far less consequence elsewhere in Europe. We must also remember that he left most Anglo-Saxon institutions intact: life did not change nearly as much as people generally suppose. As for Pope Urban II, he might have triggered the crusading movement and encouraged the Reconquista, but in Europe, the Crusades were more important for their symbolism than their achievements. And in Spain, the kings of Navarre and León hardly needed any encouragement to fight the crumbling caliphate of Córdoba. Both William I and Urban II were certainly significant characters, but if we are to consider all of eleventh-century Europe, they are both dwarfed by a figure whom Michael Hart ignored – Hildebrand, otherwise known as Pope Gregory VII.

Even before he was pope, as archdeacon of the Roman Church Hildebrand played a major role in imposing papal primacy over the Holy Roman Emperor. He was the proponent of the Gregorian Reforms that came to define Catholic priesthood. His vision of the clergy as a body separate from the secular world, and his drive to impose the authority of the papacy over rulers and subjects, changed Christendom. Can you imagine a medieval Europe in which the pope was a mere imperial appointee and the Church was without political influence? That Urban II received such an overwhelming response when he preached at Clermont in 1095 may well have been down to his oratory and religious zeal, as well as the attraction of the opportunities for conquest he outlined, but he had to thank Gregory VII for giving him such a platform. It was Gregory who had first mooted the idea of an armed expedition to help the Eastern Christians, in 1074. Urban II must therefore take second place to Gregory VII. Towards the end of his pontificate, Gregory was ousted from Rome by the emperor, dying in exile a year later, in 1085, but this does not diminish his achievement. Not all great lives end well, and certainly the manner of a man's death should not make us discount his successes in life. Gregory turned the papacy into the single most important voice in Christendom and raised the standing of clergymen above that of those who fought and those who worked, thereby empowering learning and debate, without which European society could not have developed as it did.

The Twelfth Century

On Christmas Eve 1144, the Crusader state of Edessa fell to the Muslim commander Zengi. All the Christian knights who were captured were slaughtered; their wives and children were rounded up and sold as slaves. It was an event that traumatised Christendom. A stunned Pope Eugenius III commissioned his old friend and mentor, Bernard of Clairvaux, to preach a new Crusade to win back God's patrimony. Bernard had started life as a Cistercian monk but subsequently proved himself a diplomat of the first order. On 31 March 1146, he read out Eugenius's papal bull in the church at Vézélay and started to speak to the assembled throng in his own inimitable manner. Soon men were crying, 'Crosses! Give us crosses!' as they swore to fight for Christ. The French king, who was in the congregation, undertook to go to the Holy Land himself. Inspired by his example and Bernard's rhetoric, many of his nobles did likewise. Over subsequent weeks, as Bernard made his journey into Germany to preach to the Holy Roman Emperor, people reported miracles everywhere he went. The fervour grew. Bernard himself wrote to the pope: 'You ordered; I obeyed . . . I spoke and at once the crusaders have multiplied to infinity. Villages and towns are now deserted. You will scarcely find one man for every seven women. Everywhere you will see widows whose husbands are still alive.' Finally, at Speyer, Bernard called on all his skills to persuade the reluctant emperor to join the Crusade. After two days' trying, he raised his arms and held out his hands as if he himself was Christ on the cross, and cried aloud before the court: 'Man, what ought I to have done for you that I have not done?' The stunned emperor bowed and swore to fight to recapture Jerusalem.

The twelfth century offers us a whole host of dramatic moments like this, as well as a glorious array of extraordinary characters. It was the century of the lovers Peter Abelard and Héloise, of the

composer-abbess Hildegard von Bingen, and of the greatest knight of
the Middle Ages, William Marshal. It witnessed such colourful characters as Frederick Barbarossa, Henry II and Thomas Becket. It saw
queens come to the fore: the Empress Matilda, Eleanor of Aquitaine
and Tamar of Georgia. It also saw a whole string of rulers with leonine
nicknames – William the Lion, Henry the Lion and Richard the
Lionheart – as well as kings with more unusual epithets, such as David
the Builder, Umberto the Blessed and Louis the Fat. The names of
the military orders, especially the Knights Templar and Hospitaller,
still have resonance. It was the first great age of chivalry, the century
that invented heraldry and tournament fighting. At the same time it
was robust and earthy in its culture too, giving us the great Latin
poets the Archpoet and Hugh Primas, as well as the troubadours, who
composed their moving poems to delight and seduce their ladies (or,
more frequently, other men's ladies).

It is striking how many stories and phrases from the period have
remained current in our culture. Most famous, perhaps, is Henry II's
declaration, 'Will no one rid me of this turbulent priest?' when he
had had enough of his chancellor, Thomas Becket, archbishop of
Canterbury. Then there is the immortal line uttered by the Master of the
Templars to the Master of the Hospitallers at the Springs of Cressen
in 1187, when the latter suggested that it might be foolish for 600
knights to attack the 14,000 troops of Saladin's army arrayed before
them: 'You love your blond head too well to want to lose it.' And
who can forget the bravado of William the Lion, King of Scotland,
as he launched himself into a completely hopeless attack on the English
at the Battle of Alnwick shouting: 'Now we'll see who among us are
good knights.' In the midst of so much bloodshed, you can understand
why the twelfth-century chronicler Roger of Hoveden noted that a
man 'is not fit for battle who has never seen his own blood flow, who
has not heard his teeth crunch under the blow of an opponent, or
felt the full weight of his adversary upon him'.[1]

These characters and stories give us an impression of the age:
bloody, brave, confident, wilful and passionate. And yet they have little
to do with the most profound changes of the period. It was humble
peasants, lawyers and scholars who had the most significant impact
on the twelfth century. You could argue that the Crusades brought
West and East into contact with one another, to the cultural enrichment of the West. That would be true to a certain extent, but the

East–West relationship was far more productively exploited in cities where Christian scholars could work on Arabic and Greek manuscripts in relative peace. And while the Crusader states of Antioch, Edessa, Tripoli and Jerusalem may have led the way in castle design that had far-reaching influence across the whole of Europe, they did little to change the castle's basic function, which was to permit garrisons to withstand sieges. The deeper changes in society were to be found elsewhere.

Population growth

The period from about 1050 saw significant economic growth in Europe. Huge areas of forest and moorland were cleared and much marshland was drained, adding considerably to the area of land under cultivation. A bird's-eye view of the continent would have seen it alter from a predominantly forested area to one dominated by fields. The clearances were the result of a marked increase in the population, the causes of which are still debated by historians. One possible reason was the gradual introduction of a harness for horses to draw ploughs. Unlike oxen, which can pull a great weight from a simple yoke, horses cannot be yoked together as the fastenings bite their necks and cut off the arterial circulation. Thus they require a much more protective harness in order to till the land. This technology, which seems not to have been used in the ancient world, started to be applied systematically in the twelfth century. It spread very slowly, however: even in the fifteenth century, about two thirds of draught animals in use in England were oxen.[2] Nevertheless, the use of horses as well as oxen in some places can only have added to the tractive energy available for clearing and ploughing the land.

A more important cause of population growth was what historians call the Medieval Warm Period. The average temperature rose very slowly in the tenth and eleventh centuries, and by the twelfth it was almost a degree warmer than it had been before 900. This may not sound like a huge difference: we barely notice if the temperature changes by one degree. But as an annual average, it has an enormous impact. As the historian Geoffrey Parker has pointed out, in temperate zones 'a fall of 0.5°C in the average spring temperature prolongs the risk of the *last* frost by ten days, while a similar fall in the average autumn temperature advances the risk of the *first* frost by the same amount. Either event suffices to kill the entire crop.'[3] It follows that the reverse, an increase of just 0.5°C, reduces these risks. Furthermore, the danger alters according to the altitude of the land. According to Parker, a drop of 0.5°C doubles the risk of a single harvest failure at low level, and increases the risk of two consecutive harvest failures at that level sixfold, but it increases the risk of consecutive harvest failures above 1,000 feet one hundred times. An increase of 0.5°C thus meant the difference between life and death for many people. Fewer severe winter days would have meant that fewer crops were lost to frost. Slightly warmer summers would have meant that there was a reduced risk of harvest failure and, over time, a greater yield from the grain planted. Consequently there was more food on average, and fewer children died.

A modest decrease in child mortality doesn't sound as if it should be a contender for one of the most significant changes in Western history, but when that phenomenon is extrapolated across Europe for the whole two and a half centuries of the Medieval Warm Period, it becomes hugely important. The surviving children had families of their own, and many of their children survived too; and they in turn cleared more land and harvested enough food to sustain a larger population still in the next generation. Without grain surpluses, there could have been no cultural expansion. There would have been no spare labourers to build the monasteries, castles and cathedrals, and scholars would have had to work in the fields rather than read books. Those initial few extra lives had an exponential effect for the simple reason that Europe was abundant in potentially fertile land. It just needed the manpower to bring it into cultivation.

The clearances of the natural landscape each began in one of two ways – individually, as a result of a peasant's initiative, or collectively, at the instigation of a manorial bailiff. In the individual cases, a man managing a smallholding of five or six acres might have realised that he would not be able to feed his growing family with so little land. Even in a good year it would not have left him with a surplus to sell at market or to store for security in case of a bad harvest in the future. Having identified one or two acres of overgrown or wooded land nearby, he would have arranged with the manorial bailiff to cut down the trees and plant the land with crops, and to hold it thenceforth in return for an additional rent. Such a development pleased everyone:

the peasant had more land to work and greater security for his family, and the manorial bailiff and his lord were happy with the extra rent. When the peasant's sons grew up, they could help to clear another four or five acres. And so on.

Collective clearance tended to be concerned with large-scale drainage and irrigation projects. The bailiff would employ the tenants of the manor to spend a specified number of days digging ditches and building dykes. When the work was done, the new land was apportioned amongst existing or new tenants. Some manors belonging to a monastic order might even be cleared by the monks themselves, labouring in the true spirit of the Rule of St Benedict. Thousands of acres of European forest were cut down and marshland drained by the Cistercians over the course of the twelfth century.

It is difficult to measure the extent of these clearances. Literacy was still so rare that lords and their clerks did not record the borders of manors on a regular basis, still less the specific strips of lands held by their tenants. Several manorial charters allowing 'assarting' – clearing the land for farming – survive, but these individual grants hardly establish the full extent of the process. Our best measure, therefore, is population growth itself. But this too is difficult to quantify. The most complete figures we have for the period are those for England, due to the unique survival of Domesday Book (1086), the only comprehensive survey of a kingdom and its wealth in the eleventh century. Estimates based on Domesday reveal the population of England to have been in the region of 1.7 million. Poll tax records from 1377 show that the population had by then grown to about 2.5 million, and it would have been much higher before the famines of 1315–22 and the Black Death of 1348–9. From these and other pieces of data, we can estimate that the population rose from about 1.8 million in 1100 to almost 3.4 million in 1200. The implication is that England's farmland in 1200 was nearly twice as productive as it had been at the start of the century. The only explanation for a population increase on this scale is that huge swathes of the kingdom had been brought under the plough for the first time. More land resulted in more food, which meant that more people felt secure enough to marry and raise a family, and their children were better fed. And every generation in turn brought more land into cultivation, leading to further population expansion.

How did the rest of Europe fare in the twelfth century? As shown in the Appendix (page 347), historians have come up with contrasting figures. One recent set of estimates by Paolo Malanima suggests that Europe's population as a whole grew by 38 per cent in the twelfth century. However, if we take the three best-documented countries – England, France and Italy – and construct a model for the core of Europe based on their population figures, by projecting back from the well-established figure of 84 million for 1500, we get a very different picture for the twelfth and thirteenth centuries, which suggests that the population rose by 49 and 48 per cent respectively to a total of over 100 million by 1300. Whatever the exact figures, there can be little doubt that the period from 1050 to 1250 saw the bulk of the clearances that enabled such growth. The popular image of the age might be that of chain-mail-clad Crusaders dealing out crunching mace blows on the helms of their adversaries, but the real powerhouse of social change in the twelfth century was their estates, wrought by hard-working peasants whose names are unknown to us and whose only memorial lies in the newly tilled fields they left behind.

The expansion of the monastic network

The very fact that Pope Eugenius III called upon Bernard of Clairvaux to preach the Second Crusade hints at another major twelfth-century development. Bernard was a monk and thus was supposed to have withdrawn from the world to live a secluded life of contemplation. Yet here we see him travelling far and wide, meeting kings and preaching to huge crowds. Moreover, wherever he went, his reputation preceded him. When a controversy over the election of the pope broke out in 1130, Bernard was asked to decide which candidate should be nominated. He chose Innocent II and then travelled around Europe for several years trying to persuade the rulers who had supported the other candidate to switch sides. In 1145, the principal reason for Eugenius being elected pope was that he was a friend of Bernard. Bernard's influence and reputation also gave a huge impetus to his religious order. Thousands of people now flocked to join the Cistercian order, which had been founded in 1098 and whose monks vowed to lead an austere life strictly governed by the Rule of St Benedict. By 1152, the order had more than 330 monasteries spread across Europe, and in the second half of the century it expanded further, into eastern

Europe, Scotland and Ireland. By the end of the century it had also added several dozen nunneries.

The Cistercians were not the only monastic order on the rise. The Carthusians opted for an even more austere existence, living in cells around the cloisters of their charterhouses. There were also several orders of religious clerics, such as the Order of Canons Regular (the Augustinians), whose lifestyle closely resembled that of monks. William of Champeaux established the Order of the Canons of St Victor (the Victorines) in 1108; Bernard's friend Norbert of Xanten founded the Premonstratensian order in 1120; and Gilbert of Sempringham set up the Gilbertines in 1148. When combined with crusading zeal, the monastic spirit led to a similar growth in the military orders, whose duties included prayer and the protection of pilgrims. The Knights Hospitaller developed in the wake of the success of the First Crusade. The Order of Knights Templar, founded in 1118, was enthusiastically supported by Bernard of Clairvaux. In Castile in the 1150s, the Cistercians set up their own military wing, the Order of Calatrava, and the Teutonic Knights were established towards the end of the century. These were just the most notable orders; many more flourished to protect pilgrims to the Holy Land.

The scale of monasticism's expansion becomes clear if we look at the figures for England and Wales. In 1100, there had been fewer than 148 religious houses, including some 15 nunneries. Over the course of just two decades, 1135–54, the number rose from about 193 to 307: an increase of 6 every year. In 1216, there were about 700 religious houses, with another 60–70 belonging to the Hospitallers and Templars.[4] The number of monks, canons and nuns increased to an even greater extent, from about 2,000 to about 12,000. If we were to extrapolate totals for the whole of Europe from these figures, there would have been 8,000–10,000 religious houses in Western Christendom at the end of the twelfth century, and about 200,000 monks, nuns and canons. However, when we take into account that England and Wales were relatively sparsely populated fringes of Christendom at this time, it becomes apparent that the actual numbers of religious men, women and houses in 1200 were much higher.

Why did this happen? What made people give away huge amounts of their wealth to support these new religious houses? To understand their motivation, we need to examine the rise of the doctrine of Purgatory – the Roman Catholic belief that the souls of the dead do not go straight to Heaven or Hell but rather to a spiritual holding bay, where they remain for a while before being sent in one direction or the other. Before this doctrine developed, those lords and ladies who founded monasteries did so in the hope that, as a result of their good works, their souls might ascend straight to Heaven when they died. If they didn't, they would be spending the rest of eternity in Hell. Around the middle of the twelfth century, however, the question of whether your soul went to Heaven or Hell became more nuanced. At what point exactly was the soul condemned to Hell? Did this occur at the moment of death, or might prayers for the dead still assist the passage to Heaven? Theologians elaborated on the ancient idea of redemption through prayer, and conveniently decided that prayers after death could indeed help the deceased. In the 1150s, Peter Lombard declared that prayer could help both the moderately wicked, by lessening their suffering, and the reasonably good, by aiding their path to Paradise.[5] People started to believe that such souls did not go straight to Heaven or Hell. By 1200, an elaborate doctrine of Purgatory had been established, and more and more people donated their wealth to monasteries and chantry chapels, hoping that prayers said for them after death might hasten their passage to eternal bliss.

You might be tempted to think that, as all these new monks and nuns were shut away in cloistered communities, they hardly affected what was going on outside – so how could they mark a major development in the history of the Western world? We need to consider the twelfth century in terms of connectedness, however. Today, in our Web-wondrous world of intercommunication, we believe that our methods of obtaining information and passing on our views are completely different from those of our forebears. Networks of information storage and transfer now exist that previous generations could not even have imagined. Monasticism, however, provided a similar connectedness. It was a layer of Christian interconnectivity – a monastic web – interwoven with the secular world of parish priests, court clerks and political bishops. From Iceland to Portugal, from Poland to Jerusalem, monks, canons and priests were crossing the boundaries of kingdoms, spreading knowledge and taking part in wider debates. Capitalising on the Latin orthodoxy introduced by Pope Gregory VII in the previous century, they did so in a common language that rendered itself internationally useful in the same way that the standard mark-up languages underpin the Internet today.

This Christendom-wide monastic web did not just spread knowledge; it was able to generate it too. Just think of all the roles that a monastery fulfilled. Building work required master masons, carvers and carpenters, so the monastic orders were major patrons of twelfth-century design and architecture, structural engineering and art. Monasteries required monks and canons to be able to read, so they spread literacy; several founded schools for the education of boys (and occasionally girls) outside the cloister, either as part of their good works or in order to raise money. In their libraries they preserved the works of earlier writers and produced new books, thus storing as well as disseminating information. The monastery of Bec in Normandy, for example, had a library of 164 books in the early twelfth century and acquired 113 more as the result of a bequest in 1164. It also opened a school for fee-paying students. In describing it, the chronicler Ordericus wrote that 'almost every monk of Bec seemed a philosopher, and even the least learned there had something to teach the frothy grammarians'.[6] When a monastery formed part of a cathedral establishment, as was common, the monks dealt with correspondence from the royal administration, which facilitated the creation of archives and the writing of chronicles. The monks themselves travelled, carrying news between monasteries across Europe. In their gardens they cultivated medicinal plants, and those monasteries that had an infirmary performed a moderate degree of physic. Some orders circulated their technologies across the continent, spreading the use of the water wheel, the heavy plough and better viticulture, and thus assisting with the exploitation of the newly cleared land.

Not every monastery in Europe had a library full of wonderful texts, of course, and not every monastery had a school. But many had both. 'A monastery without a library is like a castle without an armoury' ran a saying at the time.[7] Monasteries opened eyes, instructed minds, and encouraged the young who attended their schools to seek further knowledge – not just in their own monastic libraries but also further afield.

The intellectual renaissance

If you were to play a game of word association with a group of medieval historians, the response to the words 'twelfth century' would undoubtedly be 'renaissance'. This does not relate to the Italian Renaissance of the mid-fourteenth to sixteenth centuries but an earlier phenomenon, identified in 1927 by the American medievalist Charles Homer Haskins. He demonstrated that the twelfth century saw an unprecedented revival in scholarship. Two strands are particularly important in our survey: first, the dialectical method, arising from the pioneering thinking of Peter Abelard and the rediscovery of the works of Aristotle; and second, the wealth of translations from the Arabic that allowed the recovery of so much knowledge from the ancient world.

Peter Abelard was the eldest son of a Breton knight who encouraged his sons to learn to read before they could wield a sword. Inspired by the few Aristotelian texts that had survived in a sixth-century translation by Boethius, Abelard advanced quickly in the study of logic. Soon he 'wielded no other weapon but words'. But this did not make him a pacifist: Abelard's words were sharper and more dangerous than most men's swords. He studied under William of Champeaux at the school of St Victor in Paris, but was soon defeating his master in debate. His fame as a scholar quickly spread and by 1115 he was lecturing at the cathedral school of Notre-Dame, where hundreds flocked to hear him speak. He was the academic sensation of the age.

It was then, at the height of his fame, that he fell in love with Héloise, the niece of Fulbert, a fellow canon of the cathedral. He seduced her and she fell pregnant. Fulbert did not take the news well: he had Abelard barbarously castrated. Humiliated, Abelard took refuge in the Abbey of Saint-Denis to the north of Paris. Here, when he was not antagonising his fellow monks, he wrote his first theological work, on the Holy Trinity. Unfortunately, it resulted in him being charged with heresy by a provincial synod at Soissons in 1121. When he was found guilty and forced to burn his book, he decided to become a hermit. He founded an oratory, the Paraclete, and withdrew from the world. But the world did not withdraw from him. Soon students set up tents around the Paraclete. Twenty years after his first condemnation for heresy, when he was in his early sixties, Abelard came up hard against Bernard of Clairvaux, who wanted to stamp on his dangerous teaching. At Abelard's instigation, in the hope of clearing his name, a debate was arranged at Sens between the two great speakers. But the night before the debate, Bernard privately addressed the bishops who formed part of the council assembled to judge proceedings. After that, Abelard refused even to speak in his own defence. He was found guilty of heresy again and died the following year, under the protection of the Abbot of Cluny.

The reason why Abelard infuriated so many churchmen was not simply because of his combative nature and his seduction of a canon's niece. Nor was it because of his application of the logic of Aristotle. It was because of his own advances in logic and dialectic, and because he used these forms of reasoning to investigate matters of faith. At the time, the consensus was that reasoning was fine – unless it was applied to religion. Abelard tackled this prejudice fearlessly. In his book *Sic et non* (*Yes and no*), he examined 158 apparent contradictions in the writings of the Church Fathers, looking at each one from two conflicting points of view and drawing out many radical points for debate. For example, his very first principle in *Sic et non* is that 'faith is built up by reason, and that it is not'. In considering whether logic supports faith or detracts from it, he directly challenged the Biblical dictum that 'without faith, there is no understanding'. To us Abelard's approach seems straightforward: we tend to believe that what we think is reasonable; conversely, we cast aspersions on those who say something is reasonable purely on the strength of their belief. But until Abelard's time, faith *itself* was the way to understanding. It was Abelard who laid down the dictum that 'doubt leads to enquiry and enquiry leads to truth'. And he gave his application of logic to religion a name – 'theology'.[8]

Sic et non shows just how fearless Abelard was, how far beyond the brink of orthodoxy he was prepared to push his theology. Using his dialectical technique – looking at a question from two opposing points of view in order to identify and resolve the contradictions between them and thereby to answer the initial question more exactly – he postulated some ideas that, for the time, were simply dangerous. For example, when he put forward the proposition that 'God can know everything', he implied that it was possible that God did *not* know everything. He similarly proposed 'that all things are possible to God, and that they are not'. Suggesting in the twelfth century that God might not be omnipotent was scandalous. There was even the suggestion in *Sic et non* 'that God might be the cause or author of evil things, or not'. Typically, Abelard did not come down decisively on the side of divine infallibility as Bernard of Clairvaux would have done; he left the matter open for people to draw their own conclusions. Indeed, he argued that *all* views, even those of the revered Church Fathers, were merely opinions, and thus fallible. This was taking rationalism too far for many of his contemporaries, for whom questioning the holy writers was touching on heresy. But Abelard did not stop there. Whereas traditionalists glossed over the problem of whether God was a divisible or indivisible trinity of beings with talk of a mystical union, he poured scorn on them. It was ridiculous to propose that God the Father might be the same being as the Son, he maintained, for how could anything give birth to itself? At a time when most commentators wanted to reconcile opposing views among the Church Fathers who had shaped medieval theology, Abelard was determined to exploit their differences.

When it came to ethics, Abelard advocated similarly dangerous ways of thinking. He argued that intention was all-important in determining culpability. In short, if you committed a wrongdoing accidentally, then you were less guilty than someone who consciously committed that act. Your (minor) guilt lay in your negligence, not in your criminal intention. Indeed, in certain circumstances, intention could be the *only* factor determining guilt or innocence. If a brother and sister who had been separated at birth and did not know of one another met again later in life, married and had a child, then although they were clearly guilty of incest, they should not be punished for they were completely unwitting of their crime. The trouble was that the principle underlying this point implied that lords, bishops and judges could not punish all crimes in the same way without themselves committing an injustice. Not only did Abelard challenge the moral codes promulgated by the Church indirectly, he did so directly too. For example, he reasoned that the pleasure of sexual intercourse was the same within marriage as it was outside. Therefore, if that pleasure was sinful outside marriage (as the Church taught), then it was sinful inside marriage too, as the act of marrying did not wipe away that sin. But as copulation within marriage was essential for the human race, surely God would not have made its survival dependent on sin? Therefore the sinfulness of extramarital sex was clearly open to question. Even more controversially, he argued that those who had crucified Christ were not sinful because they had had no way of knowing about Christ's divinity and were only acting in accordance with what they believed was right. You can see why he got into trouble.

Abelard was not the only one seeking new truths. Across southern

Europe, scholars were becoming aware that a treasure trove of know-ledge from the ancient world had not perished with the Roman Empire, as they had previously believed, but remained locked up in the Arabic libraries of Spain and North Africa. Slowly the Reconquista was retaking territory from the Muslims and winning back access to the literature and knowledge of the distant past. Toledo fell to the Christians in 1085 and Saragossa in 1118. Soon a small army of trans-lators from all over Europe, working in the cities of Spain and southern France, was seeking the truths hidden in Arabic literature – with all the fire of a bunch of grave robbers ransacking treasure-laden tombs. There were Adelard of Bath, Robert of Ketton and Robert of Chester from England; Gerard of Cremona and Plato of Tivoli from Italy; Hermann of Carinthia from Austria and Rudolf of Bruges from the Low Countries; as well as many Spanish Jews who facilitated the work. Encouraged by Raymond, bishop of Toledo, and Michael, bishop of Tarazona, they translated whole libraries of philosophical, astro-nomical, geographical, medical and mathematical works. As we have already seen, once translated into Latin, these texts could be copied and read by scholars throughout the West. Along with the thought of the ancient world, they also gave Christendom the works of the great Islamic mathematicians. In 1126, Adelard of Bath translated al-Khwarizmi's *Zij al-Sindhind*, which introduced Arabic numerals, the decimal point and trigonometry to the West. In 1145, Robert of Chester translated the same author's *Kitab al-Jabr wa-l-Muqabala*, rendering it in Latin as *Liber algebrae et almucabola*, thereby introducing the term 'algebra' and the means of solving quadratic equations. Foremost of all the translators was Gerard of Cremona, who by the time of his death in 1187 had rendered at least 71 ancient texts into Latin. These included Ptolemy's *Almagest*, Euclid's *Elements*, Theodosius's *Spherics* and many philosophical and medical works by Aristotle, Avicenna, Galen and Hippocrates.[9]

Alongside the Spanish and southern French cities where these trans-lators were hard at work, two other centres made long-lost texts available to European scholars. In Constantinople, many of the ancient works had survived in their original Greek forms. It was there, in 1136, that James of Venice translated the *Posterior analytics* of Aristotle; the 'new logic' – so named in order to distinguish it from the 'old logic' that Boethius had translated centuries before. In the Norman kingdom of Sicily, Greek scrolls were found from the days when the Byzantine

Empire had controlled the region. Arabic books were found there too, from the time when Sicily had been under Muslim control. To please the intellectual kings of Sicily, Roger II and his son William I, court translators in Palermo produced Latin versions of Plato's *Meno* and *Phaedo*, Aristotle's *Meteorology*, various works by Euclid, and Ptolemy's *Optics* and *Almagest*. They also translated Mohammed Al-Idrisi's great geographical compendium, which included a world map stretching from Iceland to Asia and North Africa.

Did all these discoveries really mark a change for Christendom as a whole? How did the intellectual advances of the twelfth century affect the proverbial peasant in central France? Not directly, perhaps, and certainly not as much as his ability to clear a few more acres of land and feed a larger family. However, to look for every change having a direct and immediate impact on the whole population would be both unrealistic and simplistic. It would be like asking whether Einstein's Special Theory of Relativity affected contemporary factory workers: it may not have had an impact in 1905, the year it was published, but it certainly shook the world in 1945, when its explosive implications brought the Second World War to an end. In the case of the intellectual renaissance of the twelfth century, Aristotle's new logic slowly filtered through to affect the whole of society. It led to a new approach to knowledge. It taught people who had hitherto compiled larger and larger encyclopedias that knowledge was not just a matter of accumulating more and more facts: the quality of those facts was equally important. Writers like John of Salisbury, who attended Abelard's lectures in 1136 and eventually became bishop of Chartres, was just one of many intellectuals of the age profoundly affected by the new reasoning. To paraphrase one of his most famous observations: it did not matter which three pilgrimage sites claimed to have the relic of the head of John the Baptist; the important thing was which church possessed the *true* head. You only have to remember that the figures we use for numbers today are Arabic to realise how much we owe to Muslim mathematicians whose works were translated in the twelfth century. Have you ever tried multiplying figures or dividing fractions in Roman numerals? Have you ever considered how you might multiply the number π (3.1415926536 . . .) using Roman numerals? Even more significantly, prior to the translation of treatises from the Arabic, there was no concept of zero. But zero is an enormous round hole out of which tumbles an endless amount of later mathematical thinking. The quest for new knowledge might have gone way over the head of the peasant in the field and only slowly seeped through to the man in the street, but without it the future of Europe would have been very different.

Medicine

One branch of twelfth-century scholarship that did have a direct effect on people was medicine. Of course, medicine itself was not new. There had been physicians in the ancient world, and medical ideas had been transmitted in numerous forms down the centuries. Anglo-Saxon 'leechbooks' and their continental equivalents were available, and so were herbals. Hrabanus Maurus included a chapter on medicine in his encyclopedia, and the seventh-century writer Isidore of Seville inserted a dozen or so texts by the second-century physician Galen into his compendium of knowledge. But there was no systematically arranged corpus of medical texts. There were very few physicians, and no surgeons as such. Nor was there any formal medical education. Moreover, there was a belief that medical intervention was an attempt to undo God's work. Early medieval Christian writers such as Gregory of Tours had stressed the immorality of medicine, as physicians sought to change God's judgement. He gave examples of men and women being rightly punished for seeking medical help – and others being cured miraculously with holy oil after physicians had failed. That such views persevered into the early twelfth century is evident from Bernard of Clairvaux's statement that 'to consult physicians and take medicines befits not religion and is contrary to purity'.[10]

The apparent harshness of Bernard's view is perhaps more under-standable if we look at some of the medical strategies adopted in the tenth and eleventh centuries and realise that superstition played a major role in medicine. Recipes for medicines, for example, often called for excrement or the body parts of animals to be included, together with incantations and charms. One example, taken from an Anglo-Saxon leechbook, must suffice here:

Against cancer, take goats' gall and honey, mingle together both of equal quantities, and apply to the wound. [Alternatively] burn a fresh hound's head to ashes, and apply them to the wound. If the wound

will not give way to that, take a man's dung, dry it thoroughly, rub it
to dust and apply it. If with this you are not able to cure him, you will
not be able to do so by any means.[11]

In this context, Bernard's statement that medicine is 'contrary to
purity' makes perfect sense.

The contribution of the twelfth century to medicine was that of
systemising knowledge, introducing more scientific approaches,
teaching medical and surgical techniques and, most of all, eradicating
many of the superstitions that had hitherto pervaded the discipline.
It is true that a large dose of astrology remained wrapped up in the
medical corpus, but this too was now systemised and treated as a
science, replacing the incantations and charms that had previously
characterised medical treatments.

There were some home-grown monastic advances in medical
knowledge in the twelfth century. The substantial compilation of
medical recipes by Abbess Hildegard von Bingen is a prime example,
even if it is not as famous as her music. However, most of the new
methods that became available in the West were imports from the
Arab world, including works both by the ancient Greek physicians
Hippocrates and Galen, and by the influential Arabic medical practitioners who had built on their work, namely Avicenna, Rhazes,
Albucasis and Johannitius. Hippocrates, the 'father of medicine', was
a fifth-century BC physician known for a corpus of medical writings;
even today the Hippocratic Oath is sworn, in a modified form, by
most newly qualified doctors. Galen, who lived in the second century
AD, extended Hippocrates' understanding of the theory of the four
humours: black bile, yellow bile, blood and phlegm. Only by keeping
these humours in balance could you maintain good health. About 17
of his minor works were available in Latin in the eleventh century,
but dozens more were translated in the twelfth.[12] Avicenna was an
eleventh-century Islamic scholar who had absorbed Galen and
Hippocrates' works and synthesised an encyclopedia of medical
writing in five volumes, entitled *The Canon of Medicine*. This was
translated by Gerard of Cremona and proved a most enduring educational textbook, providing the basis of medical understanding at the
famous Montpellier medical school until 1650. Rhazes, a Persian physician
who died in the tenth century, was responsible for a huge number of
medical books, including two important medical encyclopedias (*The*

book of medicine dedicated to Mansur and *The comprehensive book of medicine*), works on specific diseases, and a general critique of Galen. Albucasis was the pre-eminent Arabic writer on surgery. Johannitius had translated 129 works by Galen into Arabic, thereby saving them for posterity. In addition, he wrote an influential introduction to the ancient physician's works, which was translated into Latin as the *Isagoge*.

These translations meshed with the development of systematic medical education. By 1100, the city of Salerno in southern Italy already had a reputation for medical teaching, thanks largely to the initiative of its bishop, Alphanus, who was able to draw on the work of Constantine the African, from Tunis, who had translated a large number of important medical treatises in the late eleventh century. In the first half of the twelfth century a syllabus for medical students was assembled in Salerno. Called the *Articella*, it consisted of the *Isagoge*, Hippocrates' *Aphorisms* and his *Prognostics*, Theophilus's *On urines* and Philaretus's *On pulses*; by 1190, Galen's *Tegni* had been added to this list. Those who qualified in Salerno could expect lucrative employment in the royal households of Europe. But lesser lords too could take advantage of the presence of medical men at court: as John of Salisbury declared, physicians 'have only two maxims that they never violate: never mind the poor and never refuse money from the rich'.[13] For the less well off, methods of maintaining good health – regimens – were soon devised and taught at Salerno, and circulated in the form of poems and advisory texts. Thus, whereas in 1100 hardly anyone was sufficiently well versed in medical texts to take responsibility for understanding and treating diseases, by 1200 there was a small but growing cadre of qualified physicians who professed to know the systems that governed the health of the body, and would treat the few who could afford their services.

The skill of the surgeon advanced in tandem with these developments in medicine. At the start of the century, Western surgery amounted to little more than letting blood, cauterising corrupt flesh with a hot iron, draining swellings, binding broken limbs with bandages, dressing wounds and burns with herbs, and amputating gangrenous and cancerous limbs. In moving beyond this, the Christian encounter with Jews and Muslims in the Holy Land was important, as soldiers and pilgrims were ministered to by physicians of all religions and nationalities. One famous account tells of a Syrian physician who treated a knight with an abscess in his leg and a woman troubled with a fever. To the knight's leg he applied a small cataplasm, and the wound started to heal. The woman was similarly recovering through his ministrations (which were largely dietary) when a Christian physician took over their care, refusing to accept that a Syrian could possibly help them. This physician now asked the knight whether he would rather live with one leg or die with two. On being given the obvious answer, he called for an axe and had the man's leg chopped off. It took a couple of blows, 'which caused the marrow to flow from the bone, and the patient died immediately'. As for the woman, after she too was treated by the Christian physician, she returned to her old diet and fell ill again, with a rising fever. When he saw her condition had deteriorated, the physician cut a cross in her scalp through to the bone and rubbed in salt. Unsurprisingly, she too died soon afterwards.[14] Whatever the truth of these cases, it is clear that in the Holy Land, where Islamic and Christian physicians and surgeons attended the same patients, the more systematic approach of the Arabs made an impression.

Surgery also made advances through education. The translations of Rhazes, Albucasis and Avicenna led to the collection by 1170 of a corpus of chirurgical, or surgical, knowledge. The works of Albucasis in particular showed the instruments necessary for the performance of operations, so that surgeons could do more than merely dress wounds and amputate limbs. By this time surgery was being taught as a distinct art not only at Salerno but also in northern Italy, most notably by Roger Frugardi. In his *Practica chirurgiae*, edited by his pupil Guido Aretino around 1180, surgery was presented for the first time by a Western writer as a systematic science.

Pharmacy, the third of the medical sciences, also came to be studied in southern Europe in the twelfth century. Some medical recipes had been prescribed since ancient times: autumn crocus, for example, had been known for centuries to give relief from gout. But the herbals of the previous centuries presented relatively few effective remedies. Once again, it was the Arab polymaths who changed this. Gerard of Cremona's translation of Avicenna's *Canon of Medicine* proved particularly valuable in laying out a theory of pharmacy, employing not only herbs but also minerals. Today we still use Arabic words such as 'alcohol', 'alkali', 'alchemy' and 'elixir', which indicate the depth of Arabic influence on our scientific common knowledge. By 1200, Latin writers were compiling their own pharmacopoeias. The *Antidotarium Nicolai* was put together about this time, probably at Salerno. This also showed a medicinal response to ailments that was stripped of superstition, incantations and charms. And because much of this knowledge had been translated from Arabic, like many of the other medical texts it was relatively free of Christian dogma too.

As we will see throughout this book, the medicalisation of Western society took place in numerous stages. In 1100, there were very few practitioners of medicine in Europe, but by 1200, there were hundreds who had the expertise to relieve sickness or injuries – or were believed to have such knowledge. Of course, only a tiny fraction of the population would have been able to afford their services; nevertheless, the twelfth century marks the start of the process by which men and women came to trust their fellow human beings rather than God with their physical salvation, and systematically employed medical strategies to cope with sickness rather than relying on prayer or magic. Overall, this has to be counted as one of the most profound changes under consideration in this book.

The rule of law

In the previous century, the law had been a patchwork across Europe. Some Italian city states preserved corrupted versions of the old civil law of the Roman Empire, otherwise known as Roman law. Others had adopted Lombard feudal law. In the north, old Frankish and Germanic tribal laws were maintained. There was no such thing as statute law, still less international law. Law codes were also subject to considerable local variation: so few people could read that the laws were not distributed in written form, and judges therefore relied on memory – and their own opinions. When England was conquered in 1066–71, the law of the land changed only in part to reflect the feudal Norman law, while some of the old 'dooms' (laws) and customs of the Anglo-Saxons were retained. In some places a jury would be empanelled to hear a case; in others the accused might have to undergo an ordeal – trial by water (being bound and thrown into a river), trial by fire (being forced to carry a red-hot iron) or trial by battle (a straightforward combat). Trial by battle was also sometimes used to decide between two opposing parties in a legal case, such as a dispute

over the ownership of land. Even the law governing the Church –
canon law – existed in regional variations, as local bishops drew up
compilations of those restrictions that they felt should apply in their
dioceses. Given this messy situation, it should not surprise us that the
law was not properly studied: there was no jurisprudence.

Things began to change at the end of the eleventh century in
northern Italy. Long-distance trade between city states required a form
of law that was robust and standardised. By 1076, a copy of the *Digest*,
a compendium of the best jurisprudence from the old Roman Empire,
had resurfaced in Bologna. It was only part of a larger body of legal
writing on Roman law, the *Corpus Juris Civilis*, which had been drawn
together under the command of Justinian, the sixth-century Byzantine
emperor. But before long, the whole *Corpus* had been identified and
was being studied. In the early twelfth century, Irnerius, a brilliant
exponent of the law at Bologna, explained to his students the meaning
of each part of the *Corpus* through a series of glosses and commentaries. In addition, he wrote a series of *Questions on the subtleties of the
law*, and thereby encouraged debate about the apparent contradictions
in the legal authorities. In the next generation, a group of men known
to historians as the Glossators carried on his work, constantly revising
the *Digest* to give it relevance to twelfth-century society. Bologna's law
school consequently grew in fame and importance, and in 1155, an
imperial decree placed it under the emperor's protection. Irnerius's
legacy was not only the establishment of Bologna as the major centre
for the study of the law but also the revival of jurisprudence
throughout the Mediterranean and the southern part of the Holy
Roman Empire. By the end of the century, Roman law was starting
to become the international legal language of continental Europe.

About the year 1140, a monk called Gratian, who had almost
certainly lectured at Bologna, compiled a definitive collection of canons
– laws on subjects that fell under the authority of the Church. The
result, formally entitled *A concord of discordant canons* but more popu-
larly called the *Decretum*, was an attempt to reconcile the differences
in major canon collections circulating in Europe. In doing so, he
adopted Peter Abelard's dialectical method, first displayed in *Sic et non*,
listing the differences and providing the arguments for and against
each interpretation. Soon the *Decretum* was widely accepted as the
basic authority on canon law, and it began to be extended by the
decretals or religious laws of the popes, especially those of Pope

Alexander III, who held the apostolic see from 1159 to 1181. The importance of this for the whole of Christendom cannot be overestimated: one set of laws now governed the whole Church and every person in it. For canon law was not only concerned with the rules of the clergy (such as simony and clerical marriage) and crimes committed by clergymen; it also regulated the moral behaviour of every Christian – his or her sexual conduct; commercial relations and work; fraud, bribery and forgery; baptism, marriage and burial; holy days, the writing of wills and the swearing of oaths. No one was outside the scope of canon law.

In addition to Roman law and canon law, the twelfth century saw the introduction of legislation. In this the papacy again provided the inspiration. If, acting like a head of state, the pope issued decretals that laid down the law for all his 'subjects', then kings could do likewise. The emperor Frederick Barbarossa issued statutes for the maintenance of the peace from 1152, and the kings of France and the counts of Flanders were similarly promulgating statutes by the end of the century. One particular result was that the death penalty became more common. The message was simple: obey the law or die.

England took a different path. Vacarius, an expert in Roman law, was brought to the country by the archbishop of Canterbury in 1143, but King Stephen felt that this new sort of law was a threat to the royal prerogative and silenced him. Thus the old Anglo-Saxon dooms and customs were maintained, supplemented by Norman law. But the law in England did not stand still for long. The next king, Henry II, revolutionised it through administrative acts of the royal council. In 1164, the Constitutions of Clarendon defined the border between the jurisdictions of temporal (that is, secular) and ecclesiastical courts. Two years later, the Assize of Clarendon was issued, requiring the sheriffs in each county to enquire as to who had committed murder, robbery and theft since the king had come to the throne in 1154, and who might be harbouring the criminals who had carried out these acts. Regardless of individual lords' privileges and local customs, these perpetrators were to be rounded up, juries were empanelled, and the accused were forced to undergo trial by water. This was the first time that England as a whole had been required to assemble juries to report on crimes. Royal judges toured the country to give judgement on cases in the assize courts. In the assize of Northampton in 1176, Henry II added further crimes for sheriffs to search out and punish:

counterfeiting, forgery and arson. He also set up a system whereby judges would travel on six 'circuits' of the kingdom and try all the felons apprehended by the local authorities, marking the origin of modern circuit judges. Royal coroners were to confiscate the chattels of the guilty on behalf of the Crown. By the end of the century sheriffs presided over courts in every county, as did lords of manors and hundreds. There were also central courts for the administration of justice at Westminster, where private citizens could take legal action against each other. This judicial revolution provided the context for Ranulph de Glanvill's *Treatise on the laws and customs of the kingdom of England* of about 1188, the first work on what would later become the common law of England and, in due course, the foundation law of the United States, Canada, Australia and New Zealand.

Conclusion

The question of change in the twelfth century reveals starkly how relative our historical perceptions are. If you were able to ask a twelfth-century man about the most important event of his time, his answer would probably be the loss of Jerusalem to Saladin in 1187. It was a landmark in the relationship between Christians and God: a crisis of confidence for those who believed God would always favour the Christian cause. But from our point of view, the loss of Jerusalem and the Third Crusade that followed are of limited significance. So too are most of the technological discoveries of the age. Mariners might have started to use the magnetic compass (first described by Alexander Neckham in this century) and the astrolabe, but no great geographical discoveries were made. The use of these instruments remained very restricted.

When reviewing the five changes identified above – population growth, the expansion of the monastic network, the intellectual renaissance, medical advances and the application of the law – there can be little doubt that the first underpinned all the others. It is true that the lives of ordinary men and women working the land in 1200 differed most obviously from those of their forebears in 1100 with regard to the codified nature of the law, and the rigour with which it was applied. Those who think of the mobile phone as one of the most significant changes in human history might like to reflect on whether it is as significant as the application of law and order. Given a choice between living in a lawless society and one with no mobile signal, which would you choose? But I think that even the matter of law and order has to take second place to population growth. On the most basic level, life in 1200 differed from 1100 because people had more land, more surpluses, and more chance that they and their children would be able to eat – and live – for another year.

The principal agent of change

No single individual can be credited with initiating the major changes of the twelfth century. Demographic growth was due to the weather and the spread of agricultural skills, not any one ruler's policy. Henry I and Henry II both had a huge impact on the application of the law in England, but they were of limited significance when we consider all of Europe. Although Irnerius played a major role in the development of legal education and the reintroduction of jurisprudence, he was just the first of many such teachers. One can say much the same for the translators of the Arabic texts that ushered in the intellectual renaissance. Gerard of Cremona stands head and shoulders above the others on account of the importance and number of his translations, but ultimately he was just one of many performing this task; the intellectual developments of the twelfth century would still have taken place without him. Gratian was essentially a compiler: if he had not brought the canons together at that juncture, someone else's collection would probably have been adopted by the Church. And while it is tempting to single out Aristotle as the principal agent of change, it would be somewhat disingenuous. If twelfth-century scholars had not picked up his writings and realised their value, then they would have languished in Arabic libraries and had none of the impact they did.

The foremost candidates for the principal agent of change are the two great rivals of the early part of the twelfth century. Bernard of Clairvaux inspired thousands of people to take up the cross and join the Second Crusade. He led thousands more to join the Cistercian order, influenced the election of popes, and assisted the development of what I have called the 'monastic network'. But ultimately the Second Crusade came to nothing, and many kings refused to accept his endorsement of Pope Innocent II. In his attempt to defeat the rationalism of Peter Abelard, he proved himself to be the very opposite of an agent of change – a determined brake on intellectual and social development. Thus the limelight falls on the brilliant but irascible, difficult and arrogant Peter Abelard. His rationality was novel and it is hard to believe that someone else would have come up with the same ideas. The real effects of his theology – which Bernard of Clairvaux called 'stupidology' – were to be felt in the next century, when Thomas Aquinas took rationalism still further, but even in the twelfth century Abelard had a powerful impact. Gratian adopted his dialectical method in his *Decretum*. Every university now established a faculty of theology, embracing Abelard's reasoning rather than Bernard's advocacy of unquestioning faith. Imagine if universities today followed Bernard's example and did not question received wisdom even when it was self-contradictory. For his influence in making Aristotle the pre-eminent philosopher for twelfth-century scholars, for his development of theology, for his ethics, for his critical method of thinking, and for his indirect impact on the moral law of the whole of Christendom through his influence on the *Decretum*, Peter Abelard seems to me to be the century's principal agent of change.

The Thirteenth Century

In 1227, Ulrich von Lichtenstein, a knight from Styria (in modern Austria), set off on a jousting tour dressed up as the goddess Venus, complete with two long plaits of golden hair. Wherever he went, from Italy to Bohemia, he challenged everyone to joust with him. He promised all his potential opponents that he would give them a gold ring if they rode against him three times; if he defeated them, they were to bow to the four corners of the Earth in honour of his lady. According to his own account, he 'broke lances' 307 times in a month. He had some mishaps along the way, such as falling from a great height when the basket in which he was being hauled up to the window of his lady's tower broke. Nevertheless, such incidents did not put him off repeating the whole adventure 13 years later. In 1240, then aged 40, he dressed up as King Arthur and took himself and six companions off on another jousting quest, allowing those who 'broke lances' with them to join their company of the Round Table. Later in life he also wrote a tale lamenting the decline of courtly love. There could hardly be a greater contrast between his jaunty tale, almost poking fun at the predilections of his own time, and the bitter life-and-death struggles of twelfth-century knights, accustomed to the sound of their own teeth shattering under the blow of an opponent's mace on their helmet.

This self-mocking, quixotic and martial but romantic figure seems to symbolise a new direction in European culture. We are reminded of the sweet harps and flutes of thirteenth-century minstrels, the jocular humour of the *Play of Daniel* (a sort of religious musical written at Beauvais Cathedral), and the playful gargoyles and misericords that decorated churches across Europe. There is something provocatively subversive in Ulrich's cross-dressing too, even though he is well known from other records to have been a responsible and effective regional administrator. Just for a moment, it is possible to see the thirteenth century as a bright summer of pleasure, with all the optimism of the famous English folk songs from the period, 'Miri it is while summer y-last' and 'Sumer is i-cumen in'. But then we have to remind ourselves that the early years of the century saw the power of the papacy at its zenith, in the uncompromising person of Innocent III, pope from 1198 to 1216. The same century witnessed no fewer than six Crusades, including the infamous Fourth Crusade, preached by Innocent III but directed by the money-hungry Venetians against the Christian city of Constantinople. Following Innocent's call for renewed fighting in Spain, the kingdoms of Aragon, Navarre, Castile and Portugal united to defeat the Almohad dynasty at the battle of Las Navas de Tolosa in 1212. Córdoba fell to Christian armies in 1236 and Seville in 1248; by 1294, all of the Iberian peninsula except Granada was in Christian hands. Innocent III also issued a bull sanctifying the Livonian Crusade, when German and Danish troops were sent into Livonia and Estonia in order to convert Europe's last pagans by force. The same pope was behind the Albigensian Crusade, in which Simon de Montfort led the armies of Christian righteousness in massacring thousands of 'heretic' Cathars in south-western France. When at Béziers in 1209 de Montfort butchered the entire population, it was reported that the papal legate accompanying him told him not to trouble distinguishing between the Cathar heretics and innocent Catholics but to slay everyone. 'God will recognise his own,' the cleric advised him. Last but certainly not least, this century also saw Genghis Khan's Mongol armies commit horrific atrocities that collectively rank as the worst act of genocide the world has ever known. We do not know how many people he slaughtered, but there are estimates of 30 million – at a time when the entire world population was no more than 400 million.

Even the recreational pursuits of this century were bloody affairs. Tournaments might have been glamorous, with heraldically resplendent knights jousting for honour, but they could still be fatal. Analogies with modern rugby and American football, which some historians suggest, are frankly inappropriate. Just look at what happened to the ruling house of Holland. In 1223, the count of Holland was killed in a tournament. His son and successor died the same way in 1234, and a younger son, the regent during the minority of his grandson, was killed in a tournament in 1238. If three generations of a European ruling house were killed on the rugby field, then the comparison might be valid, but the dangers of thirteenth-century jousting far exceeded those of any modern pastime. In 1241, the year after Ulrich's jaunt as King Arthur, a single tournament at Neuss in Germany saw more than 80 knights killed.[1] The more you think about the thirteenth century, therefore, the harder it is to reconcile the bloodshed with the frivolity of chivalric romances, happy songs about the arrival of summer, and duelling in drag.

When you stand back, however, you realise that these irreconcilable extremes are like bookends enclosing a wider range of social phenomena. The cult of the knight, chivalry, had been on a rising trajectory since the invention of heraldry in the third quarter of the previous century. Ulrich's style of composition follows the genre of courtly romances in the Arthurian tradition, but it also has something in common with the autobiographical troubadour poetry pioneered by William, count of Aquitaine, which reached its apogee in the work of Bernart de Ventadorn at the end of the twelfth century. At the other end of the cultural spectrum, the power of the Church to command Christians to fight against Muslims was on the wane. When the Holy Roman Emperor Frederick II led the Sixth Crusade in 1228, he did not look on it as an opportunity to kill members of another faith. The emperor had entertained Muslim intellectuals at his court all his life; in fact he had a higher opinion of them than he did of the pope. He simply negotiated the return of Jerusalem to Christian control, bribing the Muslims rather than fighting them. Thus the Sixth Crusade was the only other expedition to the Holy Land, after the First Crusade, to reach its goal of returning Jerusalem into Christian hands. It remained Christian until 1244 – and, in marked contrast to the First Crusade, no blood was shed. The Ninth Crusade of 1271–2 was the West's final military push to reclaim the Crusader states. After the loss of the last Christian stronghold, Acre, in 1291, the crusading spirit was little more than a figure of speech. Ulrich was therefore writing at a time when the rise of secular romances and the popularity of travel intersected with the decline of crusading zeal and the weakening of the feudal system. While the frivolity of his account stands out, and his long blond plaits distinguish him as unique, what is really striking is that he was a free man, choosing his own destiny, and expressing the fulfilment of his dreams in his own words.

Commerce

We saw in the last chapter how the population of Europe rose consistently from 1050: by the end of the thirteenth century it stood in excess of 110 million. Widespread famines in 1225–6, 1243, 1258 and 1270–71 did not significantly reduce the overall growth. In England, the figures indicate that the population rose at its fastest rate around the year 1200, at 0.83 per cent per year. This continued until the 1220s, when it reached four million. Thereafter it rose more slowly, at an annual average of about 0.25 per cent, reaching a peak of about 4.5 million in 1290. In some parts of the country we can confirm this level of growth using independent evidence. The bishop of Winchester levied a customary poll tax of one penny on males over the age of 12 on his manor of Taunton in Somerset: this shows an increase from 612 males in 1209 to 1,488 males in 1311 (an increment of 0.85 per cent per year).[2]

It all sounds very positive: more children surviving and having families of their own. However, the population of England was nearing its maximum potential. After two centuries of clearances, all the available land had been claimed. Lords could hardly allow all their forests to be turned into farmland, particularly when houses were predominantly made of timber. The lack of land started to have a braking effect on population growth. Once again the poor did not have enough food to feed a large number of children. As their numbers grew, the per capita wealth of the poor declined. The meat-eating wealthy added to the problem as animal husbandry is a very inefficient use of land if you are trying to sustain a large population. Land that might have been turned over to crops was retained as grassland for cattle or flocks of sheep. The inevitable result was a curtailment of how much the population could grow in rural areas and a migration of landless peasants to the market towns, where they hoped to make a living for themselves.

With their guilds and markets, schools, merchants' houses, Gothic-arched churches, high town walls and great gatehouses, towns were the big news story of the thirteenth century. In both northern Italy and Flanders about 18 per cent of the population lived in communities of 10,000 people or more by 1300. However, the proportion of French people who lived in towns was much smaller and there were only four or five towns of this size in the whole of England – and none in Scotland, Wales or Scandinavia. Indeed, the vast majority of the market towns founded across northern Europe in the thirteenth century were small: they were little more than villages laid out around a market square by lords hopeful of attracting business. However, these too have to be considered in the picture of urbanisation, for the importance of a town as a commercial centre is not solely dependent on the number of its inhabitants but also on the number of people who visit to buy and sell things in the market.

In the thirteenth century about 1,400 new markets were founded in England, in addition to the 300 that already existed. Not all of these new foundations took root; in fact, the majority failed. But 345 of them were still going strong in 1600, when they accounted for over half of the 675 markets then in existence.[3] The thirteenth century was thus the time when England shifted permanently to becoming a market-based economy. Similar processes of urbanisation were under way all across Europe. In Westphalia, for instance, where only six towns existed before 1180, the number had grown to 138 by 1300.[4] Overall, the number of towns in Europe increased from about 100 in the tenth century (half of which were in Italy) to about 5,000 by 1300.[5] Whereas European lords had founded monasteries in the twelfth century for the benefit of their souls, in the thirteenth century they established markets for the benefit of their purses.

These new markets did not just further enrich the wealthy lords, they also had a wider social benefit. By the end of the thirteenth century almost everywhere in England was within seven miles of a market town; the only exceptions were sparsely populated areas. The

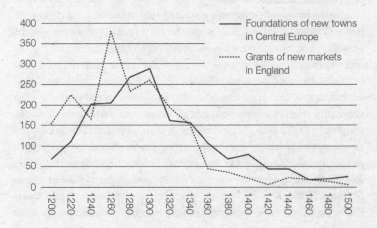

Foundation of new towns in Central Europe and grants of new markets in England, 20-year periods, 1200–1500[6]

average distance was three miles, about an hour's journey on foot each way. Whether the market was held once a week, twice a week or every day, people could come into town and sell their surpluses, or buy fresh produce. Cattle, goats and sheep were brought into the market square on the hoof; chickens, geese and other birds were transported in wooden cages on packhorses; and flitches of bacon were conveyed on carts. Here people could also buy things that were difficult to make or simply uneconomical to manufacture in small quantities, such as belt buckles, leather purses, knives, ladles, pots, kettles, nails, riding harnesses, horseshoes and stirrups. They could obtain fish and cheeses – important sources of protein when there were ecclesiastical rules against eating meat on three days of the week and over the whole of both Advent and Lent. Some men became rich carting a single commodity like eggs or eels to market. Others trapped squirrels, hares, rabbits, cats and foxes and sold their pelts to merchants, who resold them at markets to townsmen and women for fur trimmings on their robes. Some people hoarded their grain all year to sell late in the season when the price was high, along with their carefully stored apples, pears and nuts. Markets encouraged every community across Europe to pool the resources of its region and offer them to those who needed them – albeit at a price.

It should not surprise us that most of the 1,400 new markets in England failed. Towns that had no nearby competitors and were supported by large hinterlands attracted more produce and commodities and drew in more customers, thus dominating a region's trade. However, towns that were in close proximity could not similarly grow large because they were drawing from the same hinterland. As the historian Fernand Braudel observed in his 'ground rules' for world economies, 'a dominant capitalist city always lies at the centre'.[7] Thus successful market towns emerged at the centres of their hinterlands, roughly 13 miles apart from their nearest neighbours. The same principle applied to large towns and cities. Places that provided a wide range of commodities and services, including administrative functions and professional expertise, such as courts and lawyers, drew in people from much further afield. The largest and most prosperous cities were those that served the widest possible hinterland, extending for 30 or 40 miles in each direction. If the said city was a port, then it still observed Braudel's rule because it was at the centre of both its inland trade (on the landward side) and its international trade (on the maritime side). It was not just on account of its English geography that London grew to be so strong economically; it was also because it was at the centre of trade with the emerging region of the Low Countries (especially Flanders), and provided a major port for the merchants of the Hanseatic League, which linked it with the trading towns of the Baltic States and those of the Rhine and the North Sea.

The main product that powered northern European international commerce was wool. Huge quantities were sent from England to the cities of Flanders, such as Bruges and Ghent. This wool-trading network in the north was mirrored in the south by the flourishing northern Italian trading cities of Genoa, Venice and Florence. Their main commodities were silks and spices, bought in Constantinople or obtained directly by Italian merchants travelling in Asia, and traded all round the Mediterranean. Venice was the dominant city for this Oriental–Mediterranean commerce, just as London and Bruges were the major centres for wool and cloth. From the Mediterranean, spices and silks were carried to Genoa and then transported north into France, where the county of Champagne provided the great nexus of business. Six fairs in the four towns of Lagny-sur-Marne, Bar-sur-Aube, Provins and Troyes, each lasting two months, facilitated a permanent cycle of international trade between northern and southern Europe. Goods arrived at these fairs from places as far away as Lübeck on the Baltic Sea, Valencia on the Mediterranean coast of Spain, Santiago de Compostela on the Atlantic, and Augsburg in southern Germany. Even merchants from Rome and Palestine made the journey.[8]

The Champagne fairs were not an invention of the thirteenth century – they were firmly established by the end of the twelfth – but they had their greatest impact after 1200. The trunk of the huge tree of European commerce, which had its roots in northern Italy and southern France, divided in Champagne: its principal boughs stretched to London, Paris, Bruges and Lubeck, supporting branches that reached northern England, Brittany, Normandy, Flanders, Denmark, Germany and Poland by 1300. Those branches gave rise to twigs that flowered and became heavy with the fruit of many small markets, with most towns holding at least one fair of their own every year. In England, about 1,500 new fairs were added to the 146 that existed in 1200.[9] Like the markets, not all these fairs survived, but a great many did. Most towns with a weekly market held at least one fair in the summer months, and many had more than one. They usually lasted for three days and were a spectacle even for those who had little or nothing to trade. Jugglers and musicians would perform, crowds would gather, business acquaintances would meet and exchange news. Rarer commodities would be sold, both retail and wholesale. At a major fair you could buy pomegranates, which were sought after by the wealthy for their stomach-settling medicinal properties, and opium, desirable for pain relief. Through this international trading network, oranges and lemons reached northern Europe. So too did sugar, cloves, pepper, silk, medicinal drugs and carpets. These luxuries had hardly been known in the region in 1200. By 1300 they were widely available in cities and major towns, albeit at a price.

All this trade brought the old days of barter to a close: you simply couldn't do business in a market if everything had to be bartered. Huge amounts of new coinage were required. Mints were reformed and enlarged, with small provincial and private operations being closed down so governments could control the money supply more effectively. The main unit of currency throughout Christendom in 1200 was the penny, written in Latin as *denarius* (hence the 'd' in '£sd'). In France, pennies were called *deniers*, in Italy *denari*, in Spain *dineros*, in Portugal *dinheros*, in Hungary *denars*, in the Low Countries *penningen*, and in Germany *pfennige*. But pennies were remnants of a system in which money was the exception, not the rule. By the time of the Champagne fairs, some transactions could involve very large numbers of coins, and it was impractical for merchants from, say, Genoa to haul sacks of silver *denari* over the Alps. As the only gold coin available – the *nomisma* of the Byzantine Empire – was in short supply, some kingdoms started minting high-denomination coins of their own, in both silver and gold. The Venetians produced silver *grossi* – 'fat ones' – from about 1200. Brindisi, in Italy, began to mint gold coins in 1232. Florence struck the first *fiorino d'oro* in 1252 and the Genoese hammered out gold *genovini* from the following year. In England, Henry III experimented with a 'gold penny', worth 20d, but the gold was worth more than the face value of the coin, so most examples were melted down straight away. The famous Venetian *ducat* was first minted in the 1280s. France and Rome introduced their own silver *grossi* in the second half of the century, and in the Low Countries, higher-denomination silver coins acquired the name *groten*. The name travelled well: in Germany it became *groschen*, and in England, Edward I's new coinage of 1279 included 'groats', worth 4d. Coins of a lower value were minted too: halfpennies and farthings had previously been made by cutting a penny in half or into quarters; now specific coins of these values were struck. By 1300, money was established throughout Europe as the normal way – and in many places the only way – of doing business.

For those attending the Champagne fairs, even bags of high-denomination coins were an encumbrance and a liability. Rather than passing over quantities of bullion every single time a deal was agreed, merchants started keeping written notes of who owed whom money. When they sat down at the end of the fair, they could pay their debts by transferring the amounts due. But even then, they realised, there were alternatives to settling every deal in coin. Bills of exchange could be drawn up and passed on by agents, promising payment by a certain company at a certain date. Herein lies the origin of banking, for these agents started to extend credit facilities to reliable merchants. They sat on benches called *banche*, which gave rise to our word 'bank'. Along with credit and bills of exchange, these banking companies introduced double-entry bookkeeping: the method whereby outgoings and incomings are separately recorded and made to balance. By 1300, they may also have invented the means for goods in transit to be insured; certainly records attest to merchants underwriting shipped goods in the next century. The major banking companies are famous – the Ricciardi (of Lucca), the Bonsignori (of Siena), and the Frescobaldi, Buonaccorsi, Scali, Bardi, Acciaiuoli and Peruzzi (all of Florence). Their businesses were international – the Frescobaldi and the Ricciardi, for instance, both loaned money to Edward I of England in the thirteenth century. Quite a lot of the sap in that commercial tree described above was driven by their ingenuity. Next time you open your wallet and pull out a credit card, have a thought for the merchants of the thirteenth century.

Education

There is a watershed in historical evidence around the year 1200. Very few systematically created records survive from before the reigns of Richard I in England (1189–99) and Philip Augustus in France (1180–1223). In England there are no bishops' registers and no local manorial documents from before 1200. Apart from the pipe rolls – the records of loans by and debts to the King of England by his sheriffs and vassals, which date from 1130 – there are no regular royal accounts. With the exception of the Italian city states, where civic record-keeping began in the twelfth century, to write the history of Europe before 1200 we are dependent on privately composed chronicles and scraps of other evidence, such as royal letters and grants, and private charters copied into monastic cartularies.

Things started to change in the 1190s. All of a sudden, it seems, there was an explosion in record-keeping. In France, Philip Augustus ordered the Grande Chancellerie to create Le Trésor des Chartes in the royal palace, where a comprehensive record of royal business began to be preserved. The English government adopted a similar policy. Copies of every letter and charter sent out in the king's name were written on to rolls – at least one roll for every year of the reign – and stored carefully. Many of these have survived. Thus on the Confirmation Rolls we have the text of every old charter confirmed by the king after 1189; and the Charter Rolls contain every new charter issued from 1199. From 1199 there are also Fine Rolls containing the texts of grants for which a fine was payable to the Crown, such as wardships, liberties, privileges and official positions. If a letter was sent out sealed in an open or 'patent' state, it was copied on the Patent Rolls, which survive from 1201; if a letter was sealed 'closed' it was copied on the Close Rolls, which are extant from 1204. From early in the reign of Henry III (1216–72) there are files of Inquisitions Post Mortem, these being the results of enquiries made into the landholding of every person who died in possession of land held directly from the king. By 1300 you also have Treaty Rolls, Liberate Rolls (recording royal payments), Roman Rolls (letters to the popes), Scotch Rolls, Norman Rolls, Gascon Rolls, Welsh Rolls, Statute Rolls, and many other series. These are just the very tip of an enormous iceberg of records. Among the records of the Exchequer you may find inventories of royal treasure and jewels, and accounts of daily expenses in the royal household. Surveys of local landholding across the whole kingdom were compiled for the purpose of granting aids or occasional taxation to the king, such as the Book of Fees (1198–1292) and the Feudal Aids (1284). In 1279, Edward I ordered a survey to find out which royal manors had been granted to lords, in the hope of stopping further encroachments on the royal domain. Over the course of the thirteenth century, therefore, central government genuinely did pass 'from memory to written record'.[10]

This revolution was not limited to central government. In dioceses across Europe bishops started preserving records of their acts in episcopal registers. Great magnates and prelates began to keep accounts too. The bishop of Winchester's pipe roll has already been mentioned: this gives details from 1209 of the expenses, loans, fines and other income due to the bishop from his vast estate. Lords of manors started to keep court rolls to preserve information as to which tenants held what land, for how long they might continue to occupy it, and what fines were due. Manorial clerks kept accounts to keep a check on the grain and animals that the tenants produced for the lord as part of their feudal service. The earliest custumnals also date from this century, recording what by-laws and practices were to be observed in the manor: whether the lord was due to pay for a breakfast for workers at harvest time, for example, or whether they were permitted to gather firewood on his land. The earliest systematically created civic records outside Italy also date from this period – rolls of names of freemen of towns, court records and borough custumnals. In secular as well as in religious life, from cities to the remotest rural manors, records started to be systematically created and preserved.

All this recording begs the obvious question: where did these literate men suddenly come from? If we assume that every monk and clergyman was literate by 1300, and that there was a clerk serving on every manor, then maybe as many as 40,000 men had learnt to read and write in England alone. Some would have been taught by the monasteries. Others – especially scions of noble families, which had preserved literacy as part of the dignity of their class – would have had a private tutor. Peter Abelard and his brothers were examples of this tradition. In 1179, the Third Lateran Council had ordered that every cathedral should run a school; this was reinforced by the Fourth Lateran Council of 1215, presided over by Innocent III. It decreed that not only should those cathedrals that still lacked a school now set one up to teach the children of townsmen to read and write in Latin, but that every church in Christendom with sufficient wherewithal should do so. This effectively stipulated that there should be a school in every town.

The real force driving education was simply a massive increase in demand for literate men. At a time of steep economic and demographic growth across Europe, lords and landholders felt that their traditional rights were under threat. Writing things down was a way of preserving information that could later be used in court to back up one's ownership of land and enjoyment of assets. If you acquired a piece of land or a franchise in the thirteenth century, you wanted your ownership recorded in some way. If you were doing business in a town, you wanted a commercial contract to be drawn up by a notary every time you reached an agreement with another merchant. It was this need for record-keeping that filled the church schools set up in the wake of Innocent III's council. Even just a dozen more schools could make a huge difference: over the course of a few years, one master could teach hundreds of young townsmen to read and write.

Literacy is one aspect of the educational revolution of the thirteenth century. Another is the establishment of the universities, which grew out of the intellectual developments of the previous century. The University of Bologna, for example, traced its origins back to Irnerius's law school. It is sometimes argued that the schools of Paris were a 'university' at the time when Abelard was teaching. The University of Salerno claims to date from the twelfth century due to the city's role in copying and distributing medical texts at that time. In reality, the concept of the university was still evolving in the twelfth century. Abelard's pupils followed him wherever he went; if he abandoned Paris for Melun, Corbeil or the Paraclete, they went with him. It was the teacher that mattered to them, not the school. The evolution of the university in the early thirteenth century allowed for quite a different form of education: an institutional one, with a code of conduct, a set of standards and examinations, and a sense of its own social value.

The first official use of the word *universitas* to describe a type of higher-education establishment dates from a decree issued by the papal legate Robert de Courçon in 1215 to resolve a dispute in the University of Paris. Disputes were not uncommon in the early days of universities: the University of Cambridge was established as a result of a walkout by almost all the members of the University of Oxford in 1208. However, despite such teething troubles, universities were firmly established as part of the educational life of Europe by 1250. Students read for the degree of Master of Arts by studying the seven liberal arts, divided into the 'trivium' (grammar, rhetoric and dialectic) and the 'quadrivium' (arithmetic, geometry, astronomy and music). From the 1230s, graduation as a Master of Arts gave them the right to teach anywhere without further licence. If they wished to specialise in law, medicine or theology, they had to carry on with their studies in a higher faculty. By 1300, universities were firmly established at Seville, Salamanca and Lérida in Spain; Lisbon in Portugal (later relocated to Coimbra); Oxford and Cambridge in England; Toulouse, Montpelier, Angers and Paris in France; and Vercelli, Bologna, Vicenza, Padua, Piacenza, Reggio, Arezzo, Siena, Naples and Salerno in Italy. By this time it had become usual for any cleric hoping to attain high office in the Church to read for a Master of Arts degree at a university. Methods of debate, scholarship and attaining knowledge had been formalised and distributed systematically across Christendom. Education had been standardised and introduced into all the courts of Europe, and a host of masters empowered to teach the next genera-tion of clerks and scholars.

Accountability

In our day and age we take the superiority of a literate society for granted, but in 1200 the value of education was less clear. It was, after all, expensive. Having things written down for you was not a cheap service either, so if central government was keeping records of grants, then why did you need a second copy for yourself? The answer lies in a widespread cautious approach to the law. Monastic institutions and noble families kept copies of royal charters granted to them in case their terms were disputed by the grantor or questioned by a third party. A similar reason underlies the creation of financial accounts. Surviving medieval accounts are more than straightforward financial documents: they often explain *why* it was necessary to spend the money, as the compiler could be held accountable. The amounts recorded on the bishop of Winchester's pipe rolls, for instance, were set down not to inform the bishop how much he had left in his coffers but so his treasurer could acquit himself of any outstanding liabilities. In short, the major reason for record-keeping was caution, distrust and a desire for greater certainty. Things were written down so that people could hold each other accountable in law.

Accountability itself was nothing new. For centuries kings had sworn solemn coronation oaths or promised to uphold certain rights. Other men had regularly sworn solemn vows. However, there was a fundamental difference between swearing on a holy relic – in which case God was the judge of whether the oath had been broken – and being held to a certain promise in law. If you had the text of a charter or a letter from the king, you could hold him to account. Previously, rulers had made laws for their people, but they were not bound by them. Now the very relationship between a king and his people was changing. In the 1190s, Philip Augustus began to call himself 'King of France' instead of using the traditional title 'King of the Franks'. At the same time, Richard I similarly called himself 'King of England' rather than 'King of the English'. These subtle changes stressed that kings ruled over all the people in their kingdom, foreigners included, and that they made the laws for *everyone* within their domain, not just their followers. It also implied that they had a duty to preserve the borders of their kingdom from incursions and took responsibility for the Church and foreigners within their realm.

It was a Holy Roman Emperor who first found himself held to account by his subjects. In the 1160s Frederick Barbarossa had tried to impose his direct control on the great cities in the north of Italy. When they resisted him, he destroyed Milan. In defiance, the Italian cities formed the Lombard League and defeated the emperor in battle at Legnano in 1176. In the subsequent treaties of Venice (1177) and Constance (1183), they then established their right to rule themselves while still remaining part of the Holy Roman Empire. The emperor had been forced to conclude treaties with his subjects and, as a consequence, could be held accountable by them.

Far more famous in the Anglophone world is Magna Carta, or the 'Great Charter', the result of King John being forced to the negotiating table. John was never a popular king. His neglect of Normandy led to its conquest by the French in 1204, with the result that most of his leading nobles lost their ancestral lands there. Moreover, his refusal to accept Stephen Langton as archbishop of Canterbury led to a direct confrontation with Pope Innocent III. In 1208, the pope placed the whole of England under an interdict, whereby he prohibited clergy throughout the kingdom from holding church services and burying people. The following year he excommunicated the king. In 1213, John backed down. He agreed to accept Langton and resigned the kingdom of England and the lordship of Ireland to Innocent, who granted them back to him as his vassal. He also undertook to pay an annual tribute to the pope of 1,000 marks (£666 13s. 4d) and promised to go on a Crusade.

Having gained papal support, John then proceeded to advance against his enemies. He built up a series of alliances with the count of Flanders and the Holy Roman Emperor, and went to France hoping to regain possession of Normandy. But like most of his strategic plans, this enterprise went terribly wrong. His allies were crushed by the French king, Philip Augustus, at the battle of Bouvines in 1214, and John was forced to return to England empty-handed. He was now more unpopular than ever, and a number of important lords withdrew their fealty to him. In June 1215, he was compelled to agree to Magna Carta, which confirmed the ancient laws of England and established a series of rights and liberties for the English people and the Church that would henceforth be a check on royal power. A council of lords was established to override the king's rule if he should fail to observe the terms of the charter.

The idea that a king could be held to account by his feudal subjects in this manner did not go down well across Europe. John's old enemy, Innocent III, voiced his horror in the most uncompromising language. He declared that Magna Carta was 'not only shameful and base but illegal and unjust', and went on:

> We refuse to pass over such shameless presumption for thereby the Apostolic See would be dishonoured, the king's right injured, the English nation shamed, and the whole plan for the Crusade seriously endangered . . . We utterly reject and condemn this settlement, and under threat of excommunication we order that the king should not dare to observe it and that the barons and their associates should not insist on it being observed. The charter, with all undertakings and guarantees, whether confirming it or resulting from it, we declare to be null and void of validity for ever.[11]

The pope's protest was in vain. From 1215, kings of England were not at liberty to do whatever they chose. Magna Carta was reissued in different forms over the next years and was rendered permanent in 1237. In 1297, Edward I placed it on the statute book.

Throughout the various versions, two important clauses remained at the heart of the charter. The first was that 'no free man shall be seized or imprisoned, or stripped of his rights or possessions, or outlawed or exiled, or deprived of his standing in any other way . . . except by the lawful judgement of his equals or by the law of the land'. The second read: 'To no one shall we sell, deny or delay right or justice.' Over the subsequent centuries English kings were constantly reminded of these provisions – especially that they were not at liberty to lock up anyone unlawfully. They continued to do so, of course, but hereafter the supporters of any wrongfully imprisoned person could invoke Magna Carta and accuse the king of acting illegally and thus tyrannically.

Most other European kingdoms did not see an equivalent to Magna Carta. But this does not mean they were not affected by it. Like the French Revolution in the eighteenth century, it did not require every country to experience its own Reign of Terror to appreciate the significance of what had happened. Instead, evidence of the widening desire to hold kings to account and for kings to involve their people more in decision-making can be found in the establishment of parliaments.

Previously kings across Europe had ruled with the advice of their royal councils, which were formed of the leading magnates and prelates in the realm. Exceptionally, in the Spanish kingdom of León, Alphonso IX called representatives of the towns, as well as his lords and clergy, to advise him at his *cortes* from 1188. In the thirteenth century this became more common. Scottish kings summoned groups of commoners to advise them from the 1230s. The kings of Portugal started to summon representatives of the towns to their *cortes* in 1254. Other Iberian legislative and tax-controlling parliaments included those of Catalonia and Aragon (which met from 1218 and 1274 respectively), Valencia (1283) and Navarre (1300). In France, the nearest equivalent, the Estates General, was first summoned in 1302. In all these assemblies representatives and lawyers sat and heard cases; grants and rights were recorded; and appeals were made to the king when men in positions of responsibility were found to be abusing their power. Particular points on which the king needed advice were debated. Discussions increasingly took place about whether he should be granted the necessary taxation to go to war. Thus Magna Carta is not an isolated case of enforced monarchical restraint but indicative of a growing mood for people to have a say in the government of the realm.

In England, Parliament developed a particularly strong role, following on from Magna Carta. In 1258, Simon de Montfort – a younger son of the instigator of the massacre at Béziers – forced Henry III to agree to the Provisions of Oxford, which called for regular parliaments to be held. These were to include elected representatives of the towns and counties as well as the lords and important clergymen. The representatives of the 'commons' thus came face to face with their king and negotiated the terms for granting extraordinary taxation, which the king usually wanted in order to fund military campaigns. Frequently Parliament asked the king to agree to new statute legislation in return. In an enormously important constitutional development in 1297, Edward I agreed that henceforth parliamentary assent would be required for *all* such grants of taxation. This meant that, in effect, Parliament could stop the king from going to war by simply refusing to approve the necessary finance. If you cast your mind back to the start of this book, and recall the powerlessness of the common man in the age of Viking invasions, you can see that society had come a long way in just three hundred years.

Friars

Have a thought for poor old Pietro Bernadone. He was a hard-working, prosperous cloth merchant from Assisi, in Umbria, who travelled regularly to the Champagne fairs, where he developed a taste for all things French. On one such trip he even acquired a French wife, a woman from a distinguished Provençal family, and brought her back to Assisi. Later, in celebration of his fondness for France, he renamed his son, who had been called Giovanni, Francesco, or Francis for short. Francis, however, turned out to be a disappointment. First he lived the high life, enjoying lavishing his father's wealth on his friends. Next he decided to be a soldier in Apulia. Then, in his early twenties, he changed course yet again. Inspired by a vision in which he was told to rebuild the dilapidated church of St Damian near Assisi, he helped himself to a bundle of cloth from his father's house and sold it, giving the money to the priest of St Damian. The cleric refused to accept the proceeds of stolen goods, leaving Francis embarrassed. When his father found out what Francis had done, he was furious. He reported his son to the civic authorities and forced him to forgo his inheritance.

Francis was living as a hermit, repairing local churches and helping lepers, when in late 1208 or early 1209 he heard someone reading out Chapter 10 of the Gospel of Matthew, in which Christ exhorted his disciples: 'as ye go, preach, saying the kingdom of Heaven is at hand. Heal the sick, cleanse the lepers, raise the dead, cast out devils: freely have ye received, freely give. Provide neither gold nor silver nor brass in your purses, nor scrip for your journey, neither two coats, neither shoes, nor yet staves . . .' At that moment he decided to live his life according to that Biblical text. He began preaching his message of absolute poverty and penance around Assisi. Before long he travelled to Rome to see Innocent III and outline his vision for an order of brothers, or *frères* – hence our word 'friars' – who would spend their lives in poverty. The pope was impressed and gave Francis his blessing. Thus the Franciscans – otherwise known as the Order of Friars Minor, or Greyfriars (after the colour of their habit) – came into being.

Dominic de Guzman gave his father and mother an easier time as he was growing up in Calaruega in northern Spain. They too were well off, described as of 'gentle birth', which in twelfth-century Spain meant that they came from a line of warriors. In the Guzmans' case, however, the whole family was deeply pious. Dominic studied at Palencia, and when a famine overtook Spain he sold his books and possessions to raise money for the poor. Twice he offered his services as a labourer in order to buy the freedom of men who had been enslaved by Muslims. His willingness to demonstrate his piety through action inspired many, and he was welcomed as an Augustinian canon in the cathedral at Osma.

He was in his early thirties when, in 1203, he joined a diplomatic mission to Denmark with the Bishop of Osma. On his return, he decided to see Innocent III to ask for permission to go as a missionary to eastern Europe. The pope, however, gave him a more pressing task: to deal with the heretics of south-west France. Dominic thus found himself in the county of Toulouse, trying to convince Cathars to return to orthodox Catholicism. He was no doubt shocked by some of the Cathars' beliefs – such as their condemnation of marriage and their refusal to accept the resurrection of the body – but at the same time he was inspired by their vows of poverty. In 1206, he established a house at Prouille where women who resented being forbidden marriage and treated as concubines could live in their own religious community. Over the following years he developed a mission that centred on both of these strands of faith: preaching against heretical beliefs and advocating poverty. Various attempts were made to give him a bishopric, but Dominic resisted, insisting that his first priority was to found a preaching order. In 1215, he was able to realise this dream when a wealthy citizen of Toulouse gave him a large house for his growing band of followers. Later that year he attended the Fourth Lateran Council and put his proposal for a new religious order before Innocent III. The pope died before he could agree, and it fell to Innocent's successor, Honorius III, to give official papal blessing to the Order of Friars Preachers, or Dominicans, otherwise known as the Blackfriars.

The Greyfriars and Blackfriars spread with astonishing rapidity. Honorius issued bulls on behalf of the Dominicans, effectively advertising the order throughout western Christendom, and approved of the Rule written by St Francis in 1223 for the Franciscans. Their success meant that they were soon joined by other orders of friars. In 1226, Honorius gave his approval to the Order of Carmelite Friars (Whitefriars). His successor Gregory IX approved the Augustinian Friars in 1231. The idea of holy men taking vows of absolute poverty, chastity and obedience, and preaching to the common people in their communities, proved extremely attractive. Associating the mendicant orders' way of life with that of Christ, thousands flocked to join them. Both the Franciscans and Dominicans catered for women too: the Franciscans with a sister order of cloistered nuns, the Poor Clares, founded by Clara of Assisi in 1212; the Dominicans with their order of nunneries, stemming from the original convent at Prouille. The friars also became recognised as important educators. Dominicans could be found at the University of Paris as early as 1217, at Bologna in 1218, at Palencia and Montpelier in 1220, and at Oxford in 1221. The Franciscans similarly established theological colleges across Europe, most notably in the universities of Paris, Oxford and Cambridge.

What was so significant about the advent of the friars? The new mendicant orders sliced through a society strictly divided between religious and secular, and created an interim body of men who had many of the virtues of the religious and all the flexibility of the secular. They were educated like monks: they could read and write and they understood the international language of Latin. They were disciplined like monks: they followed a set of rules and answered to an ecclesiastical hierarchy. They carried the good name, trustworthiness and integrity of holy men. Unlike monks and other regular clergy, however, they could roam free: they were not bound to remain in any one house, community or parish. They lived in towns, mixing with people, and they were cheap – they did not require tithes or prebends to endow their prayers and furnish their habits. If the monastic orders of the previous century constituted a network that created, stored and disseminated knowledge, then the friars allowed that network to penetrate further, deeper and faster than it had previously done. Friars became the diplomats of choice for secular rulers and Church leaders alike. As educated messengers travelling and reaching out to people in the name of God, they made excellent administrators and negotiators. And they also made good inquisitors: popes and bishops increasingly relied upon Dominicans to interrogate heretics and, after 1252, even to torture them.

The friars demonstrated that the Church was able to move with the times – in the sense that they could travel freely, preach to merchants as well as to lords, and tackle new questions of faith – while retaining the spirit of humility and poverty that characterised early Christianity. If the Church had not been able to avail itself of this versatile body of holy men, heresy would undoubtedly have spiralled out of control. Perhaps, in addition to the Albigensian Crusade, there would have been English and German Crusades – and the Spanish Inquisition would have been established well before the fifteenth century. As it was, the three centuries after 1215 saw only minor and localised forms of heretical practice in Europe. It seems that the friars, especially the Dominicans, played a major role in securing for the pope another 300 years of spiritual authority over Christendom. That is not something to be sniffed at.

Finally, we must bear in mind the intellectual impact of individual friars. As the mendicant orders eschewed personal wealth but prized learning, they appealed to a great many unworldly men keen to engage intellectually with the principal debates of the time. Among the Franciscans was a string of great theologians – Alexander of Hales, Bonaventure, Duns Scotus and William of Ockham are the most prominent. But the foremost intellectual Franciscan of the thirteenth century was undoubtedly the remarkable scientist and philosopher Roger Bacon, who lectured on Aristotle at the universities of Oxford and Paris, studied Greek and Arabic works on optics, argued for the introduction of scientific teaching in universities, and wrote a substantial compendium of scientific knowledge, philosophy, theology, languages, mathematics, optics and experimental science. He was the first person in the West to describe gunpowder; he provided the first description of spectacles; he theorised that a copper balloon filled with 'liquid fire' might fly; and he was blessed with the most extraordinary open-mindedness. He believed, for instance, that it was possible to build colossal ships that had no oars but could be sailed by a single man; that vehicles could be invented that would travel at incalculable speed without having to be drawn by a draught animal; that men might propel themselves in machines 'like a bird in flight'; that suspension bridges might cross wide rivers 'without a pier or prop', and that divers could explore the seabed in special underwater suits.

The Dominicans had a number of intellectual heavyweights too. There was the mystic and theologian Master Eckhart; the scientist, philosopher and theologian Albert of Cologne (known as Albertus Magnus); and the master theologian of them all, Thomas Aquinas. It was Aquinas who followed Abelard in applying Aristotelian logic to religion and modified his maxim 'doubt leads to enquiry and enquiry leads to truth' to 'wonder leads to enquiry and enquiry leads to knowledge'. Whereas Abelard was prepared to accept that some things, such as the nature of God, were above and beyond rational enquiry, Aquinas felt that everything should be subject to investigation and rationalisation. He deduced the existence of God from nature by arguing that because everything that is moving has been set in motion by something else, there must have been a prime mover at the beginning of that chain. Another of his arguments is still rehearsed regularly today: the fact that the world displays order and constantly rejuvenates itself demonstrates God's role as an intelligent designer. We may suspect that Aquinas would have produced the works that made him the most important theologian of the Middle Ages even if he had not been a friar, but there can be no doubt that he and many others were supported in their learning by the resources and networks of the mendicant orders, and that they were inspired to even greater achievements by their intellectual curiosity.

Travel

There is a widespread belief today that people did not travel far before the advent of the railways in the nineteenth century. This way of thinking is superficially backed up by our family trees: prior to the mid nineteenth century, our ancestors usually married people from the same parish or a neighbouring one. But there is an obvious weakness in this argument: just because we wouldn't consider walking long distances today does not mean that people in the past were not prepared to do so. And just because people married someone from their own community, where they could be sure of support and perhaps an inheritance, this does not mean that they never went anywhere else.

There is a frequently repeated counter-argument that is also deeply flawed. The ocean-going Vikings, the sea-crossing Saxons and the road-building Romans all demonstrate that people did travel long distances. The bluestones at Stonehenge, for instance, come from the Preseli Hills in south Wales, 250 miles from where they now stand. Thus people have always been able to travel if they wanted to. To which the scholar must respond: *people*, yes, but not *most* people. Evidence for long-distance travel before the thirteenth century relates predominantly to politically influential individuals and groups that were able to protect themselves, such as Roman armies, Vikings and Crusaders. Royal officials were guarded by the king's men wherever they went. Great lords travelled with large households, accompanied by many armed men. Ordinary people, however, did not regularly travel much further than the distance to their manorial court, their animal pastures, or their church. As we saw in the chapter on the eleventh century, travelling away from home was just too dangerous. You had to have a very good reason to risk it.

In the thirteenth century, however, many people *did* have good reasons to travel: to attend fairs and larger markets further away from home; to obtain a grant from the king or a judicial decision from a central court, or to attend parliament. You might travel to ask for the attendance of a physician or surgeon. You might go right across Europe to attend a university, or walk a few miles every week to school. If you were a friar, travelling was part of your vocation. Pilgrimages became extremely popular in the thirteenth century, with thousands of people walking for a day or so to visit a local religious attraction or travelling to one of Christendom's three principal destinations – Santiago de Compostela, Rome and Jerusalem. For Italian merchants, travel was part of daily life, both by land and sea. The Venetians had established trading posts throughout the Mediterranean and the Genoese had set up a number of fortified outposts in the Crimea. The criminal law also required more people to travel. Judges sent out to administer the law or hold inquiries went on long journeys. Criminals had to be taken to towns where they would be judged. Jurors sworn in to hear a case would have to go wherever the court sat. After Gratian, canon law stipulated that archdeacons and bishops had to supervise the moral life of their flocks, and people now had to attend consistory or archdeaconry courts. This might be on account of a moral wrongdoing, such as bigamy or adultery, or a case of heresy; alternatively it could be a routine matter, such as the need to prove a will. And the more people travelled and the longer they were away from home, the greater the need for messengers to tell them of important news in their home town. Travel led to more travel.

As more people took to the roads, it became easier to do so. The bonds that bound serfs to their manors, meaning that they were unable to leave without the permission of their lords, started to break down as they journeyed to markets and fairs. Inns and monastic guest houses sprang up to cater for travellers. Wooden bridges were rebuilt in stone, securing river crossings. Roads were cleared of undergrowth on either side to prevent thieves from lurking there. The very fact that there were more travellers gave them greater security. For example, people would gather at an inn and set out together in a group, so that they could defend themselves if they were attacked. Obviously the further you travelled from home the greater the risks. But even merchants passing through foreign kingdoms could feel safer. The more thorough application of the law, discussed above, meant that they could take legal action if the worst came to the worst and they were robbed, assaulted or defrauded on their journey.

We cannot do justice to thirteenth-century travel fully without looking at the voyages to the Far East. Earlier geographical knowledge had been limited to what was preserved in the stories of Alexander the Great's travels to India, the natural histories of Pliny and Solinus, and by the geographers of the classical world. The Hereford world map illustrates the knowledge of the world in this century, showing little detail east of Jerusalem apart from the Red Sea (with a passage marking Moses's route through it), the Tigris, the Euphrates and the Ganges. Strange beasts adorn its further edges. However, this charming naivety masks a deep concern about the threat from the East. In the late 1230s Ogedai, the son of Genghis Khan, led the Mongol armies into Russia; soon afterwards, his kinsmen added parts of Germany, Poland and Hungary to their list of conquests. In 1243 the newly

elected Pope Innocent IV decided to make contact with the Mongol
leaders in an effort to convert them to Christianity. Two missions set
out two years later, one led by Giovanni da Pian del Carpine, a
Franciscan, and the other by the Dominican Ascelin of Cremona.
Ogedai's son, Guyuk Khan, responded by asking the pope to come
and swear allegiance to him. Understandably, the pope declined, but
two further missions to the East followed, led by the Dominican friar
Andrew of Longjumeau and the Franciscan William of Rubruck. In
1254 William entered the Mongol capital of Karakorum – only to find
the Hungarian-born son of an Englishman already ensconced there,
as well as a Hungarian-born Frenchwoman and the nephew of a
Norman bishop, hinting at all those unrecorded long-distance journeys
that were undertaken but are hardly known to us.[12]

Ten years after these expeditions, the Venetian merchants Niccoló
and Maffeo Polo set out on their first great venture into the Far East.
On their second journey they took with them Niccoló's son, Marco.
But perhaps the most significant thirteenth-century expedition was
that of the Franciscan friar Giovanni de Montecorvino, who was sent
as an envoy to Kublai Khan, the conqueror of China, in 1289. He
reached Peking in 1294, just after Kublai Khan's death. He remained
there and began to convert the Chinese to Christianity, becoming the
first archbishop of Peking in 1307.

The thirteenth century was not only when travel became common
for the majority of people; it was also when Christian travellers reached
places that before had only been touched by ancient legends. By 1300,
the furthest outposts of the Christian world were almost 5,400 miles
apart – from Garðar on Greenland in the west to Peking in the east.
Neither of these would prove permanent: the Christians were ousted
from Peking in the fourteenth century and the worsening weather
put an end to Garðar in the early fifteenth. But the change in the
Western imagination would endure, and it is best exemplified in the
travels of Marco Polo. Dictated in jail to a fellow prisoner, Polo's book
included eye-popping information about the massive populations and
wealth of the cities in China and Indonesia. His vivid descriptions of
their customs, so different from Western practices, left Christians agog.
When these stories accompanied the silks and spices that could increasingly be found in the markets and fairs of Europe, people began to
wonder about Asia and the rest of the world in an entirely new way.
Travelling to the Far East when you believe it is full of dragons and

monsters is unwise; going there when you know it to be full of riches, with a Christian archbishop already in place in Peking, is a much more alluring prospect.

Conclusion

There can be little doubt that the great shift of the thirteenth century is that of trade. Everyone took one step up the cosmopolitan ladder: the peasant became more familiar with the market town; the burgher from the market town visited the city more regularly, and the prosperous merchants from the major cities travelled further and more frequently, even to international fairs. It leaves me with an image of a humble man sitting in his cottage, carving a wooden bowl. Suddenly he hears a knock at the door. When he opens it, a merchant is standing there, offering him silver in return for his bowl. Another merchant is coming up behind the first, dragging a goat with him, which he wants to sell for the silver our humble man has just earned. There is a third merchant behind the second, carrying a metal bowl that is far more attractive than the wooden ones that our humble man has carved all his life; this he offers to exchange for the goat. Suddenly our humble man is inundated with people all offering goods to sell or asking to buy his wares. There is rabble and noise where before there was only silence. Above it all a preacher's voice rings out, urging our man to relinquish his money-making ways and to follow the path of penury and self-denial. A clerk also arrives, presenting him with a list of his rights on the manor, and an account of the corn he has set aside against the winter. The outside world has intruded on his life and, very clearly, it is here to stay.

What is striking about these changes is their universality. Even in Moreton, everyone's lives would have been affected. In 1207, the lord's demesne land next to the church was laid out as a market square, with a dozen burgage plots located around its perimeter. The ancient house I now live in stands on a road that once marked the southern edge of this market square. In the 1290s it was inhabited by a chaplain, Adam de Moreton, who no doubt strolled out among the stalls with his servant to buy eggs and meat, cloth and candles. People would have come in from the surrounding villages on market days, and from further afield when the five-day fair was held here, on the feast of St

Margaret. Friars would have preached in the marketplace. In 1300, after Adam de Moreton had been appointed vicar of St Marychurch, which is 19 miles away, he sold this house to a local man, Henry Suter. A deed was drawn up conveying the property, and although Suter was probably illiterate, his family kept that deed carefully until his descendants sold the house in 1525. In the twelfth century, such a transaction would not even have been written down, let alone kept.

The proceedings of the manorial courts in Moreton similarly started to be written down on court rolls from the 1280s, as they were in the adjacent manor, Doccombe (whose court rolls still exist). Moretonians guilty of moral crimes would have been summoned to make the 24-mile journey to Totnes to attend the archdeaconry court, taking a number of witnesses with them to swear an oath to attest to their innocence. Those guilty of felonies would have been forcibly marched to Exeter Castle to await trial. The bishop of Exeter would have visited the parish as he went around his diocese dedicating the altars of the churches, newly rebuilt with the proceeds of the wool trade. Even in such a quiet, remote place as Moreton, where no one would have heard of William of Rubruck or Marco Polo, daily life changed for ever.

The principal agent of change

If this book was about world history, there would be no doubt who was the principal agent of change. Genghis Khan's reputation, floating on the ocean of blood that stretched from China to the Caspian Sea, has no rival. The devastation he caused, continued by his son and grandsons to the detriment of eastern Europe, led directly to the West forging links with the East for the first time. His genocides destroyed the ability of the East to take the initiative in bringing new trade to the West. He therefore inadvertently broadened the horizons of Christendom as well as creating new opportunities for its merchants. However, in our story of change in the West, Genghis Khan is only a peripheral player. A more influential figure lurks at the heart of Europe in the person of Lotario dei Conti, otherwise known as Pope Innocent III.

Innocent III was one of the last popes to influence rulers and decision-makers at every level of society throughout Christendom. At the top end, he not only held firm against John of England; he forcibly annulled several European royal marriages and issued the decree

'Venerabilem' that finally established the relationship between the Holy Roman Emperor and the pope. At the lower end, he understood the inclinations of the masses over whom he spiritually ruled. His reception of Francis of Assisi is a good example: had he decided that the impoverished man was a ne'er-do-well, he could have sent him packing and thereby labelled a group of fervent preachers heretics. He would thus have deprived the Catholic Church of the Franciscan friars, who became one of its most powerful tools. Much the same can be said for his encouragement of Dominic de Guzman. Dominic had originally wanted to convert pagans on the fringes of Europe; it was Innocent who directed him to undertake the correction of heretics and thereby gave the Blackfriars one of their key purposes. Innocent's exhortations to the Spanish kings to collaborate in the Reconquista bore fruit in the victory at Las Navas de Tolosa, which paved the way for the reconquest of almost all of Spain by the end of the century. His determination to make every major church in Christendom teach men to read and write further accelerated the spread of literacy. There can be no doubt that he was the most influential Christian of the century. On the other hand, in some respects he was hugely conservative: he was opposed to the erosion of royal absolutism, as his condemnation of Magna Carta reveals. But as we have already seen with Bernard of Clairvaux, a determination to resist change can end up causing it – a point to which we will return at the end of this book. Innocent's role with regard to the Cathars is particularly telling in this regard: his firm stance against heretics in general and the Cathars in particular not only maintained the hegemony of the Catholic Church; it also led to the Inquisition. For all these reasons, therefore, I cannot see a rival to Innocent III as the principal agent of change for the thirteenth century.

1301–1400

The Fourteenth Century

Medieval people did not understand social history. When artists depicted scenes from the Bible or from Roman times in stained-glass windows, carved sculptures and illuminated manuscripts, they showed people wearing medieval clothes, living in medieval houses and sailing medieval ships. However, just suppose that an exceedingly well-informed and imaginative monk was consulting all the historical sources available to him in his cloister in the year 1300. This would no doubt have led him to believe that mankind had been fortunate over the last three centuries – or, to put it in the religious framework of his own time: that God had been good to Christendom. Economically, western Europe had gone from strength to strength. The population had increased enormously. Towns whose residents had once lived in fear of attack were now well defended, and rural areas were no longer under threat. Such a monk could quite reasonably have seen the Church as the principal architect and facilitator of these developments. The Church had acted as an agent for peace both by extending the borders of Christendom and by attempting to calm the warring factions within them, concentrating Christian violence on specific targets at the periphery and beyond. It had taught many people to read and write. It had developed a moral code for all Christians to follow, and encouraged courts to impose punishments should that moral law be broken. Our imaginary monk could have been confident that God was acting for the well-being of the whole of Christendom in almost every way possible. He would not have had such confidence a hundred years later.

By 1300, the situation was already changing. A succession of poor harvests in the last decade of the thirteenth century led to severe food shortages in northern France and the Low Countries. In 1309, excessive rainfall caused a major famine throughout Europe. And things then went from bad to worse. There had been many famines in the past, of course, but the population had always bounced back rapidly because of the high crop yields attainable. These were no longer the norm. Decades of intensive cultivation had steadily depleted the nitrogen in the soil to the extent that simply leaving a field fallow was no longer enough to reinstate its fertility. Wheat yields that in 1200 had been as high as 6:1 (six seeds harvested for every seed planted) had fallen to 2:1 by 1300; barley and rye yields had fallen from 4:1 to 2:1.[1] When the productiveness of the land was so low it was not possible for the population to recover quickly. Suppose a peasant had the use of 25 acres – a large amount of land for a manorial tenant in early-fourteenth-century England. Let us say that he had sufficient corn from the previous year's good harvest to plant 50 bushels and achieved a yield of 5:1, so he reaped 250 bushels from his 25 acres. At the end of that year, presuming he felt confident of another yield of 5:1 the following year, he could have set aside 50 bushels for seed corn, the same for his draught animals and 75 bushels for himself and his family: that would have left him 75 bushels to take to market.[2] If the next year's harvest turned out to be a bad one, however, yielding a return of just 3:1, he would have harvested only 150 bushels. That would have left him with nothing to take to market. After he had fed himself, his family and his animals, he would have had only 25 bushels of seed corn – half of what he needed to plant his land. Even a good harvest in the third year at a yield of 5:1 would have left him only just enough to feed his family and livestock, with no seed corn for the following year.

In reality, conditions were even harder than the above example suggests. Most peasants had less than 25 acres, yields dropped below 3:1, and the weather could cause far more damage than one slightly bad harvest in three years. When harvests failed completely due to early or late frosts – and the average temperature was decreasing at this time, with dire consequences for upland areas – people starved to death in their thousands. And when there were consecutive harvest failures, it was nothing short of a catastrophe. Europe's population was thus suffering even before the terrible famines of 1315–19. It has been estimated that 10 per cent of the population was killed by these natural adversities, which means that more than 10 million people across Christendom either starved to death or died of illnesses arising from nutritional deficiency.[3] It marked the end of three centuries of almost uninterrupted demographic and commercial expansion.

But that was nothing compared with what was to follow.

The Black Death

It is difficult to convey just what a shattering event the Black Death was. When I used to give lectures on fourteenth-century England, and stressed how catastrophic the years 1348–9 were, there would always be someone who insisted that it could not have been as terrible as the First World War, or as frightening as the Blitz. I would explain that the British mortality rate in the First World War was 1.55 per cent of the population over four years: an average rate of 0.4 per cent per year. The Black Death killed roughly 45 per cent of the population of England over a period of about seven months as it passed over the country like a wave: an annual mortality rate of 77 per cent.[4] Thus the mortality rate in 1348–9 was about 200 times that of the First World War. Or, to put it in a way that compares with the bombing of the Second World War: to replicate the plague's intensity of killing you would have had to drop not just two atomic bombs on Japan (each one killing about 70,000 people or 0.1 per cent of the population) but 450 such devices. That's two atomic bombs *every day* on a different city over a period of seven months. Had that happened, no one would have doubted that it was the worst calamity in human history. But the plague struck such a long time ago, and we are so far removed from the culture of the victims, that we cannot appreciate death on such a scale. We find it much easier to comprehend the trauma of parents who lost their much-loved sons in the First World War than the fate of whole communities that were wiped out in the fourteenth century.

The Black Death was the first wave of the second pandemic of an enzootic disease, often called bubonic plague on account of the black buboes that grew in the groin and armpits of infected victims. Its pathogen is a bacillus, *Yersinia pestis*, which is carried by fleas that normally live on rodents but which can also be spread by human fleas. In certain circumstances it can spread via the breath of infected victims too. The current thinking is that if pneumonia occurs in the course of the disease's development, the bacilli are exhaled and the airborne disease is communicated directly from human to human. In this form it is not correctly described as bubonic plague but rather becomes the far more dangerous pneumonic plague.

The first pandemic had taken place 800 years previously, breaking out in 541. That forerunner of the Black Death retained its virulence

throughout the sixth century but gradually weakened as the years
went by. It finally disappeared in the 760s. By 1347, no one had seen
plague in Europe for almost 600 years. Thus no one was prepared for
the consequences of its reappearance, which was first noted in China
in 1331. Carried by the merchants travelling along the Silk Road, it
arrived in the Crimea in the autumn of 1347, where its victims boarded
the Genoese ships bound for Constantinople. From there it spread to
Sicily, Greece, Egypt and North Africa, Syria and the Holy Land. By
the end of 1347 it had arrived in the commercial heart of Christendom
– the trading cities of Venice, Pisa and Genoa – in its most dangerous
pneumonic form. The cities affected quickly saw the bodies mount
up: mortality figures in excess of 40 per cent were normal.

News of the plague spread even faster than the infection itself.
Forewarned towns barred their gates against travellers, but nothing
could keep out something as small as a flea when the gates had to be
opened occasionally to allow in food and supplies. The plague spared
no one: rich and poor, women and children, Christian and Muslim all
perished. In Tunis, Ibn Khaldun wrote that 'it was as if the voice of
existence in the world had called out for oblivion . . . and the world
responded to its call'. Agnolo di Tura described his experience in
Siena:

> The mortality began in May. It was a cruel and horrible thing, and I
> do not know where to begin to tell of its cruelty and pitiless ways . . .
> Father abandoned child, wife husband, one brother another, for this
> illness seemed to strike through breath and sight . . . Members of a
> household brought their dead to a ditch as best they could, without
> priest, without divine offices . . . And I . . . buried my five children
> with my own hands . . . So many died that everyone believed it was
> the end of the world.

Florence was among the worst-affected cities in Europe: about 60 per
cent of the population died there. One eye-witness in the city observed
that:

> All the citizens did little else except to carry dead bodies to be buried;
> many died who did not confess or receive the last rites; and many died
> by themselves and many died of hunger . . . At every church they dug
> deep pits down to the water table; and thus those who were poor who

died during the night were bundled up quickly and thrown into the pit. In the morning when a large number of bodies was found in the pit, they took some earth and shovelled it down on top of them; and later others were placed on top of them and then another layer of earth, just as one makes lasagne with layers of pasta and cheese.[5]

The poet Giovanni Boccaccio was struck by the treatment of the deceased. He noted that 'it was common practice of most of the neighbours, moved no less by fear of contamination from the putrefying bodies than by charity towards the deceased, to drag the corpses out of the houses with their hands and to lay them in front of the doors, where anyone who made the rounds might see them'.[6] The Florentine writer Giovanni Villani himself fell victim to the plague. The last words of his chronicle were: 'and the plague lasted until . . .' The pestilence cupped its black hand over his mouth before he could fill in the date.

By January 1348, the disease had arrived in the French port of Marseilles. From there it moved northwards through France and westwards into Spain. Its deadliness did not diminish. In Perpignan, 80 of the 125 notaries died, 16 of the 18 barber-surgeons, and 8 of the 9 physicians.[7] The town's thriving money-lending business ceased to function altogether. In Avignon in France, where the popes had resided since Clement V had moved there in 1309, a third of the cardinals died. In Languedoc and Provence, half the population perished. And still the disease marched on, spreading in every direction. Givry in Burgundy, which uniquely has parish registers dating from 1334, saw burials rise from an average of 23 per year to 626 in just four months – suggesting a mortality figure of about 50 per cent. In England, every diocese saw more than 40 per cent of its priests perish: the diocese of Exeter lost over half its clergy.[8] Death rates among peasants in rural Worcestershire averaged 42 per cent, but that bald figure masks a range from lucky places like Hartlebury (19 per cent) to terribly affected manors such as Aston (80 per cent). England's two largest cities, London and Norwich, both saw mortality figures of 40 per cent. In early July 1349, a ship from London was found drifting in the Norwegian port of Bergen. When the authorities boarded it, they found the entire crew dead. They withdrew in horror from the vessel and headed back to shore. But it was too late: one of them had caught the disease. Thus the plague arrived in Norway.[9]

How many died in the Black Death? Papal officers calculated the figure to be almost 24 million Christians, which they thought amounted to a third of Christendom. Recent research has indicated that the toll may well have been much higher: 60 per cent of the population in most parts of France; possibly slightly more than 60 per cent in England, Catalonia and Navarre; and between 50 and 60 per cent in Italy.[10] Obviously, death on this scale left people traumatised, as those who performed essential daily tasks – priests, servants, cooks, cowherds, harvest workers and mothers of young children – were simply removed from the equation of daily life. In 1340, few could have imagined mortality as high as that of the famine of 1316–19, let alone anything worse. From 1347, however, Europeans had to prepare for death. In fact they had to do so again and again, as the Black Death was only the first wave of the pandemic: it returned in 1361–2, 1369 and 1374–5, and then on average every eight to twelve years for the next three centuries. And although subsequent outbreaks were not as severe as the first one, they still killed millions of people. The outbreak of 1361–2, for instance, wiped out another 10 per cent of the population of England in less than a year. A century later, the major outbreak of 1478–80 similarly claimed between 10 and 15 per cent of the population. Even 300 years after the Black Death, plague outbreaks could still kill 15 per cent of the inhabitants of a medium-sized town; in large cities it could be far worse. Over 20 per cent of the population died in London in 1563, and even larger proportions succumbed in Venice in 1576, Seville in 1649, Naples in 1656 and Marseilles in 1720–1. The fourteenth century thus heralds an age of fear. People went to bed aware that every night might be their last.

In the context of this book, however, the plague's deadliness is not its most significant feature. It is important to realise that society did not collapse. The deaths of over half the population did not mean that people threw away the rules of property ownership or abandoned the cycles of sowing and harvesting. The breakdown of law and order that occurred in some places was short-lived. In Florence, the *becchini* (grave diggers) robbed empty houses, extorted money from victims too scared to leave their homes and took advantage of defenceless women, but the lawlessness only lasted a few weeks. Although many prelates and magnates died of plague, they were quickly replaced. And Europe's rulers put on a brave face. In England, Edward III announced publicly that he would travel to France while the plague was raging there, and he did actually go, albeit only for a short time.

He also held a well-attended tournament at Windsor in April 1349, at which he completed his foundation of the Order of the Garter, while England itself was reeling under the plague. His message was simple: he believed that he enjoyed God's protection. Moreover, he was determined to display his confidence in divine approbation to his people – a brave bit of posturing, given that one of his daughters had already become a victim.

It was the long-term consequences of the Black Death that mattered most, for both secular and spiritual reasons. Medieval society had been exceptionally rigid, with people seeing their positions assigned to them by God. A manorial lord was a man bred to bear arms and command followers in battle. A shoemaker was a shoemaker and nothing more, nothing less. An unfree tenant of eight acres of his lord's land was just that. These roles were what God wanted for them. The massive population drop now caused huge cracks in this inflexible structure. Most significantly, there was a severe shortage of manpower. Working men whose families had died no longer had to accept a life of servitude: with nothing left to lose, they could simply walk to the nearest town and sell their labour. A ploughman whose children were starving no longer had to be content with farming a handful of strips of land for his lord if a neighbouring landowner was offering good wages for workers. If his lord wished to retain his services, he had to pay him better wages or reward him with more land.

Nothing separates the late Middle Ages from the earlier period as clearly as the Black Death. Although the famines mentioned at the start of this chapter meant that the optimism of the thirteenth century had fallen flat well before 1347, the plague shook the very roots of people's understanding of their place on Earth. Some had to come to terms with the almost complete annihilation of their communities, and quite reasonably they asked why God had treated them so harshly – especially when the next village might have seen far fewer victims. Could it be assumed that God sought the best for mankind if he killed babies in their cradles with this agonising, terrifying disease? In smashing the brittle material of society, the plague raised profound questions about the causes of disease. Many started to reflect on the decline of the papacy since the accession of Boniface VIII, who reigned from 1294 to 1303. From his time, the senior ranks of the Church had increasingly been connected with venality and profiteering. As the popes had fallen under the influence of a secular monarch, the king of France, their standing in the eyes of Christendom had diminished to a pale shadow of its former self. People began to doubt that the Church of Rome was leading them in the right direction. The plague, some suspected, was divine punishment on the whole of mankind for the corruption of their religious leaders.

The plague also changed people's perception of death. You might think that death is one of the few great constants in human life, but it is actually subject to quite radical alteration. Death in itself does not exist – it does not have substance – therefore it only has meaning in the minds of the living: in the absence of life and in the belief that there is some altered form of life thereafter. The latter is where the changes are to be found. All over Europe there was a profound engagement with death in the fourteenth century. Literary culture was tinged with devils, Purgatory and the afterlife. The skull motif was increasingly employed in religious paintings and sculpture. In England, the religious sect of the Lollards advocated a more intense, more spiritual way of life. The wills of Lollard knights and prelates at the end of the century increasingly stressed the loathsome, sinful condition of the testators' human flesh. *Mementi mori* – physical reminders of the rotting corpses that we will all one day become – were carved in stone and set up in churches and cathedrals. There was an intensification of the endowment of chantry chapels and the establishment of pious foundations – the building of bridges, schools, almshouses and hospitals for travellers. Beneath these individual acts of piety and expressions of self-loathing we can detect something even more profound: the unsettling question of mankind's standing in the eyes of God. What if God decided to destroy humanity entirely? After 1348, the annihilation of the human race seemed a real possibility.

For some of the survivors, however, the Black Death opened up a world of opportunities. As we have seen, the beneficiaries included those peasants who found they could sell their labour for more money than they earned on their original manor. Both England and France passed legislation to prevent the free market dictating wage rates, but these measures had little effect. Peasants realised that their labour was of value to their masters; they could insist on being treated with more dignity than previously. If they weren't, they could rebel. The peasantry had previously shown little appetite for rebellion, but as a result of the plague they acquired a sense of self-worth. This led to a number of uprisings, such as the Jacquerie in Paris (1358), the Ciompi in Florence

(1378), and the Peasants' Revolt in England (1381). Indeed, it is noticeable that throughout history, mass mortality accentuates the importance of the working man and woman, both in their own eyes and in the eyes of those who govern them.

Aspects of society that we would never have associated with the plague were profoundly affected. Marriage rights are a good example. In 1332, a bondswoman living in Doccombe called Agnes of Smallridge wanted to marry Roger the Shearman, a freeman of Moreton. The request was taken to the prior of Canterbury (the priory was lord of the manor of Doccombe). The prior refused, as Agnes would be freed from her service to the manor by such a marriage. But while in 1332 serfdom could still dictate the path of people's lives and happiness, by 1400 that system had broken down almost everywhere in western Europe. The rents that peasants had to pay to their lords declined after the fourth epidemic of plague in 1374–5, as there were fewer tenants but plenty of land. Some lords who had borrowed heavily now found themselves in debt and were forced to lease or sell whole manors to enterprising townsmen. Women like Agnes of Smallridge were now free to marry whom they chose on payment of a fine to their lord. The feudal ties that had bound workers to the land were replaced by financial obligations. Money took the place of enforced loyalty. Capitalism began to replace feudalism in the countryside, having already triumphed in the towns.

We can only touch here on a few of the major changes arising from the Black Death. Nevertheless, in terms of assessing the period for the purpose of this book, the years from 1347 to 1352 were perhaps the most formative in our history. Arguably the only years that compare are those of the two world wars, because of the rapid social changes and technological developments that accompanied them both. But even they pale into insignificance if we imagine a time when every second person suddenly died in agony – and no one understood why.

Projectile warfare

Visiting the modern gift shop of a medieval castle might leave you with the impression that medieval knights were itching to attack each other at every opportunity. Some contemporary sources support this view. The chronicles of fourteenth-century Gascony reveal fighting on a seasonal basis. The annals of medieval Ireland are filled with tale after tale of which lord stole which other lord's cattle or ambushed and murdered his son in reprisal for a similar attack the previous year. The reality, however, was that the military might of two kingdoms rarely met in battle. The whole business was just too risky. Most of the time armies only clashed when they had to – and generally speaking that only was when one king believed he had such a significant advantage over his enemy that he could not possibly lose. Knowing that your enemy was ill-equipped, tired, hungry, lacking in numbers, suffering from disease, low in morale or vulnerable to a surprise attack might sway such a decision, but even this knowledge did not ensure that a full-scale battle would ensue. Kings knew that everything depended on their own personal survival. For a king to be killed or captured in battle would not only mean a rout; it would be interpreted as God supporting the enemy and thus the entire cause would be lost. Even a demoralised and ill-equipped army could win a battle if they were lucky and managed to kill the enemy commander.

What really mattered in 1300 was having better-armed and better-trained men than the enemy – and more of them. Hence the importance of knights, and especially their massed charge, which had been a battle-winner since the eleventh century. Groups of highly trained knights with lances would gallop in formation across a battlefield on specially bred large horses, called *destriers*, sweeping everything before them: a tsunami of fast-moving hardened spear points ripping apart everything in its path. The only occasions when such a tactic did not prove decisive were when a strategic blunder rendered it counterproductive – such as when the charging knights pursued their enemy too far and became cut off from their infantry, or when the ground was so sodden that the horses slowed up and the charge lost impetus. At the end of the thirteenth century, however, the Welsh and Scots armies opposing Edward I of England found an effective defence against the massed cavalry charge. This was to arrange their troops in schiltroms: circles of men armed with spears with their backs to each other. Jamming their spears into the ground, they tilted them outwards, so that any charging horses would impale themselves or shy away. At Bannockburn in 1314, the self-proclaimed king of the Scots, Robert Bruce, neutralised the massed charge of English knights by equipping his men with sixteen-foot-long pikes and arranging them in schiltroms.

Combined with the dampness of the ground, the strategy worked. The flower of English chivalry ended up using their horses to escape from the butchery that ensued.

Bannockburn proved a decisive Scottish victory but ironically it also sowed the seed for English military dominance for the rest of the century. This was because it inspired a desire for revenge in those northern English lords who had held land in Scotland. One of them was Henry Beaumont, an experienced soldier and capable leader. Another was Edward Balliol, a claimant to the throne of Scotland who was eager to press his claim after Robert Bruce's death in 1329. In 1332, these two men led a group of English knights – known to historians as 'the Disinherited' – to Scotland to reclaim their lost ancestral lands. Beaumont had been at Bannockburn and he knew exactly what the English should have done there. Since schiltroms were slow-moving, poorly armoured and vulnerable to archers, he made sure he took 1,000 bowmen with him.

Soon after landing at Kinghorn, in Fife, the Disinherited found themselves trapped by a massive army. Scottish chroniclers state that they faced 40,000 men; English writers say 30,000.[11] Beaumont and Balliol had no more than 3,000 men in total, including their archers. On 10 August 1332, on Dupplin Moor, they fought a last-ditch stand against odds of about ten to one. Stationed equally on each flank, the archers not only broke up the Scottish schiltroms but also impeded the charge of the advancing Scots by trapping them in a killing zone between the flanks. The Scots could not retreat because their comrades were pressing them forward from behind; all they could do was clamber over the bodies of the fallen, exposing themselves to the deadly arrows. A northern chronicler noted that 'one most marvellous thing happened that day, such as was never seen or heard in any previous battle, namely that the pile of dead was greater in height from the earth towards the sky than one whole spear length'. And, remember, in 1332, a spear was 16 feet long.[12]

The strategy that Henry Beaumont employed that day became the basic principle of modern warfare. You do not wait to engage enemy troops in hand-to-hand combat; you shoot them before they get close enough to bash you over the head or stab you in the guts. The key element in this strategy is the *massed* use of projectile weapons, which produces much the same effect as a machine gun. A few dozen archers might have broken up a schiltrom but only the coordinated bowmanship of a thousand could stop the charge of well-equipped cavalry. Although a longbow was only accurate over a distance of about 200 yards, and could only penetrate armour within about half of that, the tactic of bringing down the leading enemy riders when they were nearly upon you meant that all your targets became stationary directly in front of you, in the killing zone. Each archer could shoot an arrow every five seconds; at Dupplin Moor, this meant a deadly onslaught of 12,000 arrows per minute. It did not matter that they were not all directly aimed; it amounted to a storm of arrows concentrated on the ground about a hundred yards ahead of the English lines. No army could advance in the face of such a deadly attack. We do not know whether Henry Beaumont had all this planned out in advance or whether it was luck. Either way, within a matter of days, news of his victory had reached the ears of Edward III of England.

At Edward's birth it had been prophesied that he would be victorious in war: the news from Dupplin Moor gave him the blueprint for fulfilling this prediction.[13] Within a year, at the age of just 20, he led a modest but well-equipped English army against the Scots at Halidon Hill and took revenge for Bannockburn. That was just the start. In the late 1330s, as the French sided with the Scots in the developing war, he used longbows to defeat them at sea and on land. Most famously of all, in 1346, in response to the exhortations of Parliament, he set sail with about 15,000 men, landed in France and marched on Paris. It was unthinkable that the king of a small country like England could take on the military might of France, the most powerful kingdom in Christendom. But at Crécy, on 26 August 1346, Edward lured the French to attack him. The arrows of his 5,000 archers first ripped through the enemy crossbowmen and then massacred the French knights and men-at-arms. The guns he had hauled across the uneven roads of Normandy opened fire on the enemy's men and horses, shooting bolts and cannonballs into their ranks. The French fought bravely but were annihilated. Crécy demonstrated to an international audience the effectiveness of the strategy employed at Dupplin Moor and Halidon Hill: if you equip your soldiers to shoot projectile weapons, you can defeat a large army with a smaller one. Moreover, you can do so again and again because your soldiers carry comparatively less risk of being injured. For better or for worse, a new chapter in the military history of the world had opened.

Edward III's greatest stroke of luck was that his kingdom was full of good bowmen. The regular Scottish incursions had led to a concentration of trained archers in the north of the country.[14] In addition, the partial breakdown of law and order around 1290–1320, as the famines took effect, caused many men to take action to defend their communities. In the north of England and the Welsh Marches there developed a very positive tradition of them doing so with longbows, which they had learned to shoot from boyhood. Whereas in other countries an archer was a man of low status, in England he was respected. This strong culture of archery also meant that large supplies of arrows were available. Edward III could simply place an order for three million arrows to be gathered; he did not have to wait for them all to be specially made. Bowmen, shooting rapidly together as a unit, proved to be his trump card in any battle.

Edward was not the sort of man simply to take this piece of good fortune for granted. He realised that what he could do with bows could also be done with guns, which had arrived from China a few decades earlier. In the 1340s, he ordered the production and stockpiling of gunpowder at the Tower of London; by 1346, he was producing two tons of gunpowder per year. His short, bulbous cannon had a range of at least three quarters of a mile. He also developed ribalds that contained multiple gun barrels. At the siege of Calais, in 1347, he used cannon to bombard the walls. Later in his long reign, at Dover Castle, Calais, and Queenborough, he built the earliest artillery fortifications and equipped them with cannon to guard the sea. By the end of the century, England and France were both using guns in their war against one another. Most other European countries did likewise. You can see the Loshult Gun, dating from the early fourteenth century, and the Mörkö Gun, from about 1390, in the Statens Historika Museum in Stockholm. Other fourteenth-century guns are in museums in Cologne and Nuremburg in Germany, Paris and Provins in France, Milan in Italy and Berne in Switzerland.

Despite the later superiority of guns over bows, it is important not to exaggerate their effectiveness in the fourteenth century. At that time, their main value lay in siege warfare. In 1405, Warkworth Castle capitulated after seven shots were fired at it from a large gun. Shortly afterwards, a single shot from a large cannon brought down the Constable Tower at Berwick, forcing the Scots to surrender. Thus cannon were just another type of siege engine: Christendom was no stranger to trebuchets and other machines that could fling heavy stones at a castle and bring down the walls. But they were slow, heavy and inaccurate: a thousand longbows were far easier to produce, maintain, transport and shoot. It was principally longbows, therefore, that marked the major change in warfare. When Henry V attempted to emulate his famous great-grandfather and engage the French in battle on their own soil, he used cannon in the siege of Harfleur but relied on longbows to defeat the French cavalry at Agincourt in 1415.

Once Edward III had demonstrated the effectiveness of massed longbows, he imbued his archers with the confidence to carry their own battles further afield. Companies of English mercenaries and renegades were to be found in France in the 1350s and 1360s. Others, like Sir John Hawkwood, took their skills to Italy and made fortunes fighting in the wars there. At the battle of Aljubarotta in 1385, a small group of English archers helped the Portuguese to defeat the French cavalry. In England itself, all men were now compelled every week to shoot at the butts, to preserve this military advantage. Only in the sixteenth century did the technology of muskets, arquebuses and handguns finally make longbows redundant – with the exception of 'Mad Jack' Churchill, the eccentric English officer who still fought with a longbow in the Second World War. The fundamental principle of one army systematically attacking another from a distance rather than engaging in hand-to-hand combat has never gone away, however. It could be described as one of the most significant dividing lines between the ancient world and the modern.

Nationalism

Most of us, including most historians, associate nationalism with the modern world – largely because of its contemporary relevance and its role as a motivating factor in some of the twentieth century's greatest atrocities. It is normally said that medieval monarchs reigned over kingdoms, not nations. Yet the roots of our concept of nationalism lie in the Middle Ages, and it emerges forcefully in the fourteenth century. To be more specific, nationalism appeared in three forms at that time. First it was an expression of identity, in the way people described themselves as a group when they were away from home or among people from different countries. In an ecclesiastical sense, the term 'nation' denoted a group of prelates from a certain part of Christendom. And in a political context, the term began to be used when a king and his people were united in pursuing interests that they had in common, rather than objectives that were purely local, aristocratic or royal. These three types of nationalism collectively represent a powerful force that arguably has continued to shape the world ever since.

From the thirteenth century, the greater regularity of travel and the increase in international trade meant that there were more people living abroad. Understandably, they wanted to surround themselves with people who spoke the same language, shared the same bonds of loyalty, and understood their customs (and jokes). When merchants of the Hanseatic League, the confederacy of German and Baltic trading cities, established an enclave in a foreign port, they banded together and were referred to as a 'nation'. Similarly, university students from the same country flocked together at the most popular international universities. In the early fourteenth century, the University of Paris had four recognised 'nations': French, Norman, Picard and English. In certain border towns, people of one kingdom used the word 'nation' to divide themselves and their friends from those belonging to the other kingdom. In 1305, for instance, the townsmen of the English 'nation' resident in the border town of Berwick submitted a petition to Edward I to banish those townsmen in Berwick who belonged to the Scottish 'nation'. Nationhood was thus used to define not only friends but, by implication, foes too.

From 1274, the many archbishops and bishops who attended the Church's ecumenical councils also gathered together in 'nations' to discuss and vote on motions. At the council of Vienne (1311–12), they divided themselves into eight nations: Germans, Spaniards, Danes, Italians, English, French, Irish and Scots. These ecclesiastical nations did not relate directly to political realms: there was no such thing as a kingdom of Spain in 1311, and the German nation similarly included prelates from many separate states. The concept of nationhood in this context had more to do with common culture, common language and travelling together than with loyalty to a secular monarch. In 1336, Pope Benedict XII reduced the number of ecclesiastical nations to just four: the French, Italians, Spaniards and Germans (with the English being placed with the Germans).[15] The rise of England's international standing after Edward III's victories over France put an end to it being subsumed in the German nation, however, and at the Council of Pisa (1409), the prelates acknowledged five nations: English, Germans, French, Italians and the absent Spaniards. At the Council of Constance (1415), the question of what constituted a nation was hotly debated. The French insisted that, as the Czechs and Hungarians were part of the German nation, the English should again be subsumed within that grouping. The English stood firm for their independence, telling the most outrageous lies to bolster their case. They declared, for instance, that the British Isles contained 110 dioceses, and that the Orkneys (which were ruled by Norway but were ecclesiastically part of the English nation), numbered 60 islands that were collectively bigger than France.[16]

The reason why ecclesiastical nationalism had become such a heated point of discussion was because it had taken on a political dimension. The kings of England, Scotland and France increasingly needed to widen their base of supporters, and every sort of allegiance was valuable to them. For example, in 1302, the argument between Philip IV of France and Pope Boniface VIII led to the pope issuing the bull *Unam Sanctam* and threatening to make himself ruler of all Christendom in temporal as well as religious matters. In response, Philip decided to summon the Estates General of France for the first time. Representatives of the lords, clergy and principal towns came from the whole ecclesiastical nation of France, regardless of whether they owed allegiance directly to the king of France or were vassals of the semi-autonomous dukes of Brittany and Burgundy. If the whole ecclesiastical nation could be enlisted to support the king's cause, it hugely bolstered his authority. In a similar vein, if the English representatives at a Church council were recognised as a nation in their own right, they had an equal voice to that of the French and could resist their traditional foe's initiatives.

In England, Edward III employed national interests for a wide variety of domestic and international purposes. He realised that it was in the national interest to maintain a constant war with Scotland, and later with France, for by doing so, he stopped his most powerful lords fighting among themselves. Thus he delivered several decades of uninterrupted domestic peace in England. Parliament approved of the policy and supported the king in his war in France, even though it incurred extra taxation. Through promoting a sense of nationhood, kings were able to foster a sense of unity: those who paid their taxes at one end of the country did so in order to defend those at the other. No matter which locality they were from, Englishmen were defined by their enmity of the French and Scots and their loyalty to all things culturally and geographically English, not just their king. Frenchmen and Scotsmen were similarly defined as nations by their opposition to the English.

Political nationalism was also affected by relations between kings and popes. In 1305, Philip IV engineered the election to the papacy of a Frenchman, who took the papal throne as Clement V. Following considerable hostility to his appointment in Rome, Clement moved the papal curia to Avignon in 1309. Thus the papacy came to be practically a French institution: all six of Clement's successors and 111 of the 134 cardinals created before 1378 were French. Inevitably the Church became drawn into the politics of the conflict between France and England. One scornful English joke after Edward III's run of victories over the French went: 'Now the pope has become French and Jesus has become English; soon we'll see who will do more: the pope or Jesus.'[17] Edward further undermined the authority of the pope by introducing legislation to restrict papal appointments and income in England. He also confiscated the revenues of monasteries that had French mother houses. English poets started to write vituperative lyrics against the French, such as the following lines by Laurence Minot, from the 1330s: 'France, womanish, pharisaic, embodiment of might, lynx-like, viperish, foxy, wolfish, a Medea, cunning, a siren, cruel, bitter, haughty: you are full of bile.'[18]

This is a long way from the days when magnates would regularly travel between their manors in England and France. It is also a long way from the idea of the pope being St Peter's successor as the bishop of Rome. National tensions arising from the papacy worsened in 1378, when Pope Gregory XI moved the Holy See back to Rome. When he died shortly afterwards and an Italian was elected to succeed him, a rival French pope was chosen by the French cardinals who remained at Avignon. Two popes now reigned over the various nations of Christendom. The English, Germans and Italians looked to the pope in Rome; the French and Spanish nations, together with Scotland, supported his French counterpart at Avignon. It was hardly surprising that the Council of Constance, which gathered in 1415 to end this schism, spent almost as much time arguing over what constituted a nation as it did matters of religion.

Elsewhere, there were varying loyalties to national interests. In Germany, the role and power of the Holy Roman Emperor implied a higher layer of duty to the empire. In reality, however, men's allegiance to their own lord or monarch took precedence. In the Iberian peninsula, cultural differences meant that allegiance to one's *corte* or *cort* highlighted the sharp distinctions between the kingdoms of Portugal, Castile-León and Aragon. The Scandinavian realms co-existed peacefully within a trading alliance, the Kalmar Union, which was made up of the kingdoms of Denmark, Sweden (which included Finland) and Norway (which included Iceland, Greenland and the Faroes, the Shetlands and the Orkneys). In such places, nationhood implied fraternity, not bitter rivalry with a neighbouring power. Of course, things were different in Italy. Italian cities and nobles had been divided since the twelfth century into two factions, Guelphs and Ghibellines, with the former supporting the pope and the latter the Holy Roman Emperor. In the fourteenth century, after the defeat of the Ghibellines, the Guelphs had divided into Black Guelphs and White Guelphs to fight among each other. Ultimately the loyalty of Italian men was not to Italy as a whole but to their city state or the kingdoms of Naples and Sicily, and beyond that to their Guelph or Ghibelline cause. Italian nationalism would not become a major force until the nineteenth century.

Despite these different shades and degrees of nationalism, it was in the fourteenth century that national interests became overtly more important than the unity of Christendom or the authority of the papacy. In 1300, men of substance were loyal to their lord in secular matters, and to their local bishop – and ultimately the pope – in religious ones. By 1400, things were no longer that simple. Loyalties were both locally and nationally aligned. Religious orthodoxy, taxation, parliamentary representation, language, laws and customs were all subsumed in the concept of nationhood. Hence you could oppose your king and still be loyal to your nation. Indeed, national priorities saw two English kings deposed by their parliaments in the course of the century: Edward II in 1327 and Richard II in 1399. There was even talk at the end of his long life of deposing Edward III, and early in the next century, there was an attempt to depose Henry IV too. Although it was not unusual before 1300 for European monarchs to lose their throne to a rival, it was very rare for a hereditary monarch to be deposed by his own people for failing to act in the national interest (the deposition of Sancho II of Portugal in 1247 was one of the very few examples). The fourteenth century changed that. As for loyalty to the pope, by 1400, national interest required people to be sceptical of the authority of one of the two popes, if not both of them. The English theologian John Wycliffe duly pondered on the weakness of the whole Church hierarchy and advocated obedience not to the pope but directly to Christ. It was a remarkable change from the previous century, when Innocent III's voice had thundered out across Christendom, and even kings quailed.

Vernacular languages

We tend to think that within our own lifetime things have changed faster than ever before. While this is perhaps true with respect to our use of electronic devices, the way we speak and write has changed slowly in recent times. In the English-speaking world today, millions of people can read Jane Austen and enjoy the language, which has hardly altered over the last two hundred years. Shakespeare's works are for the most part intelligible to us after more than four hundred years, even if a few words here and there have shifted their meaning and some of the grammar is difficult. In the Middle Ages, however, language changed rapidly. You may understand a good chunk of Geoffrey Chaucer's late-fourteenth-century *Canterbury Tales*, but it is unlikely that you will comprehend much of the Middle English verse written by his predecessors a century earlier. The same thing can be said for French, which saw a rapid development in the early fourteenth century from Old French (or *langue d'oil*) to Middle French, as the language lost its declension system. German also saw considerable development, as Middle High German shifted into the modern language. Later on, printing would stabilise words and syntax and set a standard for each language, but before print became common in the sixteenth century, there were no linguistic anchors, so languages changed with every generation. The truism here is that if things can be standardised, they have a much better chance of lasting for centuries – whether they are units of measurement or the words we use.

What matters about vernacular languages in the context of this book, however, is not primarily their internal linguistic changes but their use – their external history, so to speak. The various vernacular languages of Europe were already old by the start of the fourteenth century. The earliest extant documents in Old French date from the ninth century, in Anglo-Saxon from the seventh and in Slavic from the late tenth century; the oldest texts in Norwegian and Icelandic are from the twelfth century, and in Swedish and Danish from the thirteenth. But throughout Europe, Latin was the more commonly used language for record-keeping and formal literary composition. In the twelfth and thirteenth centuries, the aristocratic troubadors wrote thousands of poems in the vernacular languages of southern Europe – Galician-Portuguese, Occitan and Provençal – but their importance was largely recreational and had little or no impact on the lives of ordinary people. The same can be said of their German-singing counterparts, the Minnesingers. What happened with the vernacular languages of Europe after about 1300 (although slightly earlier in Castile) is that they came to be allied with the forms of nationalism we encountered above, and were regarded by rulers as the principal languages of their kingdoms. Latin was increasingly marginalised as the language of scholarship and the Church. Just as the influence of the pope waned and national interests grew, so the importance of the common tongue of the people rose in every region.

That link between national pride and the vernacular emerges clearly from the English evidence of the period. In 1346, in order to secure the support of Parliament for extra taxation, a Franco-Norman invasion plan dating back to 1338 was shown to the members, being described as 'an ordinance . . . to destroy and ruin the whole English nation and language'. This is a remarkable statement: almost no members of the nobility and gentry had spoken English in 1300, yet barely four decades later its preservation was being presented as crucial to the survival of the English nation. In 1362, the king confirmed in the Statute of Pleadings that men should be allowed to plead in English, confirming it as 'the tongue of the country'. Shortly afterwards, his chancellor began to use the vernacular for his speeches when opening Parliament. In 1382, another parliamentary record connects national interests with the English language:

This kingdom has never been in as much danger as it is in now, both within and without, as will be apparent to all who possess either reason or judgment: so that if God does not bestow his grace on the land and the inhabitants do not strive to defend themselves, this kingdom will be on the verge of being conquered, which God forbid, and made subject to its enemies; and as a consequence, the language and nation of England will be completely destroyed: so that now, we are faced with only two choices, to surrender or to defend ourselves.[19]

By the end of the century, English had become the dominant tongue, and most of the royal family spoke it. Edward III composed several mottos in the language. Henry IV swore his coronation oath in English in 1399. For John Wycliffe and his supporters, it was imperative that the Bible be available in English – to encourage that direct allegiance to Christ, rather than the pope, which we noted above. Geoffrey Chaucer chose to write in English rather than French, retaining the structural forms of French poetry but communicating in the tongue of the nation. The fourteenth century saw the flowering of the English language as an element of national pride.

Other kingdoms across Europe were following a similar path. Portugal and Galicia shared a language at the start of the fourteenth century, Galician-Portuguese. It was one of the languages of choice for the troubadors. Yet this vibrant vernacular fell apart in the four-teenth century as Portuguese and Galician speakers went their separate ways. In the late thirteenth century, Castilian was standardised at Toledo under the personal influence of the king of Castile, Alphonso the Wise. He commissioned many works of law, history, astrology and geology and insisted that they be made available in Castilian so his people were able to understand them. What he started continued in the fourteenth century with the works of his nephew, Juan Manuel, Prince of Villena, and Juan Ruiz, 'the Spanish Chaucer'. By the end of the century, Castilian had replaced Galician-Portuguese as the Iberian language of choice for lyric poetry. It was also the language in which the nobleman Pedro Lopez de Ayala wrote his many works, including chronicles, a satire on society, and a book of falconry. A similar attempt was made to standardise Aragonese and present it as a national language. Juan Fernández de Heredia, grand master of the Knights Hospitaller, created a corpus of Aragonese literature and brought about a golden age for the language in the late fourteenth century.

Old French was already a prestigious language by 1300: Marco Polo's book of his marvellous travels was written in French, not Venetian. Nevertheless, this language too saw considerable change in its impor-tance. Outside France it lost ground to other vernaculars (such as

English, Italian and Castilian), but within the borders of France it slowly established itself as the tongue of the nation, edging out the twenty or thirty regional dialects. In the north of France, the last writer of stature to use the Picard dialect was Jean Froissart, a late-fourteenth-century chronicler and poet. By the end of the century, Middle French was making inroads into areas where Occitan and Provençal was spoken. In the cities and towns of the Holy Roman Empire, letters, wills and chronicles were more frequently written in German. Among the Slavic languages, Polish and Czech speakers produced literary texts for the first time. Hungarian writing also first appeared in the fourteenth century. Everywhere across Europe there was a great linguistic shift in education and composition, from Latin to the vernacular languages, encouraged more often than not by a new sense of national pride and patronised by the monarch.

And then we come to Italy, the exception to almost every generalisation about medieval Europe. Only here did the rise of the vernacular lack the concomitant nationalism that we notice elsewhere. The Italians were late in breaking away from Latin, no doubt for the simple reason that Italy had been the birthplace of the language and it was where the influence of the Roman Church remained strongest. The earliest extant document written entirely in an Italian vernacular dates from about 960 but examples of Italian from before 1200 are rare. In the thirteenth century a number of Italian poets chose to write in Provençal, and Marco Polo's amanuensis was not the only Italian to use French: Brunetto Latini, Dante Alighieri's teacher, did likewise. This wide range of Romance languages in use in Italy in 1300 was described by Dante in his study of the nobility of vernacular tongues, *De Vulgari Eloquentia* (ironically also written in Latin). His own great composition, *The Divine Comedy*, was composed in Tuscan, the language of his native Florence. This work commanded such respect throughout Italy that soon after it was published it became a benchmark of Italian culture, a demonstration of what could be achieved in the vernacular. A host of Florentine writers took up the challenge of widening cultural horizons through the use of Dante's Tuscan form of Italian. The Florentine Giovanni Villani wrote his chronicle in the vernacular and praised Dante in its pages. Soon after Dante's death, Boccaccio wrote the first biography of him – in Italian, of course – and a little later, Petrarch produced his enduring poetic models for the Italian tongue. In Italy as elsewhere in Europe, by 1400 the language of the people had become, for rich and poor, literate and illiterate, the language of choice.

Conclusion

Two of the four changes picked out here are imbued with death and tragedy. But beneath the dark clouds of plague and war, many smaller things glitter. In Italy, at the start of the century, Giotto was painting expressive faces that told of human pain and suffering – the first artist to do so with a degree of depth and perspective. By the end of the century, Italian art was in demand all across Europe, especially in the form of altarpieces. On a more mundane level, buttons came into use at the courts of England and France in the 1330s, allowing clothing to be tailored elegantly to fit the human body rather than having to hang from the shoulders, as it had in previous centuries. A golden rose made by Minucchio da Siena at Avignon, and now in La Musée de la Moyen Age in Paris, shows just how exquisite the art of the goldsmith had become. Culturally speaking, the fourteenth century was an age of brilliance. Enamelled gold cups abounded at the courts of Europe; kings and courtiers listened spellbound to their minstrels; and some of the finest poetry ever written was composed. But this book is not about artistic masterpieces; rather it is about society as a whole, and few peasants ever saw Giotto's art. For the vast majority of people the century was characterised by famine, plague, war and conquest. All four horsemen of the apocalypse rode into town, and everyone shuddered. The gleaming treasures and bright tunics of noblemen might serve to remind us of the sophistication of medieval taste, but fourteenth-century people were more concerned with the proximity of death than they were with cultural innovations and Earthly delights.

The principal agent of change

The plague caused more change than anything or anybody else in the fourteenth century. But if we have to select an individual who consciously changed his world more than anyone else, it has to be Edward III of England.

Despite the fact that Edward is the only king to be reckoned a 'principal agent of change' in this book, these days he is almost forgotten. When the BBC carried out a poll in 2002 to find 'the 100 greatest Britons' of all time, many far less important monarchs featured, but Edward failed to make the list. It marks an extraordinary downturn in his reputation. His epitaph in Westminster Abbey describes him as 'the glory of the English, the flower of kings past, the pattern for kings to come, a merciful king, the bringer of peace to his people . . . the undefeated warrior, a second Maccabeus . . .' Even 300 years after his death, a Cambridge scholar called him 'one of the greatest kings that perhaps the world ever saw'.[20] The reason he has come to be neglected in recent times is because priorities change and, as time goes by, we take more and more things for granted. Few of us today stop to think about how English came to be the tongue of the English nation, or how ordinary people rather than the knightly class came to dominate the battlefield. In addition, Edward's achievements are not the sort of things we like to celebrate. He demonstrated the effective use of projectile weapons on the battlefields of Europe, and did more for militant nationalism than anyone else of his time. But in order to judge him fairly, we have to remember that nationalism was a very different thing in the fourteenth century. Forging a nation in which the king and Parliament had to negotiate with one another was a remarkably enlightened initiative in the Middle Ages, preceding as it did the absolutist monarchies of later centuries. Whether you admire Edward or not, he has to be singled out as the principal agent of change on account of his contribution to the development of methods of war, for the impetus he gave to nationalism in England and France, for his role in promoting the vernacular, and for starting the conflict that later became known as the Hundred Years War, described by one modern military historian as 'perhaps the most important war in European history'.[21]

The Fifteenth Century

You might recall the quotation from Francis Bacon at the start of this book: 'Printing, gunpowder and the compass – these three have changed the whole face and state of things throughout the world.' All three of his catalysts for change developed in the fifteenth century. The printing of texts was introduced to the West in grand fashion when Johannes Gutenberg produced a complete edition of the Latin Bible in 1455. Although gunpowder itself had been known for over a hundred years, the casting of cannon grew considerably more sophisticated. The Dardanelles Gun, for example, cast in bronze in 1464, weighs 37,037 pounds, is 17 feet in length and can blast a cannon ball 2 feet in diameter over a mile. Such guns were used by the Turks to bring down the walls of Constantinople in 1453. The compass similarly came into its own in this century as explorers crossed the Atlantic and Indian Oceans. Finally, although Francis Bacon failed to mention it, there was the small matter of the Renaissance, which represents a profound and energetic change in human awareness and thinking. On the face of things, the fifteenth century has a good claim to be the one that saw the most change over the last millennium.

The overwhelming characteristic of the century, however, is war. The rise of the Ottoman Empire dealt several savage blows to Christendom. Constantinople, the capital of the once-great Byzantine Empire, fell to the Turks, the last emperor dying in the desperate defence of his city alongside his soldiers. The Turks also seized Serbia, Albania, Bosnia, Bulgaria and much of Greece; they overran the Genoese trading outposts in the Black Sea and a number of Venetian possessions in the Mediterranean. For Italians, these losses were only one of their worries. This was the golden age of the *condottieri*, the mercenary war captains who sold their services to whichever city would employ them, and they were much in demand. Padua was defeated by Venice in 1405, the same year that Florence conquered Pisa. Venice waged a war against Milan for 21 years which finally ended in 1454. Genoa succumbed to the armies of Milan in 1464. Florence's long-standing conflict with Milan was finally resolved in 1440, allowing the Florentines to concentrate on wars with Naples and Venice. The Neapolitans sacked Rome in 1413. In the 1490s, the French invaded Italy, defeated the Florentines and overran Rome on the way to attack Naples. It becomes almost comical to describe the seemingly unending appetite of the Italian cities for fighting each other.

Nor can we say that the Italians were unusual: every country in fifteenth-century Europe experienced war, and many of them were torn apart by civil wars – the least noble and most desperate sort of conflict. In 1400, and again in 1500, England was ruled by a King Henry who had taken the throne by force from a King Richard and killed him, and who then suffered a number of rebellions before passing the throne to his son, also called Henry. Between those two successions there was much bloodshed. Henry IV had an ongoing conflict with the dissident Welsh lord Owen Glendower. Henry V aggressively restarted the war with France to prove his dynasty's legitimacy in 1415. After his death in 1422, his heirs had to demonstrate their right to the thrones of both France and England through repeated victories. Talk about a poisoned chalice! After the English were finally thrown out of France in 1453, the war they had been fighting simply moved on to English soil.[1] That phase of the conflict, the Wars of the Roses, continued intermittently from 1455 until the Battle of Stoke in 1487. Almost every landed family in England lost men or lands in the Wars of the Roses.

In Spain, too, we can find a whole assortment of military struggles. There was a war with the Hanseatic League (1419–43), the civil war of the mid 1440s and the War of the Castilian Succession (1475–9). These were followed by a 10-year-long invasion of Granada, ending with the completion of the Reconquista in 1492. The Dutch also fought the Hanseatic League (1438–41) and coped with two civil wars (1470–4 and 1481–3). In eastern Europe, the Lithuanians had a civil war over the succession (1431–5); the Teutonic Knights were finally destroyed by the Poles in 1466; the Hungarians and their allies fought the Turks until they were crushed at the battle of Varna (1444); and there were four crusades against the followers of Jan Hus, in Bohemia (1419–34), not to forget a 10-year war between Bohemia and Hungary (1468–78). And this is just the tip of the military iceberg: there were many other local and less prominent conflicts.

You cannot help but wonder what would have happened if there had been greater peace in Europe in the fifteenth century. Would there have been less change or more? Indeed, here we confront a profound historical question. In the modern world, conflicts undoubtedly speed up technological progress, as states compete against one another, and they can have a positive effect on social development. But was this also the case in the fifteenth century? The wars in Italy offered a wealth of opportunities to Renaissance artists, whose painterly skills were useful in the propaganda battles waged between competing families and rival city states. Engineers, whose skills could be employed in building walls and bridges, were similarly empowered. But at the same time, militarisation reduced the money available for the patronage of artists, scientists and writers. Violence and uncertainty inhibited trade and thus reduced the vitality of the towns and ports whose livelihood was threatened by enemies on land and sea. Indeed, many towns shrank in size. As a result, it is probably fair to say that war triggered change in some respects and inhibited it in others in the century of Gutenberg and Columbus.

The age of discovery

One of the most profound changes of the last thousand years has been the expansion of the West beyond the borders of Europe. It was not the compass that brought this about, despite Francis Bacon's assertion. That instrument had been invented more than two hundred years earlier, as we saw in the chapter on the twelfth century, but it had had little effect. Early-fourteenth-century navigators had reached the Canary Islands, and although the news of that discovery spread throughout Europe with the papal appointment of a 'prince of the Fortunate Islands' (as the Canaries were then called), it did not lead to many further voyages of discovery. As so often was the case, it was not the technological innovation that mattered but money and the political will to explore, which were frequently entwined with each other. The technology just facilitated the realisation of this combined, intensified ambition.

A crucial element behind the sudden desire to discover new lands was the circulation of manuscripts describing the wealth of foreign kingdoms. Marco Polo's work was popular, with its stories of populous cities, strange cultures and rich treasures. The equally bejewelled and almost entirely fictitious *Travels of Sir John de Mandeville*, written in the fourteenth century, was even more widely read. The retelling of such stories no doubt suggested to many poor sailors that voyages beyond the horizon might be the way to secure a fortune. In 1406, Giacomo de Scarperia's translation into Latin of Ptolemy's eight-volume *Geography*, written in Greek in the second century AD, triggered debate as to what lay beyond the known borders of the world, and how it might be systematically plotted using latitude and longitude. However, exploration was exceedingly dangerous, and, generally speaking, the armchair travellers who could read *Geography* in Latin were not the sort of people who could lead pioneering expeditions. The adventurers in society were not motivated by curiosity but by gold. However, when a highly educated prince, who had boundless curiosity and could afford to equip an exploratory expedition every year, met with a crew of treasure-hungry sailors, the world changed.

The prince in question was Henry the Navigator (1394–1460), the third son of King John I of Portugal and the great-grandson of Edward III of England who, by then, was famed as the greatest chivalric monarch Christendom had known since Charlemagne. In 1415 the youthful Henry, ardent for some glory of his own, persuaded his father to sail with a large army and lay siege to the strategic port of Ceuta, opposite Gibraltar, at the very tip of North Africa. The expedition was a success. Ceuta fell and Portugal achieved its first bridgehead in what is now Morocco. But that was just the start. From 1419, Prince Henry sent ships every year to explore more of Africa. Repeatedly they failed to get further than Cape Bojador on the western coast of the Sahara, which was notorious for its fogs, violent waves and strong currents. Sailors claimed that the 'Green Sea of Darkness' (as the Arabs called it) would wreck them if they sailed beyond that point.[2] The plain fact was that, after sailing for miles along the desert coast, they could find precious little reason to risk their lives going further. However, in 1434, Gil Eanes, one of Prince Henry's captains, rounded Cape Bojador and returned. That old Green Sea of Darkness excuse simply wouldn't wash any more.

In 1441–3 two captains, Afonso Gonçalves and Nuno Tristao,

independently reached Cape Bianco, where the Saharan coast ends. Their return put the exploration of Africa on to a new footing, for they brought back black slaves and gold dust. All fears of the dangers of the Atlantic evaporated in the excitement of imminent wealth. In 1455, Prince Henry employed the Venetian sailor Cadamosto to sail to the Guinea coast and push the boundary of knowledge even further southwards. By then, the Portuguese had begun trading horses for slaves in that part of Africa: one horse for nine or ten slaves, in case you're wondering about the exchange rate. To facilitate such a trade, Portuguese shipwrights adapted traditional ship designs to the conditions of the Atlantic, producing the lateen-rigged caravel, which could sail much closer to the wind than anything previously built. The Portuguese Crown kept up the momentum of Henry's expeditions by granting him a fifth of all the profits from southern exploration and the sole right to authorise voyages, effectively giving him the right to subcontract the exploration business, including the slave trade.

The Portuguese Empire may have been born out of chivalry and crusading zeal, but by the time of Prince Henry's death in 1460, it was driven by profit. The more money expedition leaders made, the easier it was for them to persuade backers to fund their voyages and the more ambitious they became in venturing south. The islands of São Tomé and Principe – in the armpit of the African continent – were discovered in the early 1470s. In 1482, King John II ordered a fort to be constructed at Elmina, on the Gold Coast, to protect and preserve Portuguese interests in the region – the first of many such forts, or 'factories', that the Portuguese built to run their seaborne empire. Two years later the king commissioned a team of experts to determine the best way of calculating latitude, by reckoning the position of the Sun. In 1485 Diogo Cão reached Cape Cross, south of the mouth of the Congo. In 1488 Bartholomew Dias discovered that, by sailing south-west – *away* from the African mainland – you could catch south-westerly winds that would speed you around the Cape of Good Hope. Thus the Portuguese discovered the route into the Indian Ocean.

Great discoveries are infectious: they encourage others to set out on their own expeditions. John II sent emissaries by land to Cairo and by sea to Calicut in India, to learn how to trade with the spice merchants there. His intention was, of course, to dominate such a trade, wresting control of it from the Arab seafarers who had hitherto traversed the Indian Ocean unopposed. In 1497 Vasco da Gama set out with four ships and sailed straight around the Cape of Good Hope and all the way to India. Two of his ships returned safely to Portugal in 1499. The news of the journey caused John II's successor, Manuel I, to send out another fleet of 13 vessels under the command of Pedro Cabral, with da Gama's navigators to help guide them. Cabral's fleet sailed westwards across the South Atlantic, hoping to pick up the winds that Dias had discovered 11 years earlier. Instead, his fleet sailed so far west that it landed on the coast of Brazil. What once had been a modest annual advance of a few miles down the African coastline now became a massive sweep of global trade from Portugal across to Brazil, from Brazil back across the South Atlantic and around the Cape of Good Hope, then up the east coast of Africa and across the Indian Ocean to Calicut, in southern India. Just 85 years had passed since Prince Henry had suggested to his father that the Portuguese take Ceuta as a bridgehead in Africa.

For many years, Portugal was the only seafaring kingdom expanding the boundaries of Western geographical knowledge, so it was natural that Italian seafarers should serve the Portuguese Crown. In 1482 a Genoese navigator called Christopher Columbus entered the service of John II and sailed to Elmina. Columbus, however, was one of those rare men: an explorer who had read Ptolemy's *Geography*. In 1485 he put a proposal before John II. If the king would provide him with three ships, sufficient supplies, the title of Admiral of the Ocean, and the right to rule any land he found, Columbus would sail *westwards* from Portugal to reach the shores of China. He calculated the distance to be less than 3,000 miles. The reason why he thought China so close was that he was following Ptolemy, who had grossly underestimated the diameter of the Earth, at a mere 18,000 miles (it is actually 24,901 miles).[3] King John put Columbus's proposal to his advisers; they were well aware of the shortcomings of Ptolemy's calculation, and knew that China was far further than Columbus imagined. They advised the king to refuse the ambitious Genoese captain. Undaunted, Columbus next sought the patronage of Castile. Queen Isabella similarly sent his suggestion to her advisers, and they agreed with their Portuguese counterparts. His native Genoa, the doge of Venice and Henry VII of England all turned him down too.

The whole episode appears faintly ridiculous in retrospect, but the scholars were right: Ptolemy had made a mistake and this Genoese captain was too entrenched in his ambition to see it. Columbus, however, was utterly determined. He returned to Portugal, where he was again courteously but firmly rejected by John II. By this time he knew that Bartholomew Dias had rounded the Cape of Good Hope and had found a sea route to India. So he went back to Castile, more desperate than ever. In 1492 Queen Isabella of Castile and her husband King Ferdinand of Aragon had just taken Granada and completed the Reconquista. Flushed with this success, they eventually agreed to Columbus's proposal, no doubt presuming he would sail off into the sunset and never be seen again.

On 12 October 1492 Columbus reached the Bahamas. He went on to visit Cuba and Hispaniola (the island today divided between Haiti and the Dominican Republic), and built a fort on Hispaniola, which he left garrisoned with 39 men. On 4 March 1493, after weathering a storm, he put in at Lisbon to repair his ship – and no doubt also to gloat about proving the Portuguese experts wrong: he was convinced he had been to the fabled Far East of Marco Polo. Fired with self-congratulation, he wrote an open letter to Ferdinand and Isabella telling them and the rest of Christendom about his discovery, exaggerating the wealth of the islands he had discovered and urging the Spanish king and queen to pay for another expedition, from which he would return with great wealth. His prime concern, evidently, was his own enrichment. Unlike the Portuguese pioneers, who had no desire to conquer large swathes of territory, Columbus was set on winning his own personal empire.

By the end of the year, his wish to return had been granted and he set sail with 17 ships and 1,200 settlers and soldiers, all keen to find their fortunes. On his return to Hispaniola, he found his fort in ruins and the garrison dead, killed by the natives. He immediately started to exact revenge – and never stopped. His governorship was characterised by the systematic destruction of the indigenous population through enforced labour in the mines, disruption of families' lives, enslavement, torture, capital punishment and disease. Bartolomé de las Casas, who later became the defender of the rights of the indigenous people of the New World, noted that Columbus's cruelty had reduced the population of Hispaniola from over three million to just 60,000 by 1508: a death rate of 98 per cent in 15 years. Those who had travelled with Columbus on that second voyage were disinclined to tolerate his destructiveness: they had failed to find the fortunes he had promised them. By 1500 word had got back to Spain about his tyranny, and he was relieved of his position as governor.

Columbus's actions are shocking. But it is perhaps unsurprising that the first man to lead an expedition across the Atlantic was merciless in his exploitation of the indigenous population. The early explorers had no wish to endure privations, miseries and dangers at sea for the sake of it; they only did so out of avarice. The greater their desire for gold, the greater the risks they were prepared to take. Columbus took the biggest risk of them all. If he and his men happened to encounter something of value before they drowned in a storm, there was every likelihood they would take it, torturing and killing those who stood in their way, and then simply sail on. Some historians have viewed the overseas expansion of the Iberian kingdoms as the extension of the Reconquista, and there is something to be said for that. But Christopher Columbus himself has more in common with the Vikings of the early eleventh century than the Crusaders of the twelfth.

The importance of Columbus's first voyage was immense. While he himself always maintained that he had discovered parts of Asia, more discerning minds realised that he had found a wholly new land, which they called the 'New World' – and that was the catalyst for many changes. In the Treaty of Tordesillas (1494), Spain and Portugal divided this new world between them, Portugal being given the right to appropriate all the lands outside Christendom but within a line 370 leagues due west of the Azores. Cabral's decision to sail as far westwards as Brazil in 1500 was no doubt inspired by the knowledge that Columbus had discovered land in that direction. In England, too, Henry VII sponsored the voyage of the Venetian navigator John Cabot, which led to the discovery of Newfoundland in 1497. In the east, the days of Arab and Venetian control of the spice trade were numbered. By 1500 the merchants of Europe could see that it was going to be less costly to ship large quantities of pepper, cinnamon and silk to Europe by way of the Portuguese sea lanes than carrying them in small packets along the Venetian-controlled land routes. The consequent investment in merchant shipping led to a shift in the balance of economic power. Portugal and Spain had formerly been on the edge of the known world; now they were at the heart of it, and as we have already seen, 'a dominant capitalist city always lies at the centre of its trading region'. The leading families and merchants in both countries became wealthy. And as the emphasis on international business shifted towards ocean travel, so English, French and Dutch ports were that much nearer to the action than Venice and Genoa.

The most important point about Columbus's discovery, however, was that he exploded the myth that everything worth knowing had already been discovered by the Greeks and Romans. That view had been summed up by Bernard of Chartres in the early twelfth century: medieval thinkers could see further than those of the ancient world, he argued, only because they 'were as dwarves standing on the shoulders of giants, so that although we perceive many more things than they, it is not because our vision is more piercing or our stature higher, but because we are carried and elevated higher thanks to their gigantic size'.

The rush for classical texts in the twelfth century in particular revealed how ancient wisdom underpinned medieval thinking; it continued to do so in the fifteenth century. Aristotle was still elevated above all other philosophers for his dialectical reasoning and scientific knowledge. Galen was still championed for his medical texts, Ptolemy for his astronomy and his geography. While some medieval people were capable of original thinking, Columbus's discovery demonstrated unequivocally to the whole of Christendom that there was nothing absolute about classical learning. His discovery, and those of Cabot and Cabral, smashed the authority of Ptolemy to smithereens, for if the greatest geographer of the ancient world could have failed to notice an entire continent, how could he be relied upon for anything else? The last decade of the fifteenth century therefore witnessed nothing less than a cognitive revolution: a sudden and completely new way of thinking about the world, unfettered by previous knowledge and, indeed, impelled to go beyond it.

Measuring time

You would have thought that the invention of the mechanical clock in the early fourteenth century might have drawn attention to the limitations of ancient thinking long before Columbus sailed. Such an instrument for measuring time certainly heralded a watershed in human history. But by the time clocks started to intrude on most people's daily lives, few individuals cared to remember that there had been a time when they had not existed. Shakespeare mentions clocks or describes time as 'o'clock' in *Macbeth* (set in the eleventh century), *King John* (early thirteenth century), *Cymbeline* (pre-Roman Britain) and *Troilus and Cressida* (Ancient Greece). His play *Julius Caesar* even has a stage direction for a clock to chime the hour in Ancient Rome. Clearly he did not realise (or it was not important to him) that mechanical clocks were unknown in the ancient world.

The earliest reference to a mechanical timepiece appears in a text of 1271, which noted that clockmakers were trying to make a wheel that would complete one revolution every day, 'but they cannot quite perfect their work'.[4] Sixty years later, the problem was solved. Richard of Wallingford, abbot of St Albans, was developing a mechanical astronomical clock called Rectangulus when Edward III visited him in 1332.[5] It not only told the time, it indicated the movements of the Sun, Moon and stars, and gave the time of high water at London Bridge. A mechanical 24-hour clock was set up in the church of St Gothard in Milan in 1335. Jacopo de Dondi installed an astronomical clock on the tower of the Palazzo Capitano in Padua in 1344. Four years later, his son Giovanni began work on the most famous medieval clock of them all: his seven-sided 'astrarium'. This had seven dials showing the time over a 24-hour cycle, as well as the positions of the Moon and the five known planets (Mercury, Venus, Mars, Jupiter and Saturn). By the time it was finished in 1368, there were public clocks striking regular hours in the Italian cities of Genoa, Florence, Bologna and Ferrara, and in the English royal palaces of Westminster, Windsor, Queenborough and Kings Langley. Charles V of France was so pleased with the hourly striking of the clock that he installed in the royal palace in Paris in 1370 that he ordered all the churches in the city to follow its lead in striking the hour, and had two more erected at the Hôtel Saint-Paul and the Chateau de Vincennes.[6]

While these references locate the invention of the clock firmly in the previous century, the majority of rural and small-town workers would not have heard one ring out before 1400. In the prologue to 'The Parson's Tale', Chaucer states that it was 'ten of the clokke' – indicating that the reckoning of time and 'the clock' were synonymous – but the speaker in that instance admits that he was not actually using a clock to tell the time; he was guessing it. In 'The Sergeant-at-law's Tale', the poet describes in detail the methods people normally used to reckon the hour in 1386: by the amount of the sky that the sun had crossed, and by estimating the proportion of the length of a shadow of a tree to its height. The clock can hardly be said to have been the default method of telling the time in Chaucer's period; hence its impact is better represented as a fifteenth-century change.

Rising demand for clocks led to their increased production, greater skill in metalworking, and a greater variety of clock design. All the known fourteenth-century examples are turret clocks or astronomical devices, but by 1400, portable clocks were on the horizon. Charles V of France owned an *orloge portative* in 1377, and the future Henry IV of England had a clock that could be transported in a basket by 1390.[7] Among Henry V's possessions at the time of his death in 1422 was a chamber clock; a spring-driven example made in about 1430 for Philip the Good, duke of Burgundy, still survives. Such was the rate of development that by 1488, the duke of Milan could place an order for three chiming pendant watches.[8] At the same time, astronomical clocks were added to civic halls and ducal palaces, and basic mechanical clocks were installed in parish churches and manor houses. At the Edgcumbe family house of Cotehele in Cornwall, a faceless household clock with a verge escapement was installed in the chapel not long after 1493; it is still there, ringing the hours.

So what? Did it matter how people told the time? Well, yes, it did. The use of a mechanical clock was a significant improvement on Chaucer's method: to divide the passage of the Sun across the sky into 12 parts and estimate how much of that time had elapsed. Two problems arise from this solar calculation. The first is that it is obviously prone to inaccuracy. The second is that the unit of time varies: a daylight 'hour' in summer can be up to twice as long as an 'hour' in winter, as there is twice as much daylight to divide into 12 equal hours. Mechanical clocks standardised units of time; hence the need to specify the ninth hour *of the clock*, or 'nine o'clock', which is how we end up with the modern term.

As noted above, many medieval clocks were astronomical. Calculating time exactly was essential for the accurate observation of the Sun, Moon and stars. We might not set much store by astrology in the twenty-first century but before 1600 a great deal of medical, geographical and scientific work depended on a precise knowledge of the movement of the heavens, and clocks significantly professionalised these lines of work. It goes without saying that the standardisation of the hour was crucial for scientific experimentation. Clocks also calibrated the social and economic use of time. People could meet at a particular hour; and set periods could be established for working times and opening hours. They could plan their working lives more efficiently. For these reasons, the mechanical clock deserves to be recognised as one of the great inventions of the Middle Ages, and its spread in the fifteenth century was one of the most important changes of the age.

There is another, more subtle respect in which the spread of clocks marks a significant shift. This is the secularisation of time. In the Middle Ages, time was dominated by the Church. The world only existed because God had created it, and time only existed because God had created the movement of things within Creation. Time was therefore a part of Creation: it filled divine space. Alongside this theological concept of time, there was another, more practical one. The yearly cycle was seen as part of a divine architecture in which God had ordained a time for sowing, a time for harvesting, a time for grazing flocks, and so on. Within this divine year, certain days were set aside for fasting – in Advent and Lent – and others for feasting. Some days were venerated as saints' days. Within every day of the year, particular hours were designated for divine services, such as Prime, Nones, Compline and Matins. Time was not just a sacred thing itself; its subdivisions too had spiritual significance. On a day-to-day basis, the Church controlled the perception of time through the ringing of church bells – marking the hour in towns, calling the religious to prayer, announcing the passing of the dead, and so on.

For all these reasons, time was not just time as we know it: it was a gift from God. Hence the medieval Church would not allow Christians to charge interest on sums of money they loaned to others: to do so was to charge money for time, which belonged to God, and no Christian had the right to sell what was God's. However, as time increasingly became subject to the measurement of man-made machines, it lost some of its semi-magical religious associations. It seemed to be under human control – a thing that clockmakers tamed, not an unfettered part of Creation. Most significantly, the man-made machine came to dictate to the Church when it should ring its bells and hold its religious services. While units of distance, weight and volume all still varied from place to place, the hour became the first internationally standardised unit of measurement, taking priority over both local customs and ecclesiastical authority.

Individualism

Polished metal and obsidian mirrors have existed from ancient times, and because of this, historians have usually passed over the introduction of the glass mirror as if it was just another variation on an old theme. But the development of glass mirrors marks a crucial shift, for they allowed people to see themselves properly for the first time, with all their unique expressions and characteristics. Polished metal mirrors of copper or bronze were very inefficient by comparison, reflecting only about 20 per cent of the light; and even silver mirrors had to be exceptionally smooth to give any meaningful reflection. These were also prohibitively expensive: most medieval people would only have glimpsed their faces darkly, reflected in a pool of water.

The convex glass mirror was a Venetian invention of about 1300, possibly connected with the development of the glass lenses used in the earliest spectacles (invented in the 1280s). By the late fourteenth century, you could find such mirrors in northern Europe. The future Henry IV of England paid 6d to have the glass of a broken mirror replaced in 1387.[9] Four years later, while travelling in Prussia, he paid £1 3s. 8d in sterling for 'two mirrors of Paris' for his own use.[10] His son, Henry V, had three mirrors in his chamber at the time of his death in 1422, two of which were together worth £1 3s. 2d.[11] Although these were still far too expensive for an average farmer or tradesman, in 1500 the prosperous city merchant could afford such an item. In this respect, the individual with disposable income differed greatly from his ancestor in 1400: he could see his own reflection and thus knew how he appeared to the rest of the world.

People's ability to appreciate their unique appearance led to a huge rise in the number of portraits commissioned, especially in the Low Countries and Italy. While almost all the oil paintings that survive from the fourteenth century are of a religious nature, the few exceptions are portraits. This trend towards portraiture grew in the fifteenth century, and came to dominate non-religious art. As important men increasingly commissioned artists to create their likenesses, the more those likenesses were viewed, encouraging other people to have their portraits painted. Portraits invited the viewer to 'Look at me!' and implied that the sitter was a man of substance, or a well-connected woman, worth portraying because of his or her status. They encouraged you to talk about these people, making them the centre of attention.

One of the most famous paintings of the century is Jan van Eyck's *Arnolfini Marriage*, painted in Bruges in about 1434. It shows a convex round mirror on the back wall, reflecting the backs of the subjects to the artist. If van Eyck's *Portrait of a Man with a Turban*, painted the previous year, is of the artist himself (as it probably is), then he also had a flat mirror by this date. We know from Brunelleschi's famous experiment on perspective (to which we will return shortly) that flat mirrors were available in Florence at that time. After van Eyck, self-portraits abound for the later fifteenth century in Italy as well as the Netherlands. Dürer painted quite a few, culminating in the image of himself as Christ at the age of 28 (1500); his introspection rivals that of Rembrandt in the seventeenth century. In such an artist's hands, the mirror became an instrument whereby a man could start to investigate how other people saw him. Hitherto artists had only portrayed other people; now they could put themselves into the picture. And anyone who saw the intense interrogation by the artist of his own face, searching for the clues to his nature, could not help but pause for thought about his or her own identity.

All this amounts to far more than just a series of attractive pictures. The very act of a person seeing himself in a mirror or being represented in a portrait as the centre of attention encouraged him to think of himself in a different way. He began to see himself as unique. Previously the parameters of individual identity had been limited to an individual's interaction with the people around him and the religious insights he had over the course of his life. Thus individuality as we understand it today did not exist: people only understood their identity in relation to groups – their household, their manor, their town or parish – and in relation to God. Occasionally individuals stood out from the crowd in the way they wrote about themselves – you only need to think of Peter Abelard's autobiographical *Historia Calamitatum* and Ulrich von Lichenstein's starring role in his own romances – but the average person saw himself only as part of a community. This is why the medieval punishments of banishment and exile were so severe. A tradesman thrown out of his home town would lose everything that gave him his identity. He would be unable to make a living, borrow money or trade goods. He would lose the trust of those who could stand up for him and protect him physically, socially and economically. He would have no one to plead his innocence or previous good behaviour in court, and he would lose the spiritual protection of any church guild or fraternity to which he belonged. What happened in the fifteenth century was not so much that this community identity broke down, but rather that people started to become aware of their unique qualities irrespective of their loyalty to their community. That old sense of collective identity was overlaid with a new sense of personal self-worth.

This new individualism had a religious dimension, too. Medieval autobiographical writing is not normally about the author himself, but about his relationship with God. Similarly, the hagiographies of early-medieval saints are archetypal moral stories of men and women who followed God's path. Even in the fourteenth century, a monk writing the chronicle of his monastery or a citizen writing about his town would incorporate God into his narrative, as the important element of the story was not the community itself so much as its relationship to God. As the fourteenth century drew to a close and people started to see themselves as *individual* members of their communities, they started to emphasise their *personal* relationships with God. You can see that transformation reflected in religious patronage. If in 1340 a wealthy man built a chantry chapel to sing Masses for his soul, he would have the interior decorated with religious paintings, such as the adoration of the Magi. By 1400, if the founder's descendant redecorated that chapel, he would have himself painted as one of the Magi. By the late fifteenth century, more often than not, just the patron's portrait would be on display, the emblems of faith that the artist included in the painting being sufficient representation of the religiosity the patron wished to project.

The new individualism also extended to the way people expressed themselves. The letters they wrote to one another were increasingly of a personal nature; previously letter-writers had restricted themselves to formalities and orders. There was now a marked trend towards writing about yourself and revealing your personal thoughts and feelings. Examples of such autobiographical writing abound in the fifteenth century: in English there is *The Book of Margery Kempe*; in Castilian, *Las Memorias de Leonora López de Córdoba*; and in Italian, Lorenzo Ghiberti's *I Commentarii*. Four of the earliest collections of English private letters – the Stonor, Plumpton, Paston and Cely letters – also date from the fifteenth century. Ordinary people started noting down the times and dates of their births, so they could use astrology to find out more about themselves in terms of their health and fortune.

The new self-awareness also led to a greater desire for privacy. In previous centuries, householders and their families had shared a dwelling entirely, often eating and sleeping in the same hall as their servants. Now they began to build private chambers for themselves and their guests, away from the hall. As with so many changes in history, people were largely unaware of the significance of what they were doing. Nevertheless, our vision of ourselves as individuals, not just members of a community, marks an important shift from the medieval world to the modern.

Realism and Renaissance naturalism

In some ways, realism is connected to the rise of individualism. Both incorporate new approaches to people in relation to their surroundings. Both emphasise an interest in humanity independent of mankind's relationship with God. But whereas individualism may be expressed in terms of a person's self-understanding and self-worth through reflection, realism is best understood in terms of scholars and artists holding up a mirror to the whole of Creation, in order to explain the world and everything in it in all its complexity.

You only have to look at the naturalism of Renaissance art to see signs of the new thinking. Lorenzo Ghiberti's sculptures on the Baptistry doors in Florence, created in 1401–22, were a marvel on account of their verisimilitude and daring use of perspective. Brunelleschi's famous optical experiment, performed about 1420, took things a stage further. Using a large flat mirror with a hole in it, he held up his own painting of the Baptistry (also with a hole in it) to face the actual building. By viewing the real building through the two holes and comparing it to the reflected image of his own painting, he was able to ascertain the geometric laws that govern perspective, thus vastly improving on the first attempts at linear representation by Giotto a century earlier. From the second quarter of the fifteenth century, Florentine painters did not need to estimate how to represent buildings; they could apply systematic rules to make sure they appeared 'realistic' to the onlooker. At the same time, realism crept into religious art. Northern European artists such as Robert Campin and Roger de Weyden began to paint large-scale, highly ostentatious religious scenes in which the characters' heads were no longer surrounded by haloes.

Italian artists such as Ghirlandaio and Leonardo similarly dropped the halo; others reduced it to a thin, almost invisible ring of light. These artistic changes may seem slight but they reflect a shift in priorities – from the symbolic representation of men and women to portraying them as they would actually have appeared.

Naturalistic representations came even more into focus in the matter of the nude. In the Middle Ages, depictions of nakedness seem to have lacked the erotic content they acquired in the Renaissance. The naked or semi-naked Christ on the cross was vulnerable; he was not an erotic figure. Naked buffoons blowing trumpets from their bottoms, depicted in the margins of psalters, were not included to stimulate the sexual appetites of educated readers; they were there to mock the pride of mankind, or to amuse the reader. Portrayals of Adam and Eve similarly emphasised their nakedness as shameful, not erotic. In the fifteenth century, however, the nude – the erotic unclothed figure – emerged. In the 1440s, Donatello portrayed the Old Testament figure of David as naked but for a hat and boots, allowing open inspection of the body in a way that contrasts strongly with the discreet figures from earlier sculpture and painting, which had no sensual or corporeal presence. Moreover, Donatello's work is a free-standing sculpture, unsupported by a niche or any other architectural device: it is proud and defiant in its nudity. It harks back to the Venuses of the ancient world, demonstrating not only that Donatello could rival any classical sculptor in his skill, but also that a man in his natural state, as God made him, was a suitable subject for the attention and admiration of the public.

By the end of the century, the male nude had become commonplace, appearing in everything from Leonardo's *Vitruvian Man* drawing to Michelangelo's *David* (1504). Female nudes, although rarer, first appeared in Botticelli's *Birth of Venus* (1484), Hans Memling's *Eve* (1485–90), one of Michelangelo's drawings for *The Entombment* (1500), and Giorgione's seductive *Sleeping Venus* (1508). In the early sixteenth century, in the hands of Giovanni Bellini and Titian, the erotic female nude became an established art form in its own right. On top of the physical nakedness, painters and sculptors also began to depict Man's emotional nakedness in more explicit ways than before, as can be seen in Michelangelo's *Pietà* (1498–9) or his *Rebellious Slave* (1513). Mankind was no longer depicted just as the humble recipient of God's mercy or wrath, but became a subject suitable for serious study in its own right.

Renaissance humanism also promoted the study of Man's inner life through providing classical models of education. The waning of the influence of Latin in the previous century and the rise of university courses that prepared young scholars for nothing but their allotted roles in academic life – teaching grammar, studying theology, and practising medicine or the law – led to a reaction among those who knew and appreciated the educational standards of the ancient world. The old university Trivium came to be replaced by the *studia humanitatis*, in which logic no longer played a part but history, ethics and poetry joined grammar and rhetoric as the key components of a good education. Greek too was revived as a means of unlocking the wisdom of the classical world. The Platonic Academy became the model for a new educational establishment in Florence, founded by Cosimo de Medici in the 1440s, with the humanist Marsilio Ficino as its head. Further impetus to the study of Greek came in 1453 after the fall of Constantinople, when many Greek-speaking scholars from the city arrived in Italy. Most importantly, the lack of distinction between arts and science in the *studia humanitatis* meant that it provided an open education, broadening horizons rather than limiting them by enforcing dogmatic obedience to ancient texts. Thus it built on the key virtue of medieval enquiry: to observe natural phenomena as aspects of God's Creation – in which everything is possible but everything happens for a reason – and to explain these phenomena accordingly.

You may wonder if this new realism and naturalism really deserves to be considered a major change. Is it not merely replacing one method of representing the world with another? And as for the 'inner realism' with which Renaissance humanists were concerned: was this not just a shift in educational priorities? After all, even understanding things in great depth does not necessarily have profound consequences. Consider Leonardo da Vinci, who is often said to be the ultimate Renaissance man and one of the greatest minds the West has ever produced. It is fair to say that within the parameters of this book, he was almost entirely inconsequential. Fifteenth-century peasants experimenting with horse-drawn ploughs had a greater impact on European life than Leonardo. His genius remained largely hidden, coalescing in the pages of his notebooks, which would amuse and amaze people in later centuries. Most of his paintings did not last, due to his fascination with new, untried and untested compounds of paint, many of which degraded. But what Leonardo *represents*, on the other hand, is of supreme importance. Although he did not have a university education he was able to turn his mind to a wide range of subjects, from how muscles work to how a bird flies. And while the fifteenth century saw just one Leonardo da Vinci, it also saw several hundred people of lesser genius but equal curiosity who were prepared to investigate the reality that surrounded them. That a few of them pursued experimentation into seemingly bizarre areas of enquiry, including numerology, astrological prognostication, angelology and dream interpretation, is also important. The reason why these things appear unscientific to us now is because the curious minds of the Renaissance eventually concluded that they were scientific dead ends. Thus fifteenth-century attempts to discover and portray the nature of reality are comparable to voyages to the New World: both permitted discoveries by breaking down previous assumptions and investigating the unknown, whether that be the far side of the ocean or the movement of a bird's wing in flight. To put it in a nutshell, the fifteenth century was when people in the West stopped their collective study of the abstract mystery of God and came to the conclusion that to understand God, they needed to study His Creation.

Conclusion

Some readers may be surprised that I have not highlighted printing as a major fifteenth-century change. This is not to lessen the importance of the printing press: it is probably the single most important invention of the last thousand years. But it is another example of something invented in one century that has a bigger impact in another one. The fact is that the Bible printed by Johannes Gutenberg in Mainz in 1455 was in Latin and thus unintelligible to the majority of people.[12] Moreover, most of the books published before 1500 were expensive, rivalling the production standards of fine manuscripts with wide borders and meticulously laid-out text. Not all those who could afford them could read, and many people who could read were unable to afford them. The illiterate majority had no interest in them at all. Thus Johannes Gutenberg did not change the world any more than the inventor of the compass did. The great changes that arose from printing were brought about by other people, who put his invention to wider use in the next century.

The most significant change of the fifteenth century can be summed up in the word 'discovery' – the discovery of the world and the discovery of the self. While the subtle changes of self-perception would have gone largely unnoticed, Columbus's discovery of the New World was tangible and a universal talking point. It is extraordinary that within just eight years, between 1492 and 1500, European navigators found two new continents (North and South America) and a seaborne route to Asia. And that came straight after the exploration of the southern half of Africa. Imagine waking up to the news on your radio and hearing that explorers have found another three continents full of fabulous wealth. The comparison is not as preposterous as it seems, for our confidence that such places do not exist closely resembles that of the scholars who advised the kings of Portugal, Castile and England to turn Columbus down. How wrong they were – but how quickly they made up for their mistake. The earliest globe, in which the world was mapped in three dimensions for the first time, dates from 1492 – the same year that Columbus sailed west.[13]

The principal agent of change

It may be obvious from the above which fifteenth-century individual had the most significant effect on the West. However, we should note that Columbus was just one of many mariners who made discoveries. English seafarers regularly sailed into Icelandic waters, fishing for cod, in the first decades of the century. Greenland remained part of Christendom until after 1409, so some sailors knew how to cross vast stretches of the Atlantic. The men of Bristol set out on two explorations of the ocean in 1480 and 1481, searching for the mythical 'Isle of Brasil'. And it seems that one of the main reasons why John Cabot sailed from Bristol in 1496 and 1497 was because these sailors *had* found land across the Atlantic, which would put them ahead of Columbus.[14] There are also tantalising clues that Cabot may have followed the American coast a long way south, which would mean his Venetian-English expedition beat the Portuguese to the discovery of South America. Whatever the truth of the matter, the spirit of discovery was clearly in the air in the 1480s and 1490s, and we should not inflate Columbus's pioneering role on account of his later fame and talent for self-promotion. Both Prince Henry the Navigator and John II of Portugal were even more persistent and determined than Columbus, and should be credited with providing the political and economic drive for European expansion. Columbus badgered the princes of Europe for seven years to sponsor his expedition, but Henry the Navigator spent fifteen years trying to persuade his mariners just to sail beyond Cape Bojador. John II's vision of a seaborne trading empire, linked by fortified factories, allowed the small country of Portugal to establish footholds in India without the expense of administering a territorial empire. It is tempting to say that these men each deserve to be regarded as the principal agent of change.

At the end of the day, however, it was Columbus's discovery that proved the biggest thunderbolt to hit Europe since the Black Death. It was his ambition for a lordship of his own that set Spain on a path of building a massive overseas empire. It is down to Columbus that Spanish is the second most widely spoken language in the world today, after Chinese. And it was Columbus's talent for self-publicity that ensured that all of Europe knew his name. It was thus down to him that all of a sudden people had to come to terms with a profound question: if the great writers of the ancient world did not know about two whole continents, what else had they missed?

1501–1600

The Sixteenth Century

In modern reckoning, the sixteenth century began on 1 January 1501. It was not that way at the time – unless you were living in Genoa, Hungary, Norway or Poland. In Venice, the new year started on 1 March 1501. In England, Florence, Naples and Pisa it began on 25 March. In Flanders, New Year's Day was Easter Day – a different day from year to year. In Russia it fell on 1 September, and in Milan, Padua, Rome and many German states, the year started on 25 December. Most confusingly of all, the new year in France began on one of four different days – Christmas Day, 1 March, 25 March or Easter Day – depending on the diocese in which you lived. Only in 1564, with the Edict of Roussillon, was the French year standardised as starting on 1 January, taking effect from 1567. If ever you thought the past was simpler than the present, this calendar problem should make you think again.

Agreement on a date for New Year's Day was by no means the limit of the complexity, however. All the aforementioned conventions were based on the Julian calendar of Ancient Rome. By this measurement of time, each period of twelve months was more than ten minutes short of an actual year as determined by the Earth's orbit of the Sun. Ten minutes per year might not seem a lot, but by the sixteenth century, 25 December was ten days adrift from Christmas Day in the presumed year of Christ's birth. In 1582, therefore, Pope Gregory XIII proposed a radical solution: to drop ten days out of the calendar and *not* to have a leap year in every year divisible by 100 unless it was also divisible by 400, thereby shortening every 400 years by three days (this is the system in use across the world today). Most of the Catholic Church adopted the new Gregorian calendar from Thursday, 4 October 1582; the following day now became Friday, 15 October. Of course this created a whole new layer of differences in date, as most Protestant countries kept the Julian calendar until the eighteenth century. Although both England and Florence celebrated New Year 1583 on 25 March, the Florentine celebrations took place ten days before the English ones. Such things reveal how intricately varied early-modern Europe was – even with regard to matters of daily routine that we take for granted.

The introduction of the Gregorian Calendar is only one of the thousands of changes in everyday life in the sixteenth century. In 1500, only the exceptionally wealthy travelled by coach; by 1600, it was said that 'the world runs away on wheels', and people were complaining of the dangers of road traffic accidents – with some justification. In marked contrast to medieval houses, which were sparsely furnished with maybe a trestle table, a pair of benches, beds, chests, utensils and little else, sixteenth-century houses contained a plethora of wooden and fabric furnishings, such as curtains, valances, carpets, cushions, cupboards and chairs. As for mealtimes, few people in northern Europe ate breakfast in 1501. The medieval two-meal rhythm of the day persisted: dinner was at about 11 a.m. and supper at about 5 p.m. But as more people moved into towns, and made their living by working long hours for other townsmen, the time at which they could have supper was pushed back into the evening. This meant that dinner, the main meal of the day, had to be eaten a couple of hours later and became lunch. It followed that you had to eat an early meal, breakfast, in order to get through to lunchtime. School also helped bring about this change, for more and more boys went to school, and the long lessons required that they eat breakfast. Hence breakfast was ubiquitous in towns by 1600.

The sixteenth century also saw the population expand again after a long period of stagnation. People started to complain of over-population. The shift towards individualism continued apace, with mirrors, or 'looking glasses', being available for just half a labourer's daily wage by 1600. The personal diary as we know it evolved, as people increasingly wrote chronicles of the events that took place in their communities, interwoven with their personal experiences and reflections. More and more wealthy people had their portraits painted, so it is much easier to say what sixteenth-century gentlemen and wealthy ladies looked like, while the faces of their medieval forebears are almost entirely lost to us. The well-off frequently incorporated panes of glass in the windows of their houses, allowing them to enjoy daylight indoors to a far greater degree than their ancestors. They started building and maintaining recreational gardens, with formal designs, fountains and classical sculptures; previously gardening had been almost wholly practical, intended to produce vegetables, herbs or medicines. For many townspeople, life was much less of a struggle than it had been for their forebears. Lifestyle choices became a subject for conversation at banquets and for consultation in advisory manuals.

It was different for those at the bottom of society, of course. It is a salutary thought that as Shakespeare's early plays were performed for the first time, thousands of people in England were dying of starvation in the great famine of 1594–7. But less wealthy people saw their lifestyles change too. According to William Harrison's *Description of England* (1577), the common folk had experienced three major upgrades in their living standards in recent years. First, there was a great proliferation of chimneys in towns. This might not sound very exciting to you but it was a great improvement for those who had previously been forced to warm themselves at an open fire in the centre of their hall. The smoke from an open fire never ceases to swirl around you, making your clothes smell and your eyes water, and blackening the inside of your lungs as well as the beams, rafters and walls of your house. Second, ordinary people now slept on flock beds with pillows and linen sheets; previously they had had to make do with a straw mattress and a blanket on the floor, with a log to serve as a pillow. Third, a great many people now ate with pewter spoons off pewter plates and drank out of pewter tankards, whereas before almost all tableware had been made of wood.

By 1600 most people followed a routine that you will probably recognise. They washed their face and hands and cleaned their teeth when they got up in the morning. They had breakfast and went to school or work for about eight o'clock. They ate lunch around midday, and came home and ate supper with metal knives and spoons off plates, warming themselves at a fireplace. They lay down to rest in sheets on a mattress on a proper bed frame, with their head on a soft pillow. If your main concern is the routine of daily life, you may well conclude that the sixteenth century saw the greatest developments of the millennium. However, the same period also saw changes of a far more profound nature.

Printed books and literacy

There were about 250 printing presses in Europe at the start of the sixteenth century; between them these had produced an estimated 27,000 editions by 1500. If every edition had a print run of 500 copies, then as many as 13 million books might have been circulated among a population of just 84 million people.[1] While such a figure is impressive, it requires some context. It was certainly not the case that 15 per cent of the population owned a book. Even the great majority of literate people did not own any printed texts, let alone the 90 per cent who could not read or write. Most books were printed in Latin and were theological in nature, which significantly reduced their appeal. Wealthy people who did collect books, on the other hand, were likely to own several. If 10 million books still survived in 1500, they were probably in the possession of about half a million owners, and many of those owners were institutions. It is safe to say that less than 1 per cent of the population of Europe owned a book. The popular media of 1500 were still the pulpit and the marketplace, not the printed word.

The key event that changed this was the publication of the Bible in the local vernacular. There simply was no other book that people were so eager to read for themselves. They wanted to study the word of God personally, without the intervention of a priest, in order to improve their standing on Earth in the eyes of their fellow men as well as God, and increase their chances of going to Heaven after death. They also wanted to understand it for the benefit of their families and friends, so they could advise them how to live a holy life. The Bible was thus the ultimate self-help book. Vernacular Bibles had existed in the medieval period, and some – such as the French *Bible Historiale* of Guyart des Moulins, the Provençal Bible attributed to Peter Waldo, and the English Bible of John Wycliffe – had been very influential. But these had only been available in manuscript, which meant they were both scarce and expensive. Printing made Bibles available in much larger numbers at a fraction of the cost. Even so, it was not the printing per se that made the difference, but the printing in the local vernacular. Learning to read in Latin is nigh on impossible without the necessary schooling, which very few people had received, so vernacular Bibles helped many people learn to read, as well as allowing them to study the word of God. It was thus the combination of three things – the

Moretonhampstead in Devon. In the Middle Ages the manor was inaccessible to wheeled transport. The incorporation of such places within the Latin world, through the construction of churches, is one of the most significant changes of the eleventh century.

Exeter Castle, constructed by William the Conqueror in 1068. William found Saxon England undefended by castles and thus relatively easy to conquer. Such fortifications secured a stronger political relationship between leadership and the land.

Speyer Cathedral, built between 1030 and 1106. A massive construction for its time, its chief purpose was to demonstrate the power of the Holy Roman Emperors at a time when they were losing authority to the pope.

The late twelfth-century mural from Chaldon Church in Surrey, depicting the judgement and torment of souls. In the twelfth century the doctrine of Purgatory emerged: people started to believe they were not sent directly to Heaven or Hell but could redeem themselves through good works and prayer.

An Arab physician performing a bleeding, *c.*1240. Over the course of the twelfth century a large number of medical texts from the ancient world were translated from the Arabic in libraries around the Mediterranean. In Salerno, translations gave rise to an early university specialising in medicine.

Part of an early thirteenth-century stained glass window in Chartres Cathedral, depicting a wine merchant transporting a barrel of wine in a cart. At this period trade was flourishing across the Continent.

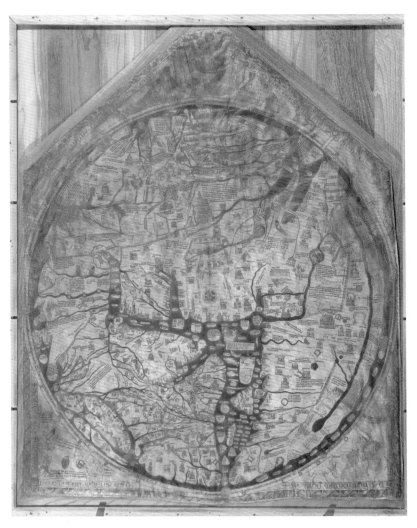

The Hereford world map, drawn by Richard of Haldingham about 1290. It is a spatial representation of knowledge rather than a true map. Jerusalem lies at the centre, the Red Sea top right, the British Isles bottom left, and the straits of Gibraltar at the bottom. Three quarters of this world – Asia and Africa – lay outside Christendom and were practically unknown to its author.

A cadaver effigy in Exeter Cathedral. The Black Death forced people to rethink their relationship with God. Many leading figures built reminders of death like this as demonstrations of their earthly humility and awareness of their sinful state.

The golden rose of Pope John XXII, made by Minucchio da Siena in 1330. Golden roses were mystical gifts given by the pope each year to a deserving prince or lord, or a favoured church. The craftsmanship and delicacy of the rose is a marked contrast to our widespread assumptions about culture on the eve of the Hundred Years War and the Black Death.

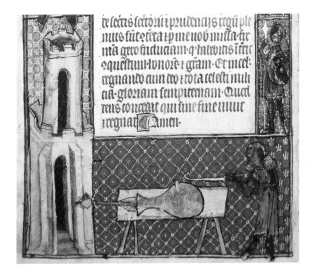

The earliest image of a cannon, from the treatise on kingship written in 1326 by Walter de Milemete for the young Edward III. Edward as a king did more than any other ruler of his age to encourage projectile warfare, including the construction of gun emplacements at his castles in southern England.

Commonly known as *Portrait of a Man in a Turban*, this is very probably a self-portrait. Johannes van Eyck dated it 21 October 1433, a year before he painted *The Arnolfini Marriage*, which openly shows a glass mirror in the background. Glass mirrors are one of the most underrated technological innovations of the Middle Ages, having an impact on everything from visual perspective to individualism.

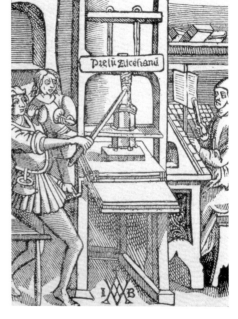

A printing press, from a book printed in 1498. In 1620 Francis Bacon declared that printing, along with gunpowder and the compass, had changed the world. By then he was right, but in its early days printing produced lavish books in Latin which few could read and even fewer could afford. It was not until the Bible was printed in the vernacular languages that the change really got under way.

The late fifteenth-century clock in the chapel at Cotehele House, Cornwall. Over the course of the late Middle Ages, time shifted from a natural, God-given state of reckoning to a secular, machine-measured one. The hour became the first internationally recognised unit of standardisation.

Portrait of Columbus by Sebastiano del Piombo. He may have been a brutal tyrant to the people of the West Indies but Columbus's legacy was of supreme importance in the history of the world, demonstrating the inadequacy of ancient knowledge as well as the wealth of the undiscovered world.

Map of the world from Abraham Ortelius's *Theatrum Orbis Terrarum* (1570). This was the first published map to use Mercator's projection. Compare it to the Hereford world map of *c*.1290 and you can see what great strides were made in the discovery and recording of the world in the fifteenth and sixteenth centuries.

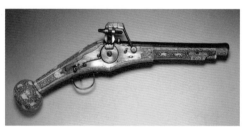

Before 1500, most guns were enormously cumbersome as well as inaccurate. Extraordinarily rapid advances in firearms technology were made over the course of the sixteenth century: this wheel-lock hunting pistol dates from 1578.

Printing was not only important for the circulation of written texts, it also was of huge importance in the dissemination of scientific knowledge in visual form. This is a hand-coloured printed (woodcut) image of an iris from Leonhart Fuchs's *De historia stirpium* (1542).

printing press, the use of the local vernacular and the spiritual significance of the Bible – that challenged the dominance of the pulpit and
the marketplace and ultimately turned Europe into a literate society.

Individual nations saw vernacular Bibles printed at varying times.
German-speaking countries were the first to receive one, when Johannes
Mentelin produced his translation in 1466. The first Italian Bible was
printed at Venice by Niccoló Malemi in 1471; the first Czech Bible
followed in 1488. A New Testament in French appeared at Lyons in
1476, and in 1487, Jean de Rély produced a printed version of Des
Moulins' *Bible Historiale*. These early versions were translated from the
Latin Vulgate; translations from the Greek only followed after the
humanist scholar Desiderius Erasmus of Rotterdam produced a Greek
edition of the New Testament in 1516. Martin Luther finished a German
version of the New Testament in 1522, based on Erasmus's Greek
edition, and he collaborated in a German translation of the Old
Testament, published in 1534. New French Bibles were produced in 1523
(New Testament) and 1528 (Old Testament). William Tyndale translated
Erasmus's New Testament into English and printed it at Worms in
1526, but fell foul of the authorities for his choice of words and was
burnt at the stake in 1536 for his supposedly heretical mistranslations.
By then he had only completed about half the Old Testament; his work
was finished by John Rogers in 1537, shortly after Miles Coverdale
printed the first complete English Bible. In 1539, the English government authorised the publication of a vernacular Bible, the 'Great Bible',
and insisted a copy be made available in every parish. The Danes and
Norwegians could own the New Testament in their own language in
1524 and the whole Bible in 1550; Swedish speakers had the New and
Old Testaments in 1526 and 1541 respectively, Spanish speakers in 1543
and 1569, Polish speakers in 1554 and 1563, and Welsh speakers in 1563
and 1588. The Finns saw the beginning of written Finnish literature
with the publication of Mikael Agricola's New Testament in 1548. Few
European communities did not have a vernacular Bible by 1600,
although both the Portuguese and the Russians had to wait until the
eighteenth century for complete Bibles in their languages.

The importance of this huge surge of people learning to read by
studying the Bible in their own language cannot be overstated. Prior to
about 1530, about half of the books published in England were in English
and half in Latin, but in the 1530s, the proportion printed in English
shot up to 76 per cent. After the publication of the Great Bible in 1539,

it passed 80 per cent. It was a snowball effect: the more books there were in the vernacular – especially Bibles – the more people learnt to read and consequently the more demand there was for new books. In England, book production increased from just over 400 titles in the first decade of the century to over 4,000 in the last decade. An Italian writer complained in 1550 that there were so many books available that he did not even have time to read their titles.[2] On top of this, individual books were read more frequently. Where once each Latin book had had one rich owner who kept it locked in his library and only shared it with trusted, highly educated friends, now most vernacular texts were passed around and read a dozen times over by different people.

With so much knowledge available in book form, the value of reading became increasingly obvious to all. Schools mushroomed in number. Universities blossomed. Printing became the natural vehicle for imparting and receiving information. It was particularly advantageous to those wishing to acquire or disseminate scientific theories. In the days before printing, scientific books had been laboriously written out by hand by copyists who often did not understand the concepts they were describing, with the result that they made many mistakes. The spread of scientific ideas was therefore flawed as well as slow. Printing allowed scientific ideas to spread more quickly and accurately, with the result that the scientific community of Europe became much more of a single body, considering each other's innovations and criticisms at the same time. This gave scientists far greater impact than they had previously enjoyed. When Nicolaus Copernicus's *De Revolutionibus Orbium Coelestium* (On the Revolutions of the Celestial Spheres) was published in 1543, the proliferation of copies meant that many astronomers simultaneously discussed its findings. In addition, the text could not be suppressed by the Church authorities, even though they wished to maintain the unquestioned truth of a planetary system in which the Earth was the centre of the universe.

It was not just the movable type of the printing press that allowed science to take several great strides forward. Just as significant was the ability to produce images. In 1542, Leonhart Fuchs published his magnificent and beautifully illustrated *De Historia Stirpium Comentarii Insignes* (Notable Commentaries on the History of Plants). A team of professional artists had engraved the blocks and painted them by hand in colour to the author's specifications. Herbals had existed for centuries but never before had they received such a scientific treatment or been so well illustrated, and certainly no scientific work had been mass-produced to such a high standard. The importance of the printed image was even greater in the case of Andreas Vesalius's *De Humani Corporis Fabrica* (On the Fabric of the Human Body), which was published the following year. In 1300, Pope Boniface VIII had forbidden the dismemberment of corpses; as a result, *Anatomia Mundini* by Mondino de Luzzi, composed in Bologna around 1315 and based heavily on Galen and Arabic works, remained the basic anatomical work throughout the later Middle Ages. A printed version appeared in 1478 and went through 40 editions, perpetuating Galen's anatomical ideas.[3] These ideas, of course, were far from perfect, as the dissection of human bodies had been frowned upon in the ancient world too: Galen's conclusions were mostly based on animal dissections. Thus, serious errors of anatomical understanding lasted for centuries. Most medical schools received only a handful of hanged criminals' corpses to cut up each year and the dissections were more ritual than experimental. On the rare occasions when a medical dissection took place a physician would read the relevant sections of Mondino's Galenic text while the surgeon made the necessary cuts. Medical students observing such dissections were told that there were three ventricles to the heart and that the liver had five lobes; in the dissecting theatres of the time they could not get close enough to the organs to question what they were told. Thus the dissection in front of them only served to reinforce the authority of the teachers who were, in effect, misleading them. Vesalius's book swept all this away and introduced the scientific study of anatomy. Many carefully drawn and engraved plates showed the dissected body in various poses to reveal its skeletal and muscular form. Such images changed attitudes to anatomy itself, leading surgeons to engage in pioneering anatomical research despite the prohibitions of the Church.

Architecture, geography and astronomy similarly benefited enormously from the printed image. Even though Andreas Palladio's *Four Books of Architecture* (1570) was only published in Italian, it had a huge impact across the whole of Europe by depicting the architectural principles of Vitruvius and other classical architects. Also in 1570, advances in image reproduction allowed Abraham Ortelius to produce the first modern atlas using Mercator's projection. Tycho Brahe's *De Nova Stella* (About the New Star), published in 1573, included charts showing where in the sky the supernova had been seen the previous year. The same author's *Astronomiae Instauratae Mechanica* (Instruments for the Restoration of Astronomy), published in 1598, depicted the technology in his observatory in great detail, revealing how he had achieved such a high level of accuracy in measuring the sky, and how others might further his work. Thus printing did not just impart knowledge, but also acted as a catalyst to scientific advancement.

The foregoing points about the printing revolution are perhaps obvious. Less obvious are its social consequences. As the spread of the *printed* word increased the number of literate people, it placed a greater emphasis on the importance of the *written* word. This in turn changed the relationships between kings and their subjects. Governments now sought to collect information on all those who dwelt within their borders. Almost every country in Europe began to keep records of baptisms, marriages and burials. England started doing so in 1538. France kept records of all baptisms from 1539 and of all marriages and burials from 1579. In Germany, where a few individual parishes kept registers from as early as the 1520s, most states did so systematically from the 1540s. In Portugal, one in twelve parishes was already keeping registers in the 1520s. The Council of Trent in 1563 recommended that baptisms, marriages and burials be recorded in every parish, and most Catholic countries that were not already doing so complied within 30 years. In Italy, for example, parish registers were ubiquitous by 1595.

Registration was just the tip of the iceberg. In England, a colossal amount of written material came to be required by the state. Every county court had to keep records of its quarter sessions. Church courts had to maintain records of their grants of probate and to keep copies of the millions of wills, inventories and accounts on which they based their decisions. The Church examined and licensed school teachers, surgeons, physicians and midwives. Magistrates licensed innkeepers and victuallers from 1552. In each parish, surveyors of highways were required to keep accounts of money collected and spent on the maintenance of roads. Churchwardens had to keep accounts of parish funds, and overseers of the poor similarly had to produce accounts of their disbursements. Local militia organisers kept records of the men trained to defend the shores and the rates levied on communities to pay for these part-time troops and their supplies.

The government discontinued its sequences of medieval rolls and formed separate departments to deal with distinct aspects of the

administration of the realm. By the end of the century these departments had started to gather statistics, assessing such things as the number
of victims of each plague outbreak and the number of inns and taverns
operating in each county, as well as centrally collecting records of the
taxes paid by individuals. The government was also suppressing the
publication of certain books. Printing outside London was permitted
only at the two university presses, and all publications had to be registered at Stationers' Hall in the capital, allowing royal officials to cast an
eye over everything put in print, censoring anything that was contrary
to its interests. The state's involvement in both controlling the new
literate culture and making use of that culture to monitor the population was unprecedented. Today we might take such intervention for
granted but the leap from a realm of unrecorded subjects in 1500 to
one of detailed state supervision by 1600 was enormous.

Another less obvious social consequence of printing in the vernacular was the changing position of women in society. In the medieval
period, very few girls had been taught to read. If a woman could
write, she knew that the vast majority of her readers would be men,
and if they did not like what they read, they could easily silence her
by destroying her manuscripts. Printing put an end to that: if enough
copies of a book were printed, it was almost impossible for her enemies
to eradicate an author's work entirely. Also, texts did not discriminate
between their readers: while many teachers would not even have
considered teaching girls, a book didn't care whether the person
reading it was male or female. Intelligent women quickly realised that
they could learn from books just as well as their male counterparts.
Moreover, women had a particular reason to want to learn to read.
For centuries they had been told that they were legally, biologically,
spiritually and socially inferior to men, and that the reason for this
was that Eve had offered Adam an apple in the Garden of Eden. Now
that they could teach themselves to read, they could interpret the
Biblical story for themselves, and express their own views on the
inequality of the sexes. Moreover, they could do so in print, and with
the confidence that their words would reach the eyes and minds of
other literate women. It is hardly surprising that in England, where
male literacy more than doubled over the course of the century, from
about 10 to 25 per cent, female literacy increased proportionally even
more, from less than 1 to about 10 per cent.[4]

Not content with merely understanding the basis of the prejudice, a number of women attempted to redress the imbalance between the sexes. In Italy, Tullia d'Aragona wrote *A Dialogue on the Infinity of Love* (1547), arguing that there was nothing morally wrong in sexual desire, and that the association of women and the sexual act with sin was itself immoral and misogynistic. Gaspara Stampa (d.1554) wrote a series of passionate and moving lyrical poems following her abandonment by a lover, which demonstrated a degree of literary skill and argumentative wit that few men could match. The relationship between the sexes became a hot topic in Italy in the last decade of the century, and the arguments of unsympathetic male writers were answered by several bright women. Lucrezia Marinella argued forcefully against misogynistic writers of the past in *The Nobility and Excellence of Women, and the Defects and Vices of Men* (1600). In Moderata Fonte's *The Worth of Women, wherein is Clearly Revealed Their Nobility and Superiority to Men* (1600), seven Venetian women discuss why men and marriage seem destined to bring women unhappiness, and how much better off they would be if they remained single.

In England, similar debates were taking place. Isabella Whitney became the first published female poet in English, expressing a heartfelt bitterness, albeit in somewhat simplistic lines, in *The Copy of a Letter, Lately Written in Meter, by a Gentlewoman to Her Unconstant Lover* (1567). Jane Anger published her brilliantly splenetic *Protection for Women* in 1589, in which she asked: 'Was there ever [anyone] so abused, so slandered, so railed upon, or so wickedly handled undeservedly as are we women?' The remarkable Emilia Lanier spoke for many when she reasoned in her poem 'Eve's apology for the defence of women' (1611) that in the Garden of Eden, the fault for the whole apple business lay with Adam. God had made him stronger to take responsibility for Eve, so if he failed in that duty, why was she solely to blame? In both England and Italy, educated women began to produce translations of classical texts. In 1613, the first original play in English written by a woman, *The Tragedy of Mariam*, was published by Elizabeth Cary, Lady Falkland.

All these were just the crest of a great wave of women's writing, published and unpublished: letters, religious tracts, diaries, memoirs and recipes. By the end of the century, affordable self-help books written by women for women were being printed and reprinted by the thousand – at a cost of no more than a skilled worker's daily wage. They helped shape women's identities and reinforced their growing sense of individuality. Printing was thus the catalyst for a whole new relationship between women and knowledge – and, by implication, between women and men.

The Reformation

It should not surprise us that the Reformation started in a German-speaking state, Saxony. By 1517 the Bible had been available in German for more than fifty years, and its various editions had provoked many people to read and discuss the word of God privately. They grew increasingly concerned. They saw that there was a huge gulf between the early Church as it was presented in the Bible and the Roman Catholic Church of their own time. For example, in the early sixteenth century you could pay an indulgence seller for a piece of paper that supposedly absolved you of some or all of your sins. The more money you paid to the Church, the greater the range of sins for which you were forgiven. But there was no basis for such a thing in the Bible. Would their sins *really* be forgiven by buying a piece of paper? Some began to suspect that unscrupulous priests simply selected passages from the Bible to suit themselves – and suppressed those that did not fit their purposes. And what about those aspects of the Church that were not in the Bible at all? Biblical texts said nothing about payments to the parish clergy beyond some vague references to tithes. There was no mention of monasteries or ecclesiastical landlords. The whole hierarchical structure of the Church had no foundation in the teachings of Christ. As for rosary beads, wedding rings, hymns and sacred vestments: where did all this religious paraphernalia come from? To the great consternation of many, it seemed that these things were superfluous to the real purpose of religion, which was to guide them to live according to the word of God on Earth.

Into this context of spiritual concern stepped Martin Luther, a monk and a doctor of divinity at the University of Wittenberg. Outraged by the demands of a papal indulgence seller who demanded large sums from unwitting citizens for a piece of paper whose ultimate purpose was to pay for building works at St Peter's in Rome, Luther set about debating the validity of the pope's actions. On 31 October 1517, so the story goes, he nailed a list of 95 theses to the door of Wittenberg's castle church. These 95 points essentially argued that the sale of indulgences was nothing more than a ploy by the pope to raise money. Luther insisted that the pope had no power to remit any penalties other than those he had imposed himself: it was down to God to forgive sins and to determine what happened to the souls of the dead in Purgatory. He asked some searing questions about the pope's authority. If the pope truly had the power to redeem souls from Purgatory, why did he not simply release all the souls of sinners therein rather than leaving them there to suffer? Why should people pay for Masses for their late relatives as well? Most of all, as the pope was so wealthy, why did he not simply build the new church he wanted out of his own funds?

There had been many challenges to the authority of the Catholic Church before 1517 –the Cathars in the thirteenth century, the Lollards in the late fourteenth and Jan Hus in the early fifteenth – but Luther's onslaught was effective because he expressed exactly what many people across Europe were thinking. Moreover, unlike the medieval heretics, his views were widely circulated with the help of the printing press. By the time he was declared a heretic in 1520, he was a respected and popular figure, and people had begun to adapt their faith and worship in accordance with his teaching. What had begun as an attempt to reform the Catholic Church from within very quickly fractured the whole ecclesiastical structure and the unity of Christendom. Reformers of many different persuasions sought to redesign the Church to suit their own spiritual and not-so-spiritual ends. Ulrich Zwingli, John Calvin, Philip Melanchthon and Thomas Müntzer all attracted followers to their various teachings, differing, among other things, on whether infant baptism was acceptable or whether the host during Mass actually contained the presence of Christ or was merely a memorial of it. Philip of Hesse, the first political leader to adopt Lutheranism as a state religion, in 1524, even sought to enlist Luther's support for reintroducing polygamy – at least in respect of his own desire to take a second wife. By 1530 the Reformation had spread beyond the German-speaking borders to the British Isles, the Low Countries, Scandinavia and eastern Europe. Further renunciations of papal authority followed. In England, Henry VIII finalised his split from Rome in the Act of Supremacy of 1534, and the Danes formally renounced the Catholic Church in favour of Lutheranism in 1536. Even in countries where the ruler did not turn against the Catholic Church, the numbers of Protestant reformers grew quickly.

Why was this so important? The answer depended to a large degree on who you were and where you lived. The Protestants themselves felt a great sense of liberation from the repression of the Catholic Church. Restrictive practices and old laws and rules were done away with. Citizens in Protestant countries were delighted not to have to pay large amounts of money to Rome for an indulgence or similar papal tax. No longer having to pretend that the bread at Mass became the actual body of Christ would have lifted a burden from the conscience of some; not having to fear ending up in Purgatory must have been a similar relief for many more. But there were down sides too. Although many Church payments did not go to Rome any more, they were not abolished; they were simply paid to secular lords and property owners. That raised new moral questions: was it right to pay ecclesiastical tithes to the lord of the manor or a university college? Many people lost their livelihood on account of the abolition of age-old rites, such as innkeepers who accommodated pilgrims travelling to the established shrines, or undertakers who furnished the great funeral services of the wealthy that had preceded the Reformation. At a deeper level, too, there was a measure of disorientation and confusion. Religion was entwined with natural philosophy, or, to use the modern term, science; thus to doubt religiously was to be plunged into scientific doubt. Consider the problem in terms of your own faith. Whatever you believe, you believe in *something*, whether it's the presence of God or some other creative force, or the random accident of chemical combinations, or something else entirely. You have an idea about how the world and everything associated with it came into being. It is not something you *choose* to believe – what you publicly say you believe could easily be quite a different thing – it is what you think is most likely to be true on the basis of such factors as your cultural background, the evidence you have seen with your own eyes and your understanding of the statements and ideas of others. Now imagine that what you believe is flatly denied by half of the known world, which will take up arms to stop you espousing your views. The result is likely either to make you question your understanding of the world or to make you defend more vigorously the truth of what you believe.

There is no doubting the divisiveness of the Reformation. But it is important to realise that it was not just a matter of setting Protestants against Catholics. Indeed, it was difficult to know exactly what the term 'Protestant' implied. However, people were sure that what they professed on Earth would affect the fate of their eternal soul. While some revelled in the iconoclastic fury that gripped Zurich in 1524, Copenhagen in 1530, England in 1540 and 1559, and the Netherlands in 1566 – when radical Protestants gleefully smashed the statues of saints, burnt the crosses that stood above the altars, and whitewashed over the paintings of Doom and the Last Judgement that decorated their churches – others were disturbed by it. Even if you resented the payment of money to the pope for spurious indulgences, it did not follow that you wanted to see the shrines of saints smashed and the relics within scattered to the winds. The question of Purgatory was particularly troublesome. In the past, people had dutifully prayed for their ancestors' souls; now they were told that those souls had either gone straight to Heaven or Hell and there was nothing they could do about it. In Protestant countries, all the friaries, chantry chapels and monasteries founded for the souls of the donors were closed and sold off by the government. Many people were deeply troubled when their family burial places were confiscated by the king and sold to rich merchants for demolition, or conversion into comfortable country houses. That was not why their ancestors had freely donated their lands and wealth to the Church rather than bequeathing them to the next generation (and ultimately to them).

Whereas in the eleventh century the Church had done much to promote peace across Europe, now it proceeded to tear itself apart. Nations were set against nations on account of whether they were Catholic or Protestant, and religious factions within those countries fought each other in civil wars that threatened everyone. There is no better example of the old adage that 'the path to hell is paved with good intentions'. Luther had only wanted to stop the corruption within the Church – a laudable aspiration by anybody's standards – but what he triggered was more than a hundred years of war in Europe, the persecution of religious minorities by European governments for the next three hundred years, and religious intolerance that in some places has lasted to the present day. Just to look at the conflicts of the sixteenth century is sobering. The first fighting broke out in the German Peasants' War of 1524–5, a widespread social uprising inspired by the teachings of Luther and Thomas Müntzer. It was brutally quashed by German princes – Protestant and Catholic alike – and the vicious persecutions that followed foreshadowed the bloodbaths that were meted out to Catholics in the Pilgrimage of Grace (1536) and the Prayerbook

Rebellion (1549) in England, and to Protestants in the Low Countries (from 1566) and in the St Bartholomew's Day Massacre in Paris (1572).

The violent tensions between nation states in the second half of the century was exacerbated by religious suspicions. Espionage by foreign agents was widely feared, and governments began to spy on their own people. In cities that had once been broadly open to newcomers, refugees were consigned to Protestant or Catholic ghettos. Those belonging to a minority sect were taxed heavily and their freedoms restricted. When English and Spanish ships met on the high seas, it was assumed on account of their religious differences that they were enemies and that they might legitimately fire on one another. In England, which had hitherto resisted the use of torture for political purposes, the government started to inflict intense pain on Catholics to force them to reveal their secrets. In Spain, the Inquisition was extended to eradicate Protestantism. The Council of Trent reaffirmed the legitimacy and orthodoxy of the Catholic Church and sought to reinvigorate it through a series of internal reforms, including the tightening of discipline among the clergy, and the prohibition of heretical literature. The new order of the Jesuits was particularly encouraged to evangelise, administer the Church more scrupulously, and stamp out heresy. The resentment between Catholics and Protestants grew ever deeper as the century progressed, and religious and political differences combined in an explosive mix that threatened people's well-being on Earth as well as their path to Heaven.

The Reformation dealt a tremendous blow to the political authority of the Church. The higher clergy had for centuries acted as a sort of unofficial opposition to rulers, not just advising but also restraining them. The best-known example in England is that of Thomas Becket, the twelfth-century archbishop of Canterbury, who opposed Henry II of England and eventually paid for doing so with his life.[5] In most countries, prelates had constituted one of the 'estates' of the realm, alongside the nobility and the commoners, and thus occupied an important part in the government. In thirteenth-century France six of the twelve peers of the realm were prelates. In medieval Germany three of the seven electors who chose the king of the Germans – who often went on to be crowned Holy Roman Emperor – were archbishops. Now, in many places, the authority of Church leaders collapsed along with the hegemony of the Church itself. In 1559, all of the English bishops, who had been appointed by the deceased Catholic queen,

Mary I, refused to accept the Elizabethan settlement, which forbade their celebration of Mass and their allegiance to the pope. Elizabeth I accordingly removed them from their positions and handed the empty bishoprics to clergymen who agreed not only to obey her but also to surrender valuable Church lands to her as soon as they were in office. Such men were not in a position to restrain their monarch. Before 1529 almost every chancellor of England, the highest post in the government, had been a bishop or an archbishop. Mary I appointed three prelates to that position in succession during her short reign (1553–8), but after her death never again was the great seal put in the hands of a clergyman. In Catholic countries like France, the higher clergy were still appointed to positions of authority – Cardinal Richelieu and Cardinal Mazarin are famous seventeenth-century examples of prelates who were also statesmen – but they were now servants of the state, not independent men who held the government to account. The Reformation thus swept away an important brake on royal power. More than that, it entwined secular strength with divine authority in investing kings and queens with the position of head of their national Church, as decreed in the maxim *cujus regio, ejus religio* (literally: 'who rules, his religion'). This was the conclusion of the Treaty of Augsburg in 1555 that brought an end to hostilities between Lutheran states in Germany and the Holy Roman Emperor, Charles V of Spain. It meant that the official religion of the state should be the faith of the ruler. Luther could never have foreseen or intended this: he had unwittingly set off a chain reaction that now gave kings absolute power – and to deny that power was not just treasonable, it could be heretical too.

Firearms

When Francis Bacon declared in 1620 that guns had 'changed the whole face and state of things throughout the world', he was not referring to fifteenth-century cannon. Such weapons were predominantly used for destroying the walls of castles and cities. They were sophisticated versions of medieval siege engines and thus of the greatest significance in large-scale conflicts rather than private squabbles. Their importance is reflected in the developments in defensive architecture. Around 1500 Italian military engineers started to experiment with the *trace italienne*: polygonal star-shaped fortifications with very thick, sloping walls that not only provided bastions capable of resisting an enemy's cannonballs but also permitted the defenders to cover every potential assault point with their own artillery. In many ways, however, large guns and the *trace italienne* represented a continuation of the medieval contest of wall against machine. It was hand-held weapons that constituted the major change. In 1500 even the more portable guns were cumbersome mini cannon that one person could barely lift, let alone shoot swiftly and accurately. While a man *could* handle and fire such a gun on his own, by the time he had loaded it, aimed it and ignited the gunpowder, his enemies would have overrun him. Thus almost all the advantages still lay with the old technology. A squadron of crossbowmen were cheaper to maintain, easier to train and quicker to transport than a squadron of gunners equipped with early arquebuses. Crossbow bolts were more accurate too. As for English longbows, arquebuses could not rival their cheapness of production, speed of shot, or portability. If a hundred gunners had run into a hundred archers in 1500, there would have been no doubt as to the outcome. After the first volley of wayward shots from the arquebuses had carried off a handful of archers, the rest of the gunners would have been mown down by the bowmen in a matter of seconds. The sixteenth century saw this imbalance reversed. If the same forces had met in 1600, it would have been the bowmen who quailed and ran for cover.

Several technological innovations led to this new precedence. The introduction of standardised shot for guns of a specific calibre allowed bullets to be cast cheaply in large numbers and shared between soldiers. The introduction of the wheel-lock firing mechanism, although expensive, permitted an efficient way of firing pistols and long-barrelled guns that did not burn miles of match. In 1584 William, prince of Orange, became the first head of state to be assassinated by a pistol, and by the end of the century over half of the troops on both sides in the conflict between Spain and the Dutch were armed with portable long-barrelled guns.[6] The Swiss mercenaries who dominated European land battles at the start of the sixteenth century increasingly had to fight with muskets after a series of defeats culminating in the battle of Pavia (1525). Another indication of the growing importance of firearms was that full armour practically vanished from the battlefield: a breastplate, backplate and helmet were all that most European soldiers now wore. Suits of heavy plate armour were more trouble than they were worth, inhibiting movement, sight and hearing. They also meant that aristocratic commanders who could afford expensive armour were easy to pick out and shoot. At close range, a bullet from an arquebus could penetrate even the best plate armour.

Some historians have spoken of a 'military revolution' in the period from 1560 to 1660. They argue that this century saw the introduction of the taxation and parliamentary representation necessary to support large armies of infantry armed with firearms, which in turn under-pinned the development of the modern nation state.[7] In reality, military technology and strategy had been in a state of flux for centuries, at least since the development of the stirrup for knightly combat in the early eighth century. It therefore makes more sense to describe the last 1,200 years as a period of military *evolution*, rather than a few intermittent revolutions.[8] In addition, the concept of a 'military revolution' at this time originated in a study of Swedish and Danish history; however, the same innovative combination of taxation, parliamentary representation and large infantry armies using projectile weapons had occurred in England two hundred years earlier.[9] Nevertheless, it is unquestionable that the development of firearms in the sixteenth century constitutes a major change from the chivalric world dominated by the nobility.

You can start to appreciate the transformation when you think of a king on the battlefield in the years 1500 and 1600. At both dates his life would have been of paramount importance: if the king was killed, the battle was lost and very probably the war too. Despite this, it was important in the medieval mind for the king to be seen to lead his army in person, not least because by being present he could most effectively demonstrate that a victory for his side was the judgement of God. In 1500, if a king wanted to reduce the risk of being killed on the battlefield, he could wear armour and stay out of the killing range of the enemy archers. Even if he charged into the melee, it was unlikely that a crossbowman would be able to pick him out and pierce his armour from a distance, and his bodyguard could protect him from enemy knights getting too close. If a king rode into a fight in 1600, however, he ran the risk of someone pulling out a pistol and shooting him at close range. Even if he lurked on the sidelines he risked a daring shot from a musketeer. Battlefields had become more dangerous places over the course of the century. Guns made the area a smoke-filled, shudderingly noisy place of terror and confusion in which the whole order of things could be upset. A commoner could

shoot a nobleman or even a king and not realise it. Thus the advent
of firearms required that kings step back from the battlefield and leave
the command to experienced professional soldiers. While in 1500 it
was still not unusual for kings to command in person – Richard III
of England fought and died at Bosworth in 1485; the future Louis XII
of France was captured in battle in 1488; Francis I likewise in 1525;
James IV of Scotland was killed while commanding at Flodden in 1513;
and King Sebastian of Portugal perished at El-Ksar El-Kebir in Morocco
in 1578 – after 1600 it was very rare for a ruling monarch to approach
closer than the edge of a battlefield. That in turn had ramifications.
When a medieval king had led his army on campaign in person, he
had no one else to blame if everything went wrong. It was God's
judgement on him. If he merely placed his trust in a general, then
defeat did not necessarily mean that he no longer enjoyed divine
support: he could simply blame the general's incompetence. People
increasingly rejected the idea that military defeat represented God's
judgement. Conflict became much more of a secular matter.

Guns had a social impact throughout society. When Edward III had
pioneered the use of massed longbows he had enjoyed a unique
advantage over his rivals. Firearms blasted this to ribbons. Soldiers did
not need to have been trained from childhood to fire an arquebus or
musket; they could learn in a matter of weeks. Thus it was once again
the larger, better-equipped army that dominated the battlefield, and
so the old rule of engagement – that the larger force is stronger – was
restored. In the 1470s the armies of France and Spain had numbered
40,000 and 20,000 men respectively. By the 1590s the French maintained
an army of 80,000 men and the Spanish one of 200,000.[10] Early in the
seventeenth century armies were larger still: the Dutch kept 100,000
men in arms, the Spanish 300,000, and the French 150,000.[11] Across
Europe, kings and princes had to tax their people heavily in order to
pay for the weapons, training and men to field a large enough army
for their military purposes. Governments dealing with the threat of
invasion recognised that they needed a better civil defence than the
old practice of country folk grabbing a sword or billhook and rallying
to the beacons' warning. Bands of militia were trained and equipped
with small arms. Gunpowder depots were organised to supply them.
Forts with low, wide walls in the style of the *trace italienne* were built
to guard harbour entrances. Large nations with the ability to tax widely
and raise enormous armies could sweep small nations off the map,

so smaller countries too had to be prepared to defend themselves. The result was an arms race across the continent. War was no longer an occasional (if frequent) event: preparing for war became an aspect of everyday life for hundreds of thousands of people as governments used the breathing spaces of peace to strengthen their defences and prepare for the next conflict.

The development of guns also had an impact beyond the continent for it allowed European nations to dominate the world's oceans and, later in the century, to make inroads into new territories. On this point we need to be quite specific: it was not firearms that allowed the Spanish to conquer South America – that was down to a combination of factors, including indigenous prophesies, superstitions, local civil wars, smallpox and the fact that stone-headed clubs and short-range bows and arrows were no match for Toledo steel blades and armour. However, the Spanish certainly relied on cannon to protect their vessels travelling to and from the New World. English and Moroccan sailors were only too keen to take advantage of poorly protected shipping laden with gold and silver. This led to a naval arms race comparable to the competition for better hand-held weapons. For their seaborne empire, the Portuguese depended on firearms to an even greater extent than the Spanish. As the sixteenth-century Portuguese writer Diogo do Couto pointed out, his compatriots in the Indian Ocean and the Far East had to contend with enemies who were armed with weapons almost as sophisticated as their own.[12] Wherever they built a trading post or 'factory' they had to defend it with fortifications, cannon and small arms. In this way European nations, which had not controlled any of the world's oceans in 1500, were masters of them all by 1600. They thus dominated the world's long-distance international trade, even if they were divided amongst themselves. And thanks to their maritime firepower, they would continue to rule the seas well into the twentieth century.

The decline of private violence

You would have thought that the development of handguns would have resulted in an increase in the murder rate; in fact, quite the reverse happened. In the Middle Ages, violence had been endemic and a part of everyday life. The homicide rate was normally around

40–45 victims per 100,000 people but at times it could be much higher. In the university town of Oxford in the 1340s, it reached 110 per 100,000 inhabitants, which was not so very far from the rate of Dodge City, one of the most dangerous towns in the American Wild West, at its gun-slinging height.[13] Often the violence was completely unpremeditated. Fights over women and arguments in taverns were common. The historian Manuel Eisner, who has made a special study of this subject, notes that two of the 145 homicides tried in London in 1278 were murders that took place following games of chess.[14] But in the fifteenth century, the number of killings started to fall, and in the sixteenth it suddenly dropped by half. As the graph below shows, there was a remarkably consistent decline in the average homicide rate – by about 50 per cent every hundred years – from 1500 until 1900.

This decline begs the question: why? What caused people across Europe to stop killing each other? There are two traditional explanations. The first is based on the work of the German social historian Norbert Elias, who postulated in his book *The Civilising Process* (1939)

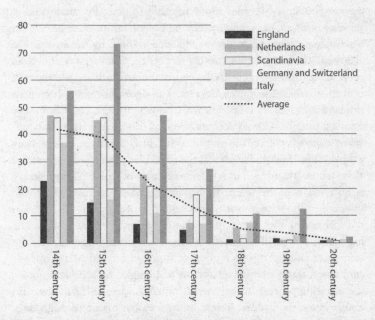

Homicide rates (per 100,000 people)[15]

that people's wilder behaviour was tamed in the early modern period through the adoption of new social rules and a greater attention to etiquette. Aristocratic violence, which once had been much more common than aggression among common folk, was brought to heel by duelling and similar codes of conduct now associated with 'gentlemen'. The rising urban classes civilised themselves, proud of a religious life that abhorred violence, and started to control the working folk through preaching morality in church. Gradually everyone came to accept the same code of civilised behaviour. The second traditional explanation is simply that governments became more efficient at punishing the perpetrators of crime, so that through the deterrent of the gallows, violent offences were checked.

A few years ago, the psychologist Steven Pinker tackled this question in *The Better Angels of Our Nature* (2011). He largely adopted the 'civilising process' theory and suggested that it was triggered by two fundamental causes: an 'economic revolution' (by which he meant the benefits enjoyed by people trading with each other), and the increasing power of the state.[16] With regard to the former, as he puts it, 'if you're trading favours or surpluses with someone, your trading partner suddenly becomes more valuable to you alive than dead'.[17] Pinker's book makes many interesting points but on close inspection his explanation of the decline of violence in the early-modern period is flawed. The nearest thing to this sort of 'economic revolution' took place in the thirteenth century. Indeed, in the sixteenth and seventeenth centuries, when the homicide rate was declining most sharply in England and Germany, per capita incomes in both countries were stagnant or falling. Gross domestic product (GDP) in relation to population fell by 6 per cent in England, and in Germany it collapsed by a third.[18] The English decline was felt particularly badly by the working class, whose income fell in real terms by almost a half.[19] There simply is no correlation between the economy and the decline of private violence in the sixteenth and seventeenth centuries.

Pinker also highlights the other traditional explanation – the increasing power of the state. Rather than simply arguing that the state was more willing to hang people, however, he argues that stronger centralised states stopped aristocrats and knights fighting and started the civilising process. As he puts it, 'a man's ticket to fortune was no longer being the baddest knight in the area but making a pilgrimage to the king's court and currying favour with him and his entourage'.[20]

But lords had curried favour with kings for centuries, and at the same time terrorised the peasants back on their estates: the one did not imply the discontinuation of the other. Moreover, as we see from the above chart, Renaissance Italy, which was the home of courtly behaviour, saw an increase in homicide at this time. While Pinker is right to point to the growing power of the state as a reason for the decline of violence his description of the mechanism by which the state brought it about is misleading. To understand what was going on, we need to go beyond the traditional explanations and ask how and why people might embrace the power of the state.

Let's say you have a wonderful orchard in which the finest apples grow – so good that a number of your neighbours regularly steal them. After a while you decide you cannot tolerate their thieving ways any longer and start to beat them with a stick when you catch them. Most desist but a few carry on stealing, so you decide to lie in wait and strike the next thief over the head with a brick. The thieves hear about this and stop taking the apples, fearful of meeting you and your brick. What does this tell us about violence? When you were tolerating the theft, there was crime but no violence. When you were taking limited action against the thieves, there was both crime and violence. The threat of the brick marked the apparent end of both crime and violence. According to Pinker, we should categorise the last stage as non-violent, for no one is beating anyone else. However, if there were no *threat* of violence, the thieves would return. The violence is still there, only it is latent. This is more or less what happened in the sixteenth century: less violence was enacted and more of it transferred to a *potential* form.

Pinker does not consider potential violence. His argument is that violence in all its forms has declined over the course of human history. This opinion has received much support in the last few years. A recent review article states that 'a number of scholars simultaneously and independently of one another [broadly agree] that war and violence in general have progressively decreased in recent times . . . and even throughout history'.[21] This judgement is sound if violence is defined purely as the enacted kind, and isolated from all other forms of force. But as my orchard story shows, to consider violence only in its enacted form is to see just half the picture. We have to consider force in all its forms if we want to understand what is happening when people refrain from attacking each other.

In its potential state, force can be transferred from one authority to another – from an aggrieved victim to the legal system, for example. What stops an aggressive man from being violent is the thought that his victim might transfer his right to seek retribution to a more powerful body, such as the state, which will exact a more dire revenge. If there is no state to undertake this, the potential perpetrator has nothing to fear; he can be as violent as he wishes, as we can see in the high levels of aggression in primitive societies or the breakdown of law and order in Florence during the Black Death, when the *becchini* robbed and raped with impunity. Where the state is willing to take on the transferred debt of violence, however, it can crush the violent oppressor, including wealthy and powerful men. This is why lords gradually ceased to terrorise and mistreat their tenants: it was nothing to do with their currying favour at court, as Pinker claims, but a growing fear that they too would be punished by the state if they broke the law. The same goes for the rest of society. When people saw that the state was prepared to punish violent offenders, they thought twice before resorting to violence.

So what made governments willing to act in the sixteenth century? It was not simply a greater readiness to hang offenders, as traditional explanations have tended to state. Ever since the twelfth century large numbers of criminals had been hanged for theft and other felonies. Rather the change was due to a phenomenon we have already discussed: increased literacy. Writing improved the lines of communication between the centralised state and its officials on the ground. As we have seen, a wealth of documentation was produced. Most importantly, records were kept of all the individuals in a locality. Over the decades, victims increasingly trusted the state to take responsibility for law and order. The wronged person who might once have felt honour-bound to draw the knife from his belt now thought, 'I do not wish to risk my neck trying to get revenge; I will call the constable and seek justice through the law.' He chose to transfer his retribution for three main reasons. First, he came to trust the legal system because it was becoming more efficient; second, he knew the legal system was more powerful than he was, and better equipped to enact revenge; and third, he himself feared falling foul of the law or starting a vendetta if he killed the original wrongdoer. At the same time the increasing efficiency of the legal system acted as a deterrent to the criminal too, by making it more likely that he would be apprehended. The would-be highway robber now decided: 'If I am caught, it won't be my pathetic victim who will enact revenge but the state.' Thus better communications in society and a more efficient legal system acted as a brake on both the victim's urge for retribution and the potential criminal's readiness to kill.

It would be simplistic to suggest that this was the only root cause of the decline of private violence. Running alongside it was an element of heightened individualism, which formed an integral part of the 'civilising process'. An enhanced sense of self can be seen in the growing tendency in the second half of the century for people to write diaries in which they expressed their internal lives. People's increased sense of themselves and their own suffering clearly facilitated their awareness of the suffering of others: we find much more empathy in the works of Shakespeare and the writers of the late sixteenth century than we do in the texts of their medieval forebears. At the same time, individualism and self-awareness were slowly changing the nature of self-respect in society. Consider that high murder statistic for Oxford in 1340: part of the reason for it was that, as in Dodge City, you had a large number of young, ambitious men with knives on their belts and friends at their sides, egging them on. The knives were still around in 1577, when William Harrison commented in his *Description of England* that almost every young man in London insisted on carrying a dagger. But Harrison was very disapproving of the practice. Men of good character were now expected to go to the law if they had a grievance. The heavily puritan society of early Elizabethan England, which frowned on violence for religious reasons, no doubt further undermined the confidence that young men felt when they went about town armed with a dagger. Where once a degree of dignity had attended the man who took revenge in person, now greater respect was accorded the man who considered such violence beneath him.

Finally we need to note that the state continued to reduce private violence directly, as it had always done, by channelling it into public enterprises, including war. Medieval kings had controlled the factions within their kingdoms by concentrating the violent tendencies of their subjects on external enemies, through crusades or wars on neighbouring kingdoms. The same phenomenon can be seen in modern times: mid-twentieth-century America saw a significant drop in killings on home soil during the Second World War.[22] Thus governments helped reduce the homicide rate by giving young men in particular a socially respectable purpose on which to expend their destructive energies. It is therefore not entirely paradoxical that the century that saw the emergence of the handgun and massive armies also saw the decline of personal violence. The state itself exercised a civilising influence, of a sort, by co-opting private force for public ends.

The foundation of European empires

The decades that followed Columbus's first voyage saw an astonishing amount of land mapped by European explorers. The Treaty of Tordesillas had divided the unknown world between Spain and Portugal, and both countries wasted no time in trying to maximise their advantage. Columbus's fourth voyage explored the eastern shores of Central America in 1502. That same year, and again in 1504, Amerigo Vespucci, a Florentine banker, and Gonçalo Coelho made two expeditions to explore Brazil on behalf of the Portuguese, sailing down the coast as far as Rio de Janeiro. Vasco Nuñez de Balboa crossed Central America and discovered the Pacific in 1513. At the same time, the Portuguese continued their exploration of the East, reaching Mauritius in 1507, Malacca in 1511 and China in 1513. Desperate not to lose their own stake in this global race, the Spanish dispatched a Portuguese captain, Ferdinand Magellan, and 270 men in five ships to seek a western passage to China in 1519. They found the southern tip of South America and sailed around it and across the Pacific to the Philippines. Magellan and almost all his men died on the journey but 18 men returned in one creaking ship under the command of Juan Sebastián Elcano in 1522, having sailed all the way around the world. In just 30 years European sailors had gone from not knowing what lay on the other side of the Atlantic to circumnavigating the globe.

The speed of discovery did not abate. Indeed, one discovery led to another. Before Elcano had returned to Spain, Hernán Cortés had brought about the collapse of the Aztec Empire. The Inca Empire soon followed. Buenos Aires was founded as a Spanish settlement in 1536. 'All has now been traversed and all is known,' declared the Spanish historian Francisco López de Gómara in 1552.[23] By 1570, when Abraham Ortelius produced the first modern atlas, the outlines of North and South America were indeed known, including much of the inland detail for the southern continent, and the eastern seaboard of North

America. Europe, Asia and Africa were shown very much as we think of them today. Australia and Antarctica were the only great land masses yet to be discovered. Ortelius's map does include an 'Australian' continent – based on the assumption that Tierra del Fuego was the northern tip of an Antarctic land mass – but it was not long before the real Australia was sighted, in 1606. The sixteenth century was thus undoubtedly the one in which Europe discovered most of the previously unknown world.

The difference this made to life in the West was colossal. For a start, it hugely expanded the geographic limits of Christendom. Previously there had been relatively sharp borders – points beyond which the Catholic Church did not exercise any influence and where travellers had to accept the rule of potentates who had no loyalty or obligations like those of Christian rulers. The Atlantic had been crossed in 1500 but it took decades for the newly discovered lands to be brought within the orbit of Christendom. By 1600, however, Latin America was ruled from Europe. The continent had European governors, used European languages for its administration, and transported gold and commodities directly to Europe. In Africa and the Far East, although sovereignty of the land remained in the hands of local rulers, the Portuguese were able to trade with many different nations. The Spanish too had set up an international trading empire, shipping silver directly from Latin America to China by way of Manila in the Philippines, which they founded in 1571. Whereas the Venetians had once seen the world come to their marketplace, the Spanish and Portuguese made their marketplace the world.

The resulting social and economic changes did not just affect Spain and Portugal. For the first time, Europe saw the effects of long-term inflation as large quantities of gold and silver flowed from Latin America into the coffers of the Spanish treasury and thence into the European economy. Tales of a fabulous *el dorado* were matched for wealth by the reality of the Potosi silver mines, discovered by Spain in 1545. Of course, such stories and discoveries encouraged more explorers and conquerors to seek their fortunes. English adventurers followed their Spanish rivals. John Hawkins of Plymouth led three triangular expeditions, buying or capturing black slaves in Africa and exchanging these for gold and silver with the Spanish in Hispaniola before sailing back to England. Francis Drake, who sailed with Hawkins on his third expedition, commanded the second circumnavigation of the world, in 1577–80, and Thomas Cavendish led the third, in 1585–8; both men returned with enormous wealth. But they were just the most successful of a huge number of fortune-hunters, who ranged from semi-official privateers to outright brigands and ruthless pirates. By the end of the century, the Portuguese, Spanish and English had been joined by the French and Dutch in attempting to exploit the world's resources.

The explorers brought back more than just gold and silver; they also returned with many specimens of the flora and fauna that they found in the countries they visited. Turkeys, potatoes, peppers, tomatoes, maize and cocoa are perhaps the most obvious new foodstuffs to have arrived, but in addition there were flavourings such as vanilla, allspice and chilli peppers. Rubber and cotton appeared in Europe, and so did dyes that were previously difficult or impossible to obtain, such as brazilwood and cochineal. The introduction of tobacco presented the previously non-smoking Europeans with an entirely new way of consuming a natural crop. The spices that had been brought to Europe from the East at enormous cost in the Middle Ages – cloves, cinnamon, pepper, nutmeg and sugar – were now shipped in such large quantities that their prices fell dramatically. Much the same can be said for other Eastern imports, such as silk, coconuts and aubergines. In 1577 William Harrison expressed astonishment at 'how many strange herbs, plants and annual fruits are daily brought unto us from the Indies, Americas, Taprobane [Sri Lanka], the Canary Isles and all parts of the world'. Kings, noblemen and wealthy merchants all collected exotic items from the furthest regions, exhibiting them in their houses for the amusement of their guests – Native American canoes, Indian daggers, Chinese porcelain, Arabian musical instruments. The traders brought people of different races to Europe too. Natives from the Americas tended to be shown as curiosities. Those from sub-Saharan Africa were traded as slaves. The Portuguese were the foremost slave-trading nation but England and Spain also took part in the business. Queen Elizabeth I herself financed one of Sir John Hawkins's slaving expeditions in the 1560s. It is ironic that the slave trade, having been declared barbarous and forbidden in the eleventh century by William the Conqueror, was supported in the sixteenth by his royal successor.

Conclusion

The sixteenth century changed what you could eat and when you might eat it. It changed what you read. It saw many people move from the country into towns. In northern Europe it introduced a whole new range of domestic comforts and saw the likelihood of you being murdered hugely reduced. And it brought to the fore some of the deepest concerns of mankind – what the interior of the human body looked like; what the place of the Earth in the universe was; how Creation was far larger than previously thought; and what you should do to save your soul. In trying to ascertain the predominant change of the century, however, it strikes me that there are two particularly strong candidates: the shift to a literate society, and the expansion of the world. Trying to decide which of those two was the more significant factor distinguishing the West in 1600 from its state in 1500 is nigh on impossible. That dilemma itself shows what an astonishing century it was.

Applying the salutary test of what happened here in Moreton-hampstead, we can see that the changes mentioned above were ubiquitous. Parish documents were created as a matter of course, and records relating to Moreton and its people were kept in Exeter and London. It is highly likely that the first printed books came to Moreton in the sixteenth century. Firearms would have made their way here for the first time too. The local militia was entirely reorganised and the muster rolls for 1569 reveal local men armed with long-barrelled guns. With regard to religion, the fear that imbued society when Henry VIII wrenched England away from the Roman Catholic Church led to the Prayerbook Rebellion of 1549, which started at Sampford Courtenay (14 miles from Moreton) and culminated in a series of battles that left more than 5,000 men dead. A vast number of men from Devon – a county with two coasts – sailed to Africa and the New World with Devonian captains such as John Hawkins, Francis Drake and Richard Grenville, returning with exotic flora and fauna, and bringing back black slaves to be pageboys and servants in the houses of the wealthy. On the domestic front, the introduction of chimneys and glass windows is evident all round this part of England. In the house in which I live, the Stoning family built a substantial fireplace in the hall, with an external chimney, showing off to the neighbours. Traces of mullioned windows, which would once have held leaded glass, also attest to the comfort they enjoyed.

Despite these things, perhaps the most significant development of all was the very realisation that society was changing. In previous centuries, people had known that war, famine, pestilence and plague altered their circumstances temporarily: these events would come and go. But at the end of the sixteenth century, people began to look back and see that life had changed fundamentally and would never be the same again. They could not 'undiscover' the New World. New books and new discoveries came to their attention every year. In the north of Europe, people could see the ruined abbeys all around them and they knew that the age of monasticism had passed. They saw the derelict castles – redundant since cannon had rendered their walls vulnerable – and understood that the age of chivalry had passed too. Towns and counties started to publish their own histories, aware of their waxing and waning fortunes. Indeed, history writing itself took off, with historians using a wide range of documents to piece together a reasoned analysis of the past, rather than just repeating snippets from old chronicles verbatim. It was not just that the sixteenth century saw enormous changes; it was that people were aware for the first time that these changes were happening. That awareness is another significant difference between the medieval and the modern mind.

The principal agent of change

The sixteenth century is simply littered with household names. It saw the world's most famous painting, the *Mona Lisa*, painted by Leonardo da Vinci in 1503. Michelangelo carved *David*, probably the world's most famous sculpture, the following year. We have encountered Magellan, Cortés, Copernicus, Erasmus, Brahe, Bacon and Vesalius, but this was also the time of Nostradamus, Machiavelli and Paracelsus. Galileo and Shakespeare were both born in 1564 and thus lived the first 36 years of their lives in this century. But whereas four of the five great changes noted above cannot be attributed to an individual, one can. The split in the Church was closely bound up with Martin Luther. Moreover, Luther did not just start the Reformation; he shaped it with his sermons, hymns, translations, theological writings and correspondence – and the example of his own life, in such things as becoming one of the first priests to marry and start a family. He was not the only Protestant reformer but whether the theology of others would have been sufficiently clear, robust and inclusive to carry a significant proportion of the population is open to question. Whether they would have had the skills to win over so many secular leaders to their cause is even more doubtful. What is certain is that Luther's stance in 1517 led to the most dramatic religious upheaval of the last thousand years. For that reason, he deserves the title of the principal agent of change for the sixteenth century.

1601–1700

The Seventeenth Century

The seventeenth century presents us with a huge paradox. On the one hand, it was the most miserable time to be alive since the Black Death. Famines killed millions. Many countries were devastated by internal conflict. Parts of Germany saw death rates of more than 50 per cent in the Thirty Years War. France lost a million citizens in the Fronde of 1648–53. England was torn apart by the civil wars of 1643–51. Several countries were locked into bitter disputes on land and at sea. And yet, despite all this conflict and devastation, most European nations today look back on the seventeenth century as their 'golden age'. The Spanish Golden Age is said to have started with the end of the Reconquista in 1492 and to have extended to the end of the Thirty Years War in 1648. In England, the years between the defeat of the Spanish Armada (1588) and the death of Shakespeare (1616) are often described as a 'golden age'. The Dutch Golden Age and the French Golden Age are regarded to be roughly coterminous with the seventeenth century. All these countries saw artistic and literary achievements of the highest order. In France, you have the palace of Versailles, the art of Poussin and Claude Lorrain, and the plays of Molière. Spain can boast the art of Velázquez, Murillo and El Greco, and the literary works of Cervantes and Lope de Vega. In the Netherlands, we find Rembrandt, Franz Hals and Vermeer, and a whole host of genre painters. In Rome, the baroque flourished to its fullest extent and Caravaggio painted his chiaroscuro masterpieces. This unlikely combination of global crisis and cultural flowering brings to mind Orson Welles's famous line in *The Third Man*: 'In Italy, for thirty years under the Borgias, they had warfare, terror, murder and bloodshed, but they produced Michelangelo, Leonardo da Vinci and the Renaissance. In Switzerland, they had brotherly love, they had five hundred years of democracy and peace – and what did that produce? The cuckoo clock.' While the idea that the Swiss enjoyed 'five hundred years of democracy and peace' is far from the truth, Welles's point about war and cultural achievement going hand in hand really can be applied to the seventeenth century.

Perhaps we can begin to understand this paradox if we reflect on the context of hardship. The prosperous might have lived in houses with chimneys, glass windows and more comfortable furniture, and they might have been better fed than their malnourished ancestors, but life expectancy at birth in late-seventeenth-century Paris was just 23. If you were the son of a bourgeois citizen of Geneva, then you might live to 30, and a daughter might make 35. Life expectancy at birth in England remained at a fairly constant 30 years – fluctuating from a low of 24.7 in a particularly bad year (1658) to a high of 35.3 in a particularly good one (1605). This was significantly lower than it had been in the previous century, when it sometimes exceeded 40 and seldom dropped below 30.[1]

A fundamental factor underpinning this dismal outlook was climate change. For the last 40 years, historians have referred to the seventeenth century as the 'Little Ice Age' but only recently has the full impact of the weather been appreciated. As we saw in the twelfth century, a drop in the average temperature of 0.5°C meant that the first and last frosts might appear 10 days sooner and later than was usual, wiping out the entire harvest.[2] The risk of consecutive harvest failures rose dramatically with even this small drop in temperature, especially at high altitude. In addition, a heavy rainfall could damage a crop, reducing it by a third or even a half. As we saw in the chapter on the fourteenth century, it did not take a complete harvest failure for a farmer to have nothing to take to market: a yield of 3:1 rather than 5:1 could render him with nothing to spare. Whether this was due to the lack of nitrogen in the over-cultivated soil or a cold and wet summer did not make much difference: if he needed 70 per cent of his corn to feed his family and his animals and to set aside as seed corn for the next year, a depletion of one single harvest by just 30 per cent would leave him with nothing at all to sell. That was the start of a chain reaction. People living in the nearest market town were deprived of grain. The price of bread increased as more people competed for what little there was. And as they had to spend more money on food, they had less to spend on non-essential things such as furnishings, tools and trinkets. As demand for these items diminished, their prices fell and the artisans who made them had less money coming in at exactly the time when they needed to spend more on food. Ultimately, those at the bottom of the food chain grew weak, fell ill and died. Such was the effect of just one poor harvest with a yield of 50 per cent. Consecutive harvest failures thus killed thousands, including the farmers and their families, who were left with nothing to plant or eat. Even without a severe frost, an average summer temperature drop of 2°C could wipe out between 30 and 50 per cent of the crop – as happened in northern Europe in the 1640s.[3]

The results of such bad weather were horrific. A French commentator in 1637 declared that 'posterity will not believe it: people lived off the plants in gardens and fields; they even sought out the carcasses of dead animals. The roads were paved with people . . . Finally, it came to cannibalism.' Unfortunately, that writer was wrong: posterity could easily believe it. An observer at Saint-Quentin in 1651 wrote: 'of the 450 sick persons whom the inhabitants were unable to relieve, 200 were turned out [of the town], and these we saw die one by one as they lay on the roadside . . .' Ten years after that, another Frenchman stated that 'the pasturage of wolves has become the food of Christians, for when they find horses, asses and other dead animals they feed off the rotting flesh'.[4] France suffered a particularly severe winter in 1692, which was followed by the Great Famine of 1693–4, resulting in about 1.3 million deaths out of a total population of 22 million. The winter of 1695–6 killed 10 per cent of Norwegians and perhaps 15 per cent of Scots. A third of the population of Finland and a tenth of that of Sweden starved in the famine of 1696–7.

On top of these food shortages, people had to contend with the changing patterns of disease. There were several calamitous outbreaks of plague in the larger European cities: Milan in 1629; Venice in 1630; Seville in 1647; Oslo in 1654; Naples and Genoa both in 1656; and Vienna in 1679. London saw a succession of major outbreaks in 1603, 1625 and 1665, and Amsterdam did likewise in 1624, 1636, 1655 and 1663–4. Smallpox – which had previously been regarded as a children's disease – took on a far more deadly form from about 1630, becoming the second-most feared illness for both adults and children. With starvation and disease both looming large, death played an even more significant part in people's lives, snatching away young siblings, parents and children, and focusing minds intently on God.

This backdrop of famine and disease goes some way to explain the paradox of the century and its simultaneous 'golden ages'. People suffered terribly, but what was later remembered of the period was not the suffering itself, but the things people did to escape it. And they were prepared to do practically anything. Men who could not feed their starving families by scratching a living from the soil left the land that had sustained their forebears for generations and moved to the city: about 6,000 people came to London every year. Large numbers emigrated, so that by 1700, the population of the American colonies stood at over a quarter of a million. One fifth of the adult male population of Scotland abandoned the country, many of them heading off to seek a better life in Poland, of all places. About a quarter of a million Portuguese men left their native soil to seek their fortunes elsewhere in the Portuguese empire.[5] For many Frenchmen and Spaniards, war was their friend. The quarter of a million men in Louis XIV's army in the 1690s might have been barely 5 foot 7 inches in height – their growth having been stunted in childhood – but, as they destroyed all the towns in the Rhineland, they were no doubt pleased not to be back in Paris, suffering the chronic bread shortages there.[6] As for the Dutch, their 'golden age' is attributable not only to their victory in the Eighty Years War against the Spanish but also to the great riches of their empire.

The extreme disparities of wealth in all these countries also enhanced cultural achievements, leading to competition between everyone – from businessmen and architects to writers and musicians – and leaving a legacy of great treasures. The artists of the age, surrounded by the empty eyes of the starving and the prim smiles of the emerging bourgeoisie, could not fail to be moved to pity and contempt. What passed down to later generations was a sense of the great intensity of life at this time. In a world where everyone was struggling to survive and advance their careers, those who had the greatest ability were forced to exploit it to the full. To paraphrase the famous words of the seventeenth-century English poet Andrew Marvell, people knew they did not have 'world enough and time'. They needed to grasp every opportunity that arose, to innovate and experiment, and thereby to help themselves.

The Scientific Revolution

In the previous century, scholars had begun to realise that what they read in the venerated texts of ancient writers was not always true. The discrepancies between Galen's anatomy and the human body have already been mentioned, as have the geographical lacunae of ancient geographers. Both of these disciplines had to go through a lengthy process of experimentation to discover the limitations of ancient knowledge. Intercontinental navigators contributed to scientific discovery because they required more sophisticated mathematical methods for determining their position, direction and speed at sea. Their explorations yielded previously unknown plants from the New World, which in turn forced botanists to produce new studies of the world's flora. These discoveries also posed new scientific questions. Those who were genuinely interested in understanding the true nature of things (as opposed to citing ancient authorities) began to adopt what quickly came to be recognised as the scientific method: to postulate a research question and identify a suitable data set that would allow a hypothesis to be advanced to answer it, and then to test this hypothesis and discard it in favour of a new one if the original proved inadequate. Such a model of research was outlined in England by Francis Bacon in his *Novum Organon* in 1620 but by that time it was already being applied by most natural philosophers across the continent. Historians generally refer to this shift as the Scientific Revolution.

Of all the observable phenomena that made men think anew, it was the stars that most gripped people's attention and forced them to apply innovative methods of investigation. The 'new star' or supernova of 1572, most famously witnessed by Tycho Brahe, did not enter the Earth's atmosphere. It followed that it was a new, movable part of the firmament. Such a phenomenon directly contradicted Aristotle's teaching, in which the stars formed a crystalline structure around the Earth and planets. Brahe built a new laboratory and charted all the stars he could in order to try to explain the form of the heavens. Shortly before his death in 1601, he was joined by the young German astronomer and mathematician Johannes Kepler, who observed a 'new star' of his own in 1604. Using Tycho's data, Kepler formulated the first two of his famous laws of planetary motion and published them in his *Astronomia Nova* (1609). These were scientifically tested theories: Kepler's data for the motion of the planet Mars allowed him to establish that it followed an elliptical orbit; this in turn gave him the means to predict its future movement. Whereas once the movement of the planets had been a thing of wonder and faith, now it was a matter of scientific knowledge and understanding.

While Kepler was nursing his manuscript through the press, a lens-maker in the Dutch town of Middelberg, Hans Lippershey, built a telescope that could enlarge images three times. In 1608, he patented the idea. Very soon, word of his invention was carried abroad. In England, the following year, Thomas Harriot built a telescope with which he observed the surface of the Moon. In Italy, Galileo constructed a telescope capable of magnifying images 33 times and used it to observe the four largest moons of Jupiter. He published the results in 1610 in *Sidereus Nuncius* (The Starry Messenger). It was an apt title: these telescopes were like ships that brought back sights and knowl-edge previously far beyond men's dreams. Kepler joined Galileo in exploring the moons of Jupiter, building an improved telescope in 1611 and publishing his findings the same year. He produced his and Brahe's compendium of the measurements of over a thousand heavenly bodies in the *Rudolphine Tables* in 1627. These allowed other astronomers to see for themselves whether the planets orbited the Sun as Kepler claimed.

What followed over the rest of the century was an outpouring of astronomical endeavour and experimentation. Observatories were constructed in Leiden (1633), Gdansk (1641), Copenhagen (1637–42), Paris (1667–71) and Greenwich (1675–6). Experimenting with refracting telescopes, Johannes Hevelius realised that the longer the instrument, the greater the detail he could see. In 1647, he built a 12-foot telescope that magnified an image 50 times. In 1673, he built one that was 150 feet long, encased in a wooden tube. The new marvel was not particu-larly practical, as it could only be used outside, suspended with ropes from a 90-foot pole, and shook in the slightest breeze, but it gives a good indication of the lengths to which astronomers were prepared to go. It required a genius to improve upon it. That genius was Isaac Newton, who invented the reflecting telescope in 1668: this produced a magnification of 40 times despite being only a foot long. With such instruments at their disposal, astronomers began the systematic explo-ration of space. Many of them are still household names: Giovanni Cassini of Genoa, who helped set up the Paris Observatory and discovered the moons of Saturn; John Flamsteed, the first English

Astronomer Royal, who catalogued three times as many stars as Brahe; and Christiaan Huygens, the Dutch polymath whose work on lenses and telescopes allowed him to see the rings of Saturn properly for the first time and to measure the distance between the Earth and the Sun at 24,000 times the Earth's radius (a margin of error of just 2.3 per cent).

So what? Space did not affect life on Earth, so was there any real value in all of this? Actually, at the start of the seventeenth century, many people believed that the stars *did* have a direct effect on life on Earth. Astrology was not just a superstition of those eager to have their horoscope read: the stars were believed to be connected to everything else in nature. If you had a disease, a physician would want to know when it started so he could check which planets were in the ascendant. Similarly, a surgeon would advise you to have your blood let when the stars seemed particularly fortuitous. The courts of European monarchs had official astrologers. Even natural philosophers took astrology seriously: one of Kepler's original reasons for studying the stars was his desire to cast more accurate horoscopes. Thus, when the old world of star-gazing collided with the new science of astronomical observation, it had an enormous impact. People could now see that the stars were spheres following predictable orbits, not the semi-magical arbiters of human fortune and suffering. They could see that the Moon was a barren lump of mountainous rock. So how could such things affect people's health and well-being? Some people began to wonder whether the further planets were inhabited by people like them. Did God's Creation extend to other worlds? And as the stars were clearly not fixed in a crystalline structure, what else had Aristotle got wrong? Where was Heaven, which was supposed to lie on the far side of the stars? The very fact that we can ask 'so what?' in relation to astronomy itself marks a considerable development in scientific understanding since 1600.

Astronomy powered scientific investigation in other areas too. It was interest in the planets that led the English physician William Gilbert to postulate in *De Magnete* (1600) that space was a vacuum, *electricitas* a force, and the Earth a giant magnet with a core of iron which revolved daily on its axis. In Italy, Galileo was as much interested in physical laws on Earth as in the night skies. As a young man, he famously observed the swinging of a chandelier in Pisa Cathedral and noted that the pendulum took the same amount of time to complete a single arc even as the distance reduced. Later his research into the properties of pendulums led him to design a pendulum-driven clock. It was never constructed but the idea passed to Christiaan Huygens, who built the first example in 1656. Far more accurate than any earlier timepiece, it became the pattern for clocks for the next 300 years. In 1675 the English natural philosopher Robert Hooke proposed that the pendulum clock might be used to measure gravity; the necessary experiment was accordingly carried out and the theory proved to be correct by Jean Richer in 1671. Huygens also worked with the German mathematician and philosopher Gottfried Wilhelm Leibniz, who designed the first mechanical calculator and, at the same time as Isaac Newton, developed the mathematical method of calculus.

These men were polymaths. They were interested not only in optics, physics and mathematics but many of them enquired into chemistry, biology and botany too. Robert Boyle, who in 1675 extended Gilbert's work on *electricitas* by showing that electrical forces could pass through a vacuum, experimented with gases and formulated Boyle's Law: that the volume of a gas varies in inverse proportion to its pressure. Advances in telescope design were matched by the development of microscopes. Galileo took the idea for the microscope from Lippershey and his colleague Zacharias Jansen, and developed a better version that he called his 'little eye'. Robert Hooke reproduced large-scale depictions of plant 'cells' (as he called them) and flies in his *Micrographia* (1665). The Dutch microbiologist Antonie van Leeuwenhoek surpassed all others in his microbiological explorations. Using microscopes that magnified up to 200 times, he discovered bacteria, sperm cells, blood cells, nematodes, algae and parasites. Where some small creatures had previously been thought to replicate themselves by merely duplicating their own forms, now it became clear that even the smallest life forms were capable of sexual reproduction. The application of magnifying lenses made an enormous difference to mankind's appreciation of the natural world.

All this pioneering work on scientific knowledge reached its apogee in Isaac Newton's *Philosophiae Naturalis Principia Mathematica* (1687). Although it took some time to be widely accepted, the book was later hailed as one of the greatest scientific achievements of all time. It outlined Newton's theory of gravity – which according to legend first occurred to him when an apple fell on his head – and thereby ended the debate about what kept the planets in their orbits. It supplied ratios by which the forces of gravity could be calculated, so they could be studied quantitatively, not just understood qualitatively. It supplied the means to measure the relative densities of the planets and the Sun, confirmed Copernicus's heliocentric theory, and explained the motion of the Moon and how the tides on Earth are affected by it, and why comets follow their paths. It also contained Newton's three famous laws of motion. Alongside his work on optics, which he began in the 1670s and finally published in 1704, these discoveries overturned much of Aristotle's erroneous reasoning about the natural world, and presented a framework for a more rigorous testing of natural phenomena.

A critical factor that makes all the above work so important is that the findings were quickly shared between natural philosophers so that one could build on the knowledge of another. With very few exceptions, these were not quiet discoveries by semi-reclusive mystics who copied them into manuscripts that slipped almost unseen into the archives of science. They were paraded in publications and talked about throughout Europe. Educated men were expected to have a knowledge of the latest scientific debates. Encyclopedias were expected to be up to date with them. A series of finely executed prints of tiny organisms could become a best-seller – as Hooke's *Micrographia* showed. The leading natural philosophers of the day also started a whole series of national scientific organisations. In Venice, the Accademia dei Lincei was founded in 1603, with Galileo as a member. The Academia Naturae Curiosorum (later the Leopoldina) was established in Bavaria by a group of natural philosophers in 1652, and received imperial recognition in 1677. The Royal Society of London was founded in 1660, and received its first royal charter in 1662. In France, the Académie des Sciences was founded by Louis XIV in 1666. These societies started circulating to their members and other subscribers regular publications of new discoveries: the Royal Society's *Philosophical Transactions* began to appear in 1665; the Leopoldina's *Ephemeriden* was first published in 1670. It became widely accepted that there was an infinite set of discoveries to be made; it was patently not the case that, after a few breakthroughs, a new stability would assert itself. From now on, scientific knowledge would forever be in a state of flux.

It goes without saying that scientific discoveries had a huge impact on the philosophy of the time. For a start, there was the empirical nature of the scientific method. Francis Bacon was not the only one to realise that empiricism nailed the lid on the coffin of theology-based science, in which the causes and meaning of natural phenomena were interpreted in accordance with holy scripture. No less significant was the rise of rationalism – the philosophy that knowledge could be attained by reason alone. The most famous exponent of this way of thinking, René Descartes, is to this day associated with the deductive formula 'I think therefore I am'. But Descartes and later rationalists, such as Leibniz, were men of science as well as philosophers. Hence there remained a close link between those doing the scientific research and those formulating the processes by which scientific knowledge could be attained and validated. This helped to maintain the links between empiricism and rationalism – people naturally wanted to test empirically any knowledge attained by rational means. Only occasionally does one find a major thinker allowing his rationalism to run away with his imagination. Christiaan Huygens's last book, *Cosmotheoros* (1698), is partly given over to discussing the living conditions on Jupiter and Saturn, and whether the inhabitants of those planets lived in houses and had water, plants, trees and animals. He reasoned that they did. With the benefit of 300 years' more research, it seems to us that his conclusion rather casts doubt on his rationalism. Nevertheless, at the time it was acceptable for well-qualified men to speculate on scientific matters, and reasonable for less highly qualified people to believe them.

This gets us to the crux of the matter. It was not only the new knowledge that marked a major change: it was the shift of authority in determining that knowledge. While in the medieval period Church leaders and purveyors of community folklore had enjoyed that authority, from the mid sixteenth century natural philosophers took over. Consider Galileo's case. In 1613, he responded to an invitation to write a letter explaining Copernicus's heliocentric theory to the Grand Duchess of Tuscany. The letter was published and Galileo was accordingly brought before the Roman Inquisition in 1616. He was informed that to talk of a heliocentric universe was both absurd and heretical, and that the idea that the Earth was revolving on a daily basis was preposterous. On that occasion he got off with a warning but in 1633 he was charged with teaching heliocentrism again, and this time Pope Urban VIII imprisoned him for life. Yet within a few decades, papal opinion on scientific matters became inconsequential: people looked to the authors of scholarly papers to advise them, not theologians. This is the real Scientific Revolution. In 1633 authority in scientific matters was still exercised by the Church; by 1670 it lay exclusively with the scientific community.

The fact that this shift in authority set religion against science has often been taken as evidence that from then on, the two began to go their separate ways. This is fundamentally incorrect. Almost all the men who made great scientific discoveries were deeply religious: they saw their collective endeavour as a constant enquiry into the nature of God's Creation. Francis Bacon wrote a scornful pamphlet attacking atheism, and in his *Théodicée*, Gottfried Wilhelm Leibniz attempted to reconcile his Christian faith and his scientific philosophy, arguing that in making this world, God had created 'the best of all possible worlds'. Isaac Newton too was a pious man, who spent his life looking for scientific truths in the Bible – including prognostications of the end of the world. This combination of a religious purpose and scientific investigation proved a heady cocktail in the seventeenth century, and the religious dimension should not be underestimated. This is particularly true in the case of those trying to understand Creation through scientific experimentation, who were keen to combat superstitions, which they thought not only false but ungodly. Whereas religion and superstition had coincided for centuries, now religion colluded with science to drive out irreligious beliefs and to teach the people of Europe the divine truth.

Scientific knowledge soon began to filter into daily life. Superstitious practices such as burying cats in the chimneys of houses fell into abeyance. People stopped using medicines that incorporated ground-up animal remains or excrement and sought those that had some demonstrable efficacy. Most interestingly, they stopped believing in witchcraft. In the sixteenth century, dozens of witches had been hanged in England and Wales and hundreds burnt elsewhere. (Only in England and Wales and, later, America, was witchcraft a secular crime, punished by hanging, and not considered heresy, which was punished by burning.) In the early seventeenth century, the numbers killed rose into the thousands, however, with a particularly severe spate of persecutions taking place in Germany in the late 1620s. Most notorious of all, the prince-bishop of Bamberg built a witch house where people were imprisoned and systematically tortured until they confessed to acts of witchcraft and implicated others in such acts. The victims would then be burnt at the stake or, if they gave all their possessions to the prince-bishop, decapitated. Across Europe, tens of thousands were horrifically tortured and killed. But the whole structure of witchcraft persecution collapsed in the late seventeenth century. The last burning of witches in France took place in 1679 (Peronne Goguillon and her daughter). The last hangings in England for witchcraft were in 1682 (the Bideford witches). The Salem witch trials of 1692 saw the end of hanging witches in America. The last mass execution for witchcraft took place in Scotland in 1697 (the Paisley witches).

Was the Scientific Revolution responsible for the discontinuation of belief in witchcraft? As one scholar has put it: 'It is hard to see exactly why Isaac Newton's thoughts on the paths of moving bodies, as set out in his *Principia* of 1687, should make assize judges less likely to convict witches, let alone why they should make villagers less willing to launch witchcraft accusations against each other.'[7] This point is all the more valid as it has been estimated that only seven people in all of Europe were qualified to understand Newton's *Principia Mathematica* when it was published. If scientific knowledge did indeed lead to the decline of superstitions generally and the belief in witchcraft in particular, how did it happen?

The answer lies in why people believed in witchcraft in the first place. Witches had been a feature of European culture for centuries, but at the end of the fifteenth century the connection between witchcraft and heresy led to an increase in witches being brought before the Church courts. The more such court hearings took place, the more the message of witchcraft spread. The news acted like propaganda, alerting people to the potential danger. Although many accusations of witchcraft were motivated by misogyny or plain hatred, the notion that witchcraft was real grew alongside the belief that there were unseen and unaccountable forces in nature. The idea of a heliocentric universe, the discovery of *electricitas* and other changes in scientific understanding had allowed people to believe that many things existed that the eye did not see. A mathematician like John Dee could still subscribe to alchemy and astrology – and even see potential in experimenting through seances to understand the will of angels. Who really knew what to believe? It seems likely that the plethora of discoveries had left people doubtful that they understood the world at all. This gave room for their fears to circulate and grow, and eventually to seize the public mind. In the mid seventeenth century, however, the scientific community managed to stabilise society's doubts and provided a new equilibrium. As we have seen, knowledge that the planets were not arbiters of human fortune but orbited the Sun in a predictable manner became orthodox, diminishing that superstition. Scientific organisations, recognised by royal charter, provided a stability that had been lacking since the early years of the previous century. You did not need to understand Newton's *Principia Mathematica* to have confidence that Newton himself and other members of the Royal Society understood it, and that they could explain many aspects of the universe that had at first been so disconcerting. With that new-found confidence, the complete lack of empirical evidence for witchcraft was not hard to interpret: the inescapable conclusion was that even self-confessed witches had been burnt or hanged for no reason.

The Medical Revolution

The medical discoveries of the seventeenth century were in many ways a subset of the Scientific Revolution. However, the implications were particularly profound. Everybody had a stake in medical knowledge because, sooner or later, ill health affected everyone. It wasn't just a case of whether society felt that physicians or Church leaders should be trusted in medical matters; it was a matter of personal faith. What did you really believe? If you or your kin fell ill, should you seek medical help or pray?

As we have seen, surgeons, physicians and apothecaries had existed since medieval times, and medical training had been available since the twelfth century. You might therefore assume that when people fell ill, they sent for the physician or, if they injured themselves, the surgeon. But it wasn't as simple as that. English probate accounts allow us to measure exactly what people did when faced with terminal illness or injury. Most people did not pay for medical help but sent for the priest for the benefit of their soul, and, if their family was unable to look after them, paid local nurses to alleviate the pain, clean their clothes and bedding, and cook for them. It turns out that in 1600, less than one in fifteen men of moderate means paid for professional

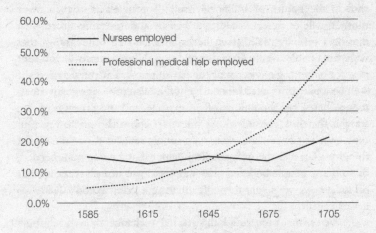

Medical and nursing help purchased by men dying in East Kent
with movable goods worth £100–£200[8]

medical expertise when they were approaching death. But by 1700, about half did so.

The chart above is based on the region for which by far the best evidence survives, but other series of accounts for Berkshire and West Sussex show that similar shifts from nursing to professional medicine took place in other parts of southern England. Another study has shown that men of property from all over the country started to pay for medicine much more regularly at about this time.[9] Even in remote Moretonhampstead, which was still inaccessible to wheeled transport in the seventeenth century, a medical man set up his practice: Joshua Smith became the town's first qualified physician when he obtained his licence in 1662.[10] Interestingly, while the quality of medical assistance available was obviously related to wealth, the regularity of medical consultation by the prosperous – those with movable goods worth more than £200 – was only a little in excess of the figures displayed in the chart above. At the other end of the spectrum, some of the poor who could not afford to pay for medical help were attended by a physician employed by their parish. It is fair to say that, one way or another, the majority of those who wanted professional medical help could obtain it by 1700.

Underpinning this change in the reliance on medical help was a shift in the nature of medicine itself. If you fell ill in 1600, your mother, wife or nurse would treat you by administering a medicinal diet or a remedy made from herbs, animals and other things that were obtainable locally. If your condition did not improve, you sent next for either the physician or the parish priest. The physician might well be a part-time practitioner. He might also prescribe costly treatments, based on the colour of your urine and the position of the stars at the time you fell ill. As the chart above shows, however, in 1600, over 90 per cent of dying men simply sent for the priest. Christ the Physician was the only medical helper that a dying man needed, or so many people believed. The emphasis was not on recovery, but on redeeming your soul through suffering – dying 'a good death', as it was called.

As the century progressed, this started to change: men increasingly sent for both the priest *and* the physician. The fundamental reason for this change was a more rigorous approach to medicine. Physicians were no longer being taught according to Galenic texts but were taking a much more scientific approach to the human body. William Harvey published his theory of the circulation of the blood, *De Motu Cordis*, in 1628 – a landmark in the understanding of the human body – and dozens of other medical texts were produced over the course of the century. But the real catalyst for change was an arrogant Swiss genius who had died in 1541. His name was Philippus Aureolus Theophrastus Bombastus von Hohenheim, but he is better known to history as Paracelsus. In the early sixteenth century he had travelled around Europe practising medicine and astrology, publishing a number of texts that were directly in contravention of accepted teaching and advocating the use of metal-based medicines and chemical compounds to treat illnesses. By the 1590s his ideas were beginning to take hold across the whole of the continent. Increasingly, metal-based medicines and natural medicinal substances were stocked by apothecaries throughout Europe. Physicians started to be called 'doctors' – as if they all had doctorates in medicine – which clearly denoted an expression of trust in their expertise. Many of them settled in small market towns so that by 1650 almost everyone was within easy reach of a physician. Thus a relatively small number of professionals provided for a massive expansion of medical demand. They dealt with a rapidly increasing number of patients by prescribing effective medicines for specific diseases rather than plotting complex treatments based on

astrological observances, superstition, the taste of the patients' blood
and the colour of their urine.

It is important to remember that dying men and women sent for
both the physician and the priest. Trust in doctors did not mean a
diminished trust in God. Indeed, trust in God was very much part of
the Medical Revolution of the seventeenth century. Licensed physicians
and surgeons had to be examined as to their morality, to make sure
that they were worthy conduits of God's healing power. Moreover,
as more exotic flora and fauna were discovered in out-of-the-way
places, the discoveries gave rise to a philosophy that God had nurtured
antidotes to all the world's diseases when he created the world. In
1608, Maria Thynne, wife of Sir Thomas Thynne of Longleat, wrote
to her sick husband, 'Remember we are bound in conscience to maintain life as long as is possible, and though God's power can work
miracles, yet we cannot build upon it that because He can, He will,
for then He would not say He made herb[s] for the use of man.'[11]
The discovery of Jesuit's bark in Peru in the early seventeenth century
seemed to confirm this faith: it provided an effective treatment for
malaria, the biggest killer in human history. Then there were those
other naturally occurring medicines – pomegranates for digestive
problems, and colchicum crocuses for gout. It seemed to be just too
great a coincidence that the antidotes for a number of diseases existed
in nature. Thus there was a religious context to both the physician
who gave you the remedy and the medicine itself. Taking a remedy
prescribed by a doctor was not exactly prayer, but it was still putting
yourself in God's hands.

In speaking of the Medical Revolution, therefore, we must not think
in terms of people suddenly switching from prayer to pills. Their
behaviour shaded from one to the other gradually. They shifted from
faith in prayer alone to a phase in which they both prayed and were
treated by a religious physician; to a phase in which the morally correct
physician both acted as the instrument of God's healing power and
prescribed all sorts of remedies; and eventually to a phase in which
the physician simply prescribed the remedies without any religious
context at all. Not everyone moved from one phase to the next at
exactly the same time; there were still some people who trusted their
prayers more than their doctor in 1700. However, by that date most
people were prepared to put themselves in the care of the medical
profession. If you think about all the temples, chapels, churches and

shrines built before that in the hope of recovering health, and all the pilgrimages undertaken and all the saints and relics to whom prayers had been offered on behalf of the unwell, you can see that this relocation of hope for human well-being from God to medical professionals is one of the most profound revolutions that society has ever experienced.[12] Indeed, it marks one of the most important stages in the transition of Europeans from the highly religious, group-thinking community members of the Middle Ages to conscientious modern individuals.

Settlement of the world

No one can look at a map of the world and fail to notice how many locations outside Europe are named after English, Dutch, French and Spanish places. New York (formerly New Amsterdam), New Hampshire, New England, New London (now Bermuda), New Holland (now Australia) and New Zealand all have precious little in common with their namesakes. The same could be said for many of the places on the east coast of North America: Boston, Yorktown, Plymouth, Jersey, Dover and Durham, to pick out just a few. Then there are those places named in honour of French and English rulers: Louisiana (for Louis XIV), Jamestown (for James I), Carolina and Charlestown in Massachusetts (for Charles I), Charleston in South Carolina (for Charles II), Williamsburg (for William III) and Maryland (for Mary II). All of the above were named in the seventeenth century. If the sixteenth century had seen Europeans discover the North American mainland, the seventeenth century saw them name it, settle it, lay claim to it with legal documents, fence it off and defend it with long-barrelled weapons.

At the time of her death in 1603, Elizabeth I of England ruled over nothing outside the British Isles. The only exceptions were a failed settlement at Roanoke, in Virginia, whose inhabitants were either all dead or lost among the Native Americans; and Newfoundland, first claimed in 1497 and claimed again in 1583 but as yet still uninhabited. Her successor James I granted a charter to the Virginia Company of London in 1606 that established the first lasting settlement, Jamestown, the following year. A supply mission supporting the settlement was wrecked in 1609 off the coast of Bermuda, where the survivors established St George's, the second English colony. In 1611, the town of Henrico was founded in Virginia, named after King James's eldest son, Prince Henry. The farming of tobacco in Virginia proved the mainstay of the troubled settlement from 1616, and continued to be so even after a massacre at the hands of Native Americans in 1622. By then, merchants from Bristol had settled at several places in Newfoundland and a community of Puritan refugees from England, known as the Pilgrim Fathers, had founded the Plymouth Colony. The volume of tobacco exported from Virginia grew rapidly: in 1628, more than 250 tons was shipped to England. That same year, another group of English settlers founded the Massachusetts Bay Colony. There might not have been any gold in the region but there were good livings to be made from fur trapping, farming and tobacco.

The New World also promised a degree of freedom from the constraints of Europe. By 1650 there were over 50,000 European settlers living on the American east coast in self-governing communities, including 1,600 black slaves bought for harvesting the tobacco. By 1700 that figure had grown to over 250,000 people, including more than 27,000 slaves.[13] Crucially for those who dreamed of liberty, vast areas of land were available. The province of Carolina, granted to eight English lords, the Lords Proprietors, in 1663, stretched from the Atlantic to the Pacific; it amounted to more than a million square miles (about 20 times the size of England). Although most of it was not occupied by settlers, by 1700 approximately 75,000 square miles of the North American east coast was under their control.[14] An even larger area was claimed to the south of Hudson's Bay and in Newfoundland. By the end of the seventeenth century the North American colonies even had two seats of higher education – Harvard (founded in 1636) and William and Mary (founded in 1693) – the same number of universities as England itself.

The English were not the only European nation to see North America as a land of opportunity and freedom. The New Netherland Company, founded in Amsterdam in 1614, received a charter from the Estates General of the Dutch Republic granting them trading rights for a period of three years. They hoped to profit from the fur trade in what is now Canada. The successor to this company, the Dutch West India Company, which received its charter in 1621, continued the expansion of the North American fur trade, but it also had rights to exploit whatever business interests it could find in the Atlantic, from

the west coast of Africa to the Arctic. Soon it had territorial possessions on the Delaware river in America, on the Gold Coast in Africa,
and in the West Indies. Political problems at home prevented the
Dutch from investing heavily in the defence of their North American
possessions, and these were surrendered to the English in 1664. At the
same time, the Spanish strengthened their grip on Florida and New
Mexico, and ventured into Texas. A series of French companies took
possession of most of eastern Canada, including parts of Newfoundland,
the area around the Great Lakes, and a central belt of land that reached
right down to the Mississippi river and Louisiana. When you add the
Spanish and Portuguese possessions in Central and Latin America,
you can see that by 1700, European states ruled over far more land in
America than they did in Europe. Long before Horace Greeley advised
a friend to 'go West, young man', the principle of expanding westwards
in search of a better life was established in the European mind.

Young European men had a choice: they could either build a new
life for themselves in North or South America, or they could join the
trading world and sail to the East. It was the Dutch who saw the most
benefit from eastwards trade. As the United Provinces (which included
the Dutch Republic and six other minor states), their whole population was only about 1.5 million; nevertheless, they made significant
dents in the Portuguese Empire. The main instrument of their success
was the Dutch East India Company, established in 1602. The following
year they founded their first permanent post in the Far East, at Java.
In 1605 they captured the Spice Islands. The Eastern headquarters was
set up at Batavia (now Jakarta) in 1611 and from there the governors
general organised the systematic domination of the silk and spice
trade. The Portuguese managed to keep hold of Brazil and some of
their African ports but in the Far East they only hung on to a few
trading outposts: Macao, the Lesser Sunda Islands and Goa. While
capturing Portuguese ports the Dutch discovered most of the rest of
the as yet unexplored world, including Australia in 1606, and Tasmania,
New Zealand, Tonga and Fiji in 1642–4. They also founded a colony
in South Africa in 1652 to supply their ships. In 1668, when their empire
was at its greatest extent, a French observer reckoned that the Dutch
had 6,000 ships under sail, not including small single-masted vessels.
A Portuguese Jesuit in 1649 had put the figure at 14,000, and added
that all these ships were bigger than those in the Portuguese fleet.[15]
Historians today generally agree that in the 1660s the total tonnage

of the Dutch fleet was about 600,000 tons – as much as the rest of Europe put together.

Finally, we must look at the English advances in the East, which centred on the East India Company, established in 1600. Nine profitable trading ventures had returned to England by 1612 and in that year the company founded a trading base at Surat, north of Bombay. Shortly afterwards, the affable English diplomat Sir Thomas Roe secured the right for the company to trade inland in India, wisely adopting a policy of never using armed force to exact trading advantages. The company also uneasily did business alongside the Dutch in Indonesia – even more uneasily after a number of Englishmen were executed for treason against the Dutch Republic in 1623. In 1661 it was given an unexpected lift when the Portuguese handed over Bombay to the English Crown as part of Catherine of Braganza's dowry. This gave the English their first sovereign territory in India and it soon became the company's headquarters. Successive royal charters granted the company sovereign rights in India: it was able to issue its own coins, exercise jurisdiction over English subjects, and form agreements with local rulers. After the English trading post at Bantam in Java was lost, the company gave up trying to compete with the Dutch in Indonesia and concentrated on the Indian trade, bringing back calico, spices, silk, indigo and saltpetre (for gunpowder). In 1684 the East India Company purchased goods worth £840,000 in India and sold them in Europe for £4,000,000.[16]

Modern readers may prefer to forget about this period of colonial rule but it cannot be denied that it transformed the world, including Europe itself. For a start, the Spanish and Portuguese territories – united under one crown from 1580 to 1640 – were the world's first 'empire on which the sun never set', with settlements on five continents. As a result of these wide networks, developments in technology spread far beyond the confines of Europe. The war between the Dutch and the Portuguese, which saw military action in Latin America, India, Africa and the Far East, and in the Atlantic, Pacific and Indian Oceans, and which drew in the Spanish, French, British and Danish as well as local rulers, was the first world war. Colonial expansion in this century also established the basis for the later British Empire, especially through acquisitions of sovereign territory in India, the West Indies and Canada. But just as importantly, the colonies provided a safety valve to relieve the pressure of population growth. Over the course of the sixteenth

century, the population of Europe had risen by over a third to about
111 million – the continent was as heavily populated as it had been in
1300. With no governments making allowances for the extra mouths
that needed feeding, the situation was ripe for Europe to turn on itself
violently. When harvests failed, people became desperate, risking the
gallows to poach livestock or steal bread. In France, lords began to
exercise authority over their lands to the point of tyranny. In Italy and
Spain, feudal lords commanded brigand armies. Revolutions and rebellions broke out throughout Europe: the Bolotnikov Rising in Russia
(1606–7), the Dosza insurrection in Hungary (1614), the peasants' war
in Upper Austria (1626), the civil wars in England (1643–9), the
Neapolitan Revolution (1647–8) and the Fronde in France (1648–53). In
addition, Catholic, Protestant and Jewish minorities were persecuted
in many countries. Thus the opportunity to make a voyage to the
New World and live freely in a self-governing community, without
fear of famine or religious persecution, was deeply attractive. It exercised a potent effect on people's minds even if the numbers of actual
emigrants at this point were still small. The next two centuries would
see the population pressure valve open more fully, as millions of
Europeans flocked to the New World, and set about building a society
that would one day rival those they had left behind.

The social contract

The existence of seemingly less sophisticated native cultures in Africa
and America raised significant new questions for European intellectuals.
Societies that did not include formalised marriage, money, writing or
land ownership seemed to support the Biblical story that mankind had
once existed in a primitive form in a 'Garden of Eden'. Europeans only
had to look at their own society – at the guns, the printed books and
the scientific and navigational achievements of the last hundred years
– to realise how rapidly they were moving away from this original
state. Above all else, the juxtaposition of the developed and primitive
caused European philosophers to reconsider matters of law and
morality. What was the natural law that governed all mankind? How
had mankind progressed morally from a state of nature, in which men
and women had presumably obeyed their own personal desires, to a
state in which they agreed common rules of social engagement?

Thomas Hobbes was the first significant commentator on the subject. In his *Leviathan* (1651), he argued that the emergence of society was entirely due to human interactions, not divine intervention. He theorised that men in a state of nature had natural rights but lived in a state of brutality. Over the years they compromised some of these rights by means of agreement with others for their mutual benefit. For example, a group of men might agree to waive the right to kill each other and choose instead to defend each other against outsiders. Their understanding of this consent formed a social contract, and that in turn provided the philosophical justification for the commonwealth or political body. Hobbes argued that there were only three types of commonwealth – monarchy, democracy and aristocracy – and that of these three, monarchy was the best. Only a strong centralised government, the 'Leviathan', could uphold peace and civil unity and protect individuals and their property. For this reason, rebellion against the monarch was never justified, even if the ruler acted contrary to the interests of the people. Moreover, no aspect of religion could be allowed to claim greater authority than the commonwealth, so no personal spiritual insight could be permitted to rival the monarch's position or the civil law.

Contemporaries of Hobbes saw things somewhat differently, especially with regard to what form of government was ideal and how far people could go in holding their rulers to account. But almost everyone agreed that there were natural rights. The concept appealed to radicals, who used it to complain against government officers and landowners abusing the natural rights of the common man. It also appealed to other philosophers. The most significant of these was John Locke, who further developed the idea in his *Two Treatises of Government* (1689–90). Locke argued that all men were equal in the state of nature and that they enjoyed three natural rights: first and foremost, the right to life; second, the right to liberty, to do what they wanted as long as it did not conflict with the first right; and third, the right to enjoy their property as long as it did not impinge on the first two rights. Locke did not agree that monarchy was necessarily the best form of government, and even suggested that if the monarch did not protect the rights of his people, they had the right to depose him. Indeed, he praised the Glorious Revolution in England (1688), which had recently enacted this very principle. He also approved of the Bill of Rights that followed, which limited the power of the king. Henceforth the monarch was unable to intervene in the law or in the workings of Parliament. He could not levy his own army or his own taxation without the authority of Parliament, and was not permitted to use or authorise 'cruel and unusual punishments'. However, Locke did follow Hobbes in arguing that no personal religious insight could override the social contract. In his view, religious intolerance was an infringement of a man's liberty. No one could prove one faith true and another false, so the reasons for religious intolerance were delusional.

In such ways, the discovery of the New World stimulated Europeans to think outside the box of a divinely ordained hierarchy, and to imagine the liberty of all people – rich and poor, Catholic and Protestant. The irony was that at the same time as they were developing these ideas, Europeans in Africa and America were busy removing the liberty of indigenous people by forcing them into slavery. Despite this tragedy, the stories told of prosperity in the New World inspired a fresh, more libertarian vision of life in Europe. This in turn gave rise to ideas that would be exported back to North and South America in the eighteenth and nineteenth centuries – prompting their respective fights for their own liberty from their colonial masters in Great Britain and Spain.

The rise of the middle classes

In 1709 the Venetian artist Marco Ricci painted three opera rehearsals in a practice room at the Queen's Theatre, on the Haymarket in London. These pictures reveal a world strikingly different from London life a century earlier. Nothing you see in them could have been depicted in 1600. The clothing worn by the singers is of the latest fashion, including long curly wigs rather than the ruffs of a century before. Portraits and landscape paintings line the walls of the practice room; prior to the career of the French artist Claude (1600–82), artists didn't paint landscapes except to illustrate a religious scene, and few people hung paintings on their walls. As for opera itself, the genre did not exist in London in 1600: the very first opera was performed in Florence in 1597, and it was not until the 1680s that Henry Purcell wrote the first English opera. One of Ricci's three pictures shows the foremost Italian castrato of the day, Il Niccolini, which hints at other developments: in 1600, singers did not travel on international tours, and nobody was castrating boys to develop their high singing voices in later life. The audience being seated on chairs was a relative novelty too, as individual chairs were relatively uncommon a century earlier. A standing member of the audience in one of the paintings is even holding a porcelain cup and saucer, drinking tea. Finally, the very subject of Ricci's painting – an indoor scene, which has nothing to do with myth or legend – is something that you would not have found in 1600. The painting speaks of a changed world, with changed people, changed tastes and changed ideas.

At the root of all these changes was the emergence of the bourgeoisie. All those silks and spices brought to Europe from the Far East were not being purchased by the poor but by the new arbiters of taste, the enriched middle classes. For centuries their numbers had been growing, principally as a result of merchants in towns buying and selling profitably and making money. But previously there had been no 'upper middle class' as a distinct social group. When merchants saw their wealth rise to the level of the gentry or aristocracy, they sold up and acquired a country estate, effectively subsuming themselves in the ranks of the upper class. In sixteenth-century England, the largest fortunes were made by lawyers and public officials but still the successful few crowned their achievement by buying a landed estate. After 1600, however, the urban middle classes and their wealth increased dramatically. Gregory King, a pioneering statistician, estimated that in 1695, apart from the nobility, higher clergy and armigerous gentry (gentlemen with a coat of arms), there were 10,000 people in England living off the fruits of an office (excluding clergymen and military officers), with a collective income of £1.8 million. In addition, there were 10,000 international merchants, whose combined income amounted to £2.4 million; 10,000 lawyers, bringing in £1.4 million; and 12,000 unarmigerous gentlemen, sharing £2.9 million. At a time when the total income of the nation was £43.5 million, the upper middle classes had almost a fifth of the whole amount – about three times as much as the aristocracy and armigerous gentry combined.[17]

Clearly not all these people could pretend to be aristocratic. However, most tried to emphasise their elevated position in society in some way or other. They dressed in the latest fashions and did all they could to be seen in public – going to the theatre or the opera, and travelling to appropriate social engagements in horse-drawn carriages. They equipped their houses with all the distinctions of modernity they could lay their hands on: large glass windows, paintings and decorative prints, musical instruments, board games with carved wood or ivory pieces, books, gilt-framed mirrors, carpets, cushions, curtains and valances, embroidered tablecloths, elaborate silver candlesticks, pendulum clocks, Venetian wine glasses, imported ceramics from the Far East, polished pewter plates engraved with the family coat of arms, and elaborately carved, turned or inlaid furniture. They also took pride in their education, and travelled widely in order to broaden their horizons. Many maintained 'cabinets of curiosities': items of a strange nature, usually reflecting life in distant or ancient places such as Ancient Egypt or the New World. They also enjoyed eating and drinking well. The seventeenth century saw the introduction of tea, coffee, chocolate, lemonade, orange juice and spirits such as brandy, aquavite and Dutch gin. They developed a taste for the fine wines newly emerging in France: Château Latour, Château Lafite, Château Margaux and Château Haut-Brion. The last mentioned was tasted by the London diarist Samuel Pepys in 1663. Sparkling champagne was also introduced to London and Parisian society at this time.[18]

In many ways, it was this urban class that created the model for modern life. They lived not in vast halls but in houses of more modest proportions. The three-storey brick terraced houses built in London after the Great Fire of 1666 provided the blueprint for urban building for the next 250 years. These houses were divided into parlour, dining room and sleeping chambers. Smaller, more efficient fireplaces were employed, and coal was increasingly used for heating. Kitchens – which had often been located in separate buildings in previous centuries – were brought indoors and placed not too far from the dining room, with a separate scullery for all the dirty jobs, such as washing plates and bowls, scrubbing saucepans and preparing meat, fish and vegetables. The ideal was that everything should be in order. Genre paintings by Dutch artists reveal the striking differences between the taverns of the ordinary people, with their dark interiors, old planks of wood hanging at angles, crumbling fireplaces, puddles on the floor, broken earthenware, and people in torn and shabby clothes, and the light, clean and tidy houses of the prosperous middle-class families.

In all those Dutch paintings, you can't help but notice how often the bourgeois individuals are portrayed with serious, concerned expressions, while most of the laughing faces are those of the red-nosed drunken poor.

Perhaps they were worried about trade or burdened with the weight of office. Or perhaps they wanted to be depicted as responsible. Social climbing was, after all, a serious business. And the ladder these people sought to climb reached very high. Although the bourgeois of seventeenth-century England, Holland and France were no longer under pressure to emulate the aristocracy, who maintained households of 40 servants or more, they aped their social superiors in almost every other respect. More and more families claimed to be entitled to bear coats of arms. Increasingly they sent their sons to universities to receive a degree. They claimed other marks of dignity too. In 1650, a French observer wrote that 'before this century it was unknown for wives of secretaries, lawyers, notaries and traders to call themselves *Madame*'.[19] In England, men increasingly insisted on being addressed as 'Mister' and women as 'Mistress' or, from the 1660s, 'Miss' if they were unmarried. New fashions became de rigueur. When Louis XIV received an ambassador from the Ottoman emperor in 1669, everyone became obsessed by all things Turkish – drinking coffee, reading Turkish stories, wearing turbans and lying on rugs and piles of cushions. For all those in subsequent centuries who have set such great store by 'keeping up with the Joneses', this is where it all began. The French drama-tist Molière wrote a scathing satire on the aspiring middle class in 1670. *Le Bourgeois Gentilhomme* tells the story of Monsieur Jourdain, the son of a cloth merchant who would do anything to be recognised as an aristocrat – and makes himself a laughing stock in the process.

It should not be supposed that the middle classes arrived fully fledged overnight. In many ways, the consumerism of the late seven-teenth century was just a prelude to the more extensive waves of social mobility in the eighteenth and nineteenth centuries. But the seventeenth century saw the structure of European society swell in the middle – and bulge distinctly above the tight belt of social control.

Conclusion

It is tempting to say that this is the century that marks the threshold between the ancient and the modern worlds, as people's hope for their well-being shifted from God to their fellow men. It reflects a change towards secular materialism that is to be found in everything from Hobbes's social contract to the conduct of war. In earlier centu-ries, the outcome of a battle was seen as indicative of God's will; in the seventeenth century, it was understood to be the result of how well or badly the commander performed with the assets available to him on the day. In other respects, too, the seventeenth century seems to have ushered in the modern world through the rapid diminution of superstitious beliefs, the commensurate rise of scientific rationality, and the continued decline of violence.

Having said that, not all the novelties of the century mark a progressive march towards modernity. Puritanism in England and America, which had begun with great zeal for moral as well as theological reform, now nurtured monstrous injustices. In 1636 the Puritan preacher of the Massachusetts Bay Colony, John Cotton, drew up a model law code according to which a married couple who had sex while the woman was menstruating should be put to death.[20] In May 1650 the Adultery Act was passed in England, which required the death penalty for sex outside marriage. It is shocking to read of a Devon woman, Susan Bounty, who was convicted of adultery. When her child was born, she was given him to hold for a few brief moments; then he was taken away and she was hanged. And yet just six years later, Charles II came to the throne – a married man who had at least eight illegitimate children by six mistresses. The mind spins. It spins even more when reading of the Salem witch trials of 1692–3, when 19 people were hanged, and one crushed to death for refusing to plead. Then you think of all the civil wars, revolutions and uprisings due to famine and social injustice. Despite scientific thinking, natural rights and bourgeois refinement, the truth is that the modern world was not born easily. It struggled into existence, kicking and screaming like a new baby, bloody and hungry. If it seems to us now that the greatest achievement of the seventeenth century was its rational approach to the world, then we need to remember that tens of thousands of people laid down their lives at the same time – in the witch houses, at the stake and on the gallows of Europe.

The principal agent of change

There are three leading contenders for the seventeenth century – Galileo, Isaac Newton and John Locke. But we should also consider the outsiders, particularly William Harvey, Christiaan Huygens and Antonie van Leeuwenhoek. Of these six, Newton is normally given the most credit for shaping the modern world. However, we have already seen that it took some time for people to come to terms with his work. The point of selecting a principal agent of change for each century is not to highlight someone who had a profound effect many years after his time. If that were the criterion, Aristotle, rather than Peter Abelard, would have been the principal candidate for the twelfth century. Therefore, I put Galileo first. He not only popularised the scientific method, but also led the way in instrument-making, basic physics, timekeeping and astronomy. He did more than anyone else of his time or any other to challenge the Church's pretence to knowledge in order to maintain its authority. In expressing a profound belief in his scientific findings, even at the cost of his own freedom, he stood for much more than a set of scientific truths. He stood for truth itself.

1701–1800

The Eighteenth Century

In 1738, a Devonshire rogue, scoundrel and rapscallion called Bampfylde Moore Carew, the self-styled 'king of the Gypsies', was arrested for vagrancy. He had been apprehended begging for money from travellers by impersonating a shipwrecked mariner. At his trial the judge demanded to know where he had been; on hearing the answer 'Denmark, Sweden, Russia, France, Spain, Portugal, Canada and Ireland', he sentenced him to be deported to Maryland in America. Carew thanked him for sending him to 'Merryland' and responded to the prospect of spending the rest of his days in servitude by escaping immediately on arrival in the New World. He was soon recaptured and fitted with a heavy iron collar. Undaunted, he escaped again, and made friends with some Native Americans, who sawed off his collar. The king of the Gypsies then set out for New York on foot, tricking and begging his way, and worked his passage incognito back to England. Not long after his return, strolling along the quay in Exeter with his long-suffering wife on his arm, he bumped into the captain of the ship that had transported him to the New World. An awkward moment ensued, as the law dictated that escapees from transportation should be hanged. Would the captain betray him? Damn right he would. But Carew was lucky: rather than being hanged, he was once more dispatched to 'Merryland'. Needless to say, he escaped again, and had many more adventures on his way home. In 1745, at the grand age of 52, he dictated his memoirs, which became a nationwide best-seller. A hundred years later, people in England were still telling stories about 'the notorious Devonshire stroller and dog-stealer', comparing him to Robin Hood.

Carew falls into the same category of eighteenth-century characters as the pirate Edward Teach, better known as Blackbeard, and Henry Every, the 'King of the Pirates', whose exploits were recounted along with those of many other felons in Charles Johnson's *A General History of the Robberies and Murders of the Most Notorious Pyrates* (1724). You could put the highwayman Dick Turpin in that category of heroic villain too; his life story was published immediately after his execution in 1739. Hundreds of such criminals were celebrated in the cheap literature of the age, and in dramatic works such as John Gay's enormously popular satire *The Beggar's Opera* (1728). All this celebration of crime might make you pause. This was, after all, the century of the Enlightenment, of political economy and scientific experiment. It was the age of elegance, harmony and order: the music of Vivaldi, Bach, Handel, Haydn and Mozart; rococo architecture; the furniture of George Hepplewhite, Thomas Chippendale and Thomas Sheraton; the landscapes of Capability Brown and Humphry Repton; Canova's exquisite sculptures; the Venetian paintings of Canaletto and Guardi; and the French works of Watteau, Fragonard and Boucher. Even the English – at long last – managed to produce painters of international importance in Joshua Reynolds, Thomas Gainsborough, Joseph Wright, George Stubbs, George Romney and William Hogarth. Above all else, this was the age of revolution – the American, the Industrial and the French. But the celebrity status of these eighteenth-century criminals is not as out of place as it first appears. Today's society sees a very similar championing of social outcasts, misfits, reprobates and iconoclasts. John Gay has his successors too: crime fiction is one of the most popular genres, and films about real criminal organisations such as the Mafia are watched by millions. The stories about Carew, Blackbeard, Every and Turpin found favour with the eighteenth-century public, which hankered after the excitement and freedom that such men represented. Indeed, there is a certain modernity to the contradictions of the eighteenth century – a mixing of order and regulation with the romantic impulse, the desire to escape. That goes for most aspects of life, from sex and crime to religion and opera. By comparison to the taste of previous centuries, which could be salty, sour, bittersweet or just plain bitter as the circumstances dictated, the taste of the eighteenth century has a certain fizz to it – like fireworks and string quartets bursting above the mere mud of human tragedy.

Transport and communications

Prior to the introduction of semaphore signalling for military purposes at the end of the century, the speed of information depended on how fast people could travel. A number of factors affected this: how much daylight there was at that time of year; the condition of the roads, which also depended on the seasons; how rich the person sending the message was, which affected the number of fresh horses that could be hired along the way; and lastly, the remoteness of the destination. If there were good roads all the way to the door, a rider with several changes of horses could travel at great speed – up to 120 miles in a day in summer. But good roads were rare in 1700. If the destination was a remote rural backwater like Moreton, the winter quagmires and boulder-strewn trackways could severely hamper progress, and it might take a day for a messenger to travel just 20 miles. The greatest distance covered by a messenger in a day I have yet come across for the centuries before 1800 was part of Sir Robert Carey's mission to inform James VI of Scotland of the death of Elizabeth I of England in March 1603. Sir Robert rode the 397 miles from Richmond to Edinburgh in less than three days, completing 162 miles on the first day and 136 on the second. The previous year, Richard Boyle had travelled from Cork to London in just two days, including the sea crossing between Dublin and Bristol, despite the state of the roads in January.[1] Most long-distance travellers, it has to be said, could never have come close to these speeds, being lucky if they could cover 30 miles in a summer's day.

The late seventeenth century saw the first serious attempts to improve European transport links. For a start, carriages with suspension were developed and new lightweight coaches built. Even more importantly, governments improved the state of the highways. In England, the old legislation that made local people responsible for the upkeep of the roads fell into abeyance and was replaced by the principle that the users of a road should pay a toll for its maintenance. Parliamentary approval had to be obtained in the form of a Turnpike Act for the construction of a specified road and the establishment of a turnpike trust to maintain it, but after that permission had been granted, the trust had a monopoly on charging tolls on that section of the highway and could spend all the funds on its upkeep. By 1750 there were about 150 such turnpike trusts in England, providing greatly improved access to much of the south-east and the Midlands. Halfway through the century there was a sudden explosion of interest in road-building. More than 550 new trusts were established between 1750 and 1800 as the rest of the country was opened up to wheeled transport. In 1770 a Turnpike Act created a new road from Exeter to Moretonhampstead and the first wheeled vehicles reached the town shortly afterwards. A road was built across Dartmoor ten years later. By 1799 a Moreton inn, the White Hart, was advertising single-horse carriages for hire, so tourists could embark on pleasure excursions over the moor. It marked quite a turnaround from 1646, when, during the English Civil War, General Fairfax's army had tried and failed to pull wheeled cannon to this remote town.

The most important aspect of this development was the improved speed of information. The post system in England had been in existence since the sixteenth century but it only covered four routes: from London to Ireland, Plymouth, Dover and Edinburgh. A cross-country route was added in 1696 when a post service from Exeter to Bristol was set up; this was followed by services from Lancashire to the south-west in 1735, and from Bristol to Salisbury in 1740.[2] Joining up the spokes of the wheel emanating from the hub of London meant letters no longer needed to be sent via the capital, so messages travelled more quickly. The gravel-covered turnpike roads allowed information to be carried to London more quickly too. A new speed record was set by Lieutenant John Richards Lapenotière, who landed at Falmouth on 4 November 1805 bearing the news that the English fleet had defeated the French in a great battle at Trafalgar. He covered the 271-mile journey to the Admiralty in London in 37 hours, changing horses 21 times – at a cost of £46 19s. 1d.

Ordinary people noticed the difference in the quality of the roads too. In the sixteenth century it could take as much as a week to travel the 215 miles from Plymouth to London. In the early nineteenth century rival coach companies were advertising in the local papers that they could get you there in 32 hours: an average speed of 6.7 m.p.h. – not far short of the 7.3 m.p.h. that Lieutenant Lapenotière achieved.[3] That stagecoach journey also cost a lot less than it had done a hundred years earlier. A yeoman travelling from Plymouth to London before 1750 had to pay for his own accommodation and subsistence, as well as stabling and fodder for his horses. For a week's travel, this amounted to a hefty sum. Over the next fifty years costs went down, comfort went up and speeds increased. The stagecoach companies that whisked travellers to London in 1800 advertised their 'extraordinary cheapness'. Coach design also improved – especially after the introduction of John Besant's patent mail coach in 1787. The coaches travelled non-stop, so there was no need for a chamber at an inn every night, and the cost of the horses was shared between the multiple occupants of the vehicle.

In France, travel was similarly transformed. The engineer Pierre-Marie-Jérôme Trésaguet developed a method of building a concave self-draining road covered in small chippings that hugely improved the quality of coach travel. The progressive French government minister Anne-Robert-Jacques Turgot reorganised the state mail coach service in 1775, slashing the time it took to pass information around the country. In 1765 it took twelve days to send a message from Paris to Marseilles and fifteen to reach Toulouse; in 1780 both places could be contacted within eight days.[4] This was a crucial improvement for the administration of a large country like France. If it took eight days for news of a problem to reach the capital it would also take eight days to send an order back directing what was to be done. Halving the time to Toulouse in each direction meant the remedy was supplied more than two weeks sooner.

The spread of information was facilitated by another eighteenth-century development – the newspaper. Occasional newsletters had been circulated in the previous century but few had gone on to appear regularly. The French *Gazette de France* was first published in 1631; the Spanish *La Gazeta* in 1661; the Italian *Gazetta di Mantova* in 1664; and the English *London Gazette* (formerly the *Oxford Gazette*) in 1665; but these were all formal weekly publications. The first British provincial paper, *The Norwich Post*, appeared in 1701, and the first daily newspaper, *The Daily Courant*, hit the London newsstands the following year. The numbers of papers then grew quickly on both sides of the Atlantic. *The Boston News-Letter*, the first regularly published paper in the New World, appeared from 1704. By 1775 42 newspapers were being sold in the American colonies. Several of these – including the *New York Journal* and the *Philadelphia Evening Post* – took a highly partisan line against the British in the struggle for independence: a French visitor to the States commented that 'without newspapers the American Revolution would never have succeeded'. By 1800 there were 178 weekly publications and 24 daily newspapers in the United States. The press proved even more important in the French Revolution: in the last six months of 1789 more than 250 newspapers were established.[5] Of course, these newspapers reached their readers via the newly improved road systems. The combination of print and transport meant that the previous centuries' slow, irregular trickle of news became a rapid flood. It marked the beginning of mass communication between governments and their people, even in the most obscure places. News of Trafalgar, for example, came to Moretonhampstead on 7 November 1805 from Crockernwell, a village through which Lapenotière had passed on 4 November. Lapenotière had reached London in time for the news to be printed in the *London Gazette* for 6 November – copies of which were delivered to Moreton on the 9th. Within three or four days of a government announcement, therefore, it could reach the whole of the British Isles. That speed of communication was markedly different from earlier centuries, when even the death of the king might not be known in some remote places until weeks after the event.[6]

What the new roads did for travel, the waterways did for goods traffic. In 1600 the easiest way to transport goods safely through France was to use the great rivers of the Loire, Seine, Saône and Rhône. The problem was that at some point, the cargoes had to be transferred from one river to another, and that was not an easy task. The 35-mile Briare Canal, with 40 locks rising 128 feet and then dropping 266 feet, which linked the Seine and the Loire, was finished in 1642. The even more ambitious 150-mile Canal du Midi was built in 1666–81 to connect the Mediterranean and the Atlantic. In Germany, the rivers Oder, Elbe and Weser were all linked by canals in the early eighteenth century. In England, the engineer James Brindley oversaw the construction of the Bridgewater Canal, which opened in 1761. It had been commissioned by the duke of Bridgewater, who, inspired by the Canal du Midi, realised the benefits of using waterways to transport his coal from Worsley to Manchester. That success proved the catalyst for a further 4,000 miles of canals to be built in England over the next 50 years. Such cheap means of transporting the fuel necessary for the expanding industries proved vitally important to Europe's economic development. With the opening of the Canal du Centre in 1784, the river systems of the Seine and Saône – and thus the Rhône – were joined, allowing heavy cargoes to be transported directly from Rouen, Paris and the English Channel to the Mediterranean.

It might come as a surprise to hear that the origins of aviation also lie in the eighteenth century. For thousands of years people had attempted to fly, and now they finally did it. On 21 November 1783 Joseph-Michel and Jacques-Étienne Montgolfier launched the first manned flight in Paris. The brave men in the basket of that hot-air balloon, which was made of sackcloth and paper, were Pilatre de Rozier and the Marquis d'Arlandes. Ten days later Jacques Charles and Nicolas-Louis Robert pioneered the first manned hydrogen balloon flight in Paris. All those men who had thrown themselves off medieval church towers with strapped-on wings, beating the air furiously as they plummeted to their deaths; all those natural philosophers who, since at least the time of Roger Bacon in the thirteenth century, had designed contraptions so that Man could fly like birds – all of them were finally vindicated in their belief that human flight was possible, albeit not in the ways they had imagined.

The whole of Europe now went balloon-crazy. The names of courageous aviators spread rapidly across the continent. In Britain, James Tytler made the first flight in Edinburgh in August 1784; Vincenzo Lunardi took to the air in London the following month. In October Jean-Pierre Blanchard travelled 70 miles in a hydrogen balloon across southern England. The editor of *The London Magazine* who reported these flights was most sceptical about the whole craze and concluded:

> However [much] such exhibitions may gratify the idle and lounging part of society, it is attended with a very serious loss to people in business. It is scarcely to be conceived what a deal of time has been trifled away, from first to last, by the various exhibitions of this bungling and mis-shapen *smoke-bag*.[7]

Shortly afterwards, on 7 January 1785, Blanchard together with an American patron, Dr John Jefferies, flew across the English Channel, reaching a height of 4,500 feet. After two and a half hours, having thrown everything out of the basket except themselves, they flew over Calais. It was an astonishing achievement. When Blanchard arrived in Paris on 11 January, there was no talk of 'bungling and mis-shapen smoke-bags'. Rather it was reported that

> [Blanchard's] appearance there had much the appearance of a triumph. Flags were displayed, guns fired, the bells set a ringing, & the magistrates went in procession to meet him and gave him as well as his companion the freedom of the city in a gold box. He was presented soon after at Versailles to the king, who . . . [granted] our hardy adventurer a bounty of 12,000 livres (£525) and an annuity of 1,200 livres (£52 10s.).[8]

Hundreds of balloon exhibitions took place in Europe and America. In 1797 André Garnerin pioneered the use of folded silk parachutes for the emergency evacuation of balloons, starting a new craze with which to wow the crowds. But that *London Magazine* editor was not so far from the mark. Balloon flight remained almost entirely a spectator event, a fairground novelty, without any practical application. It is hugely ironic that after so many centuries of dreaming of flight, all mankind could do with this new invention was to stare up at it, and gawp.

The Agricultural Revolution

Any reader who has laboured through this book this far will be aware by now that the greatest ongoing challenge facing our ancestors was the unreliable and inadequate supply of food. The eighteenth century did not solve this problem but it did see significant advances in agricultural management that led to higher grain and livestock yields, which went a long way to removing the fear of starvation.

The Agricultural Revolution is traditionally said to have originated in England with a handful of clever innovators. There was Jethro Tull, an inventor of farm machinery such as the seed drill, first outlined in his book *Horse-hoeing Husbandry* in 1733. There was his contemporary, Lord Townshend, the pioneer of the 'Norfolk Rotation' system, which involved planting turnips, clover, wheat and barley on the land in succession – hence his nickname, 'Turnip Townshend'. And then there was Robert Bakewell and the brothers Charles and Robert Collings, who advocated the selective breeding of livestock. It all seems to add up to a pretty picture of progressive landowners pioneering improvements – too pretty, in fact, for today's academic historians. According to one of them, it is all a 'grossly misleading caricature'. Bakewell is dismissed because one variety of his livestock has subsequently died out, and Lord Townshend because he clearly did not personally introduce the use of turnips for enriching the soil. The same historian disregards another famous agricultural reformer, Thomas Coke, earl of Leicester, for being just a 'great publicist (especially of his own achievements)'.[9]

The agricultural reformers have perhaps been over-praised in the past but they deserve more credit than these revisionist historians now give them. For a start, 'great publicists' were exactly what was needed for a nationwide change in long-established agricultural practices. And while Jethro Tull's book did not cause farmers to rush out and commission machines for sowing crops – the editors admitted as much in their preface to the fourth edition in 1762 – it did make people aware that mechanical improvement was *possible*. 'Turnip' Townshend may well have been stretching the truth in claiming that he introduced turnips to Norfolk but the fact that a peer of the realm became identified with such a humble way of improving land was excellent publicity that helped to spread the practice among the land-owning classes and tenant farmers alike. In short, the Agricultural Revolution only came about because a whole series of reformers changed perceptions of how profitable farming could be. If you consider Robert Bakewell's prize sheep and the Collingses' prize cattle in this light, it does not matter that one of their breeds died out. It was not so much their particular advances in breeding that made the difference but that farmers began to realise that animals did not have to remain the same size and shape as they had been since Noah's Ark struck dry land. Why rear scrawny sheep with little meat when you could produce large, fat ones, which sold for a lot more? When Bakewell started charging 80 guineas (£84) or more for the hire of his prize ram, the entire farming community talked about it. What marvellous publicity for the improvement of agricultural methods was that!

Like the explorers of the sixteenth century and the natural philosophers of the seventeenth, the agricultural reformers shared their discoveries with others. Indeed, they boasted of them. Why did they reveal their trade secrets? Many of them saw themselves as scientists; quite a few were elected as Fellows of the Royal Society. We might suspect that some who had made their money in trade and bought a country estate dedicated themselves to improving farming methods in order to be able to take their place more easily among the landed gentry. One agricultural improver who deserves to be mentioned in this respect is the businessman-turned-landowner John Mortimer (no relation to the author), who bought an estate in Essex and set about improving it. He was elected to the Fellowship of the Royal Society in December 1705, five months before Lord Townshend. Mortimer's two-volume work, *The Whole Art of Husbandry or the Way of Managing and Improving Land*, appeared in 1707 and went into a fourth edition in 1716. Among his many observations on how to improve farming he mentions the benefits of turnips as a winter crop for cattle, and clover for the soil, as well as the general efficiency of mixed husbandry.[10]

He recommends potatoes as an easy crop to grow, and especially promotes them for rearing pigs. He goes into great detail about the best way of improving the land in each county with dung, clover and ray-grass. It was systematic, scientifically minded landowners like Mortimer who helped bring about the Agricultural Revolution – by communicating their personal breakthroughs to large numbers of other landowners and lesser farmers.

Another reason for giving the reformers more credit is that they introduced the idea of enterprise to farming. Hitherto land had delivered a stable income but you could not gain great wealth from it; it was rather a means of consolidating wealth earned elsewhere. The reforming landowners now wanted improved returns from their land – and they were prepared to invest to get them. They were not mere dilettantes. Consider the low-lying Romney Marshes in south-east England. For centuries these had been little more than breeding grounds for mosquitoes, and the local people were blighted by malaria. The landlords of the region now started to turn what had been a dank and morbid deathtrap into some of the richest pasture in England. They did not do this to save the local inhabitants from malaria; they did it for profit. The hope of increased yields drove other landowners to try out 'Turnip' Townshend's four-field Norfolk Rotation system. With this method there was no longer any need to leave fields fallow to restore the ground's depleted nutrients as had been necessary in the past. The clover now added nitrates to the soil and was good for the cattle to eat, and the turnips similarly provided animal feed throughout the winter. The profit motive also drove yeomen farmers to adopt new plough designs at the end of the century.[11] Money might not always be the best incentive to do something but in the eighteenth century it had the universally welcome consequence of producing more food.

For tenant farmers, profit was not just to be reckoned in pounds, shillings and pence but also in terms of security. In this context, the humble potato played an increasingly significant role. It became a staple crop for workers in the north-west of England in the late seventeenth century and spread slowly south until it took off in the late eighteenth. In the fields around Moreton it was used for breaking up the soil and preparing the way for corn to be sown, as well as providing nutrition in its own right. It proved excellent insurance against the failure of a grain harvest. Not only was it cheap and easy to grow, it provided two and a half times as many kilocalories as wheat. When you are living on the breadline and trying to feed a growing family, *that* is an innovation worth developing.[12]

The significance of the reformers' new methods should not be judged by the reputation of any one reformer but by the agricultural yields subsequently produced. In medieval England, 10.5 million acres had provided 49.5 million bushels of wheat every year; in 1800, 11.5 million acres yielded 140 million bushels.[13] In addition, the land sustained 133 per cent more cattle, 33 per cent extra sheep and 50 per cent more pigs. All these animals were far larger than their medieval counterparts. The average cow that had provided just 168 pounds of meat in the Middle Ages now yielded 600 pounds; the average sheep that had provided 22 pounds of meat now gave 70 pounds; and the amount of meat from pigs rose from 64 pounds to 100 pounds.[14] Extra wool and leather were by-products, as was the extra dung that was returned to the soil to maintain the agricultural cycle. Selectively bred animals matured more quickly and so produced more meat at a faster rate. Every country fair now awarded prizes for the largest cow, pig and sheep. The paintings that landowners commissioned show how proud they were of their prize animals, the results of their own breeding programmes. The enclosure of common land further added to the efficiencies of farming when it was taken over by a landlord keen to intensify productivity. The resulting improved supply of food allowed the population of England to rise from 5.21 million to 8.67 million over the course of the century – an increase of about 80 per cent.

The Agricultural Revolution was not just confined to England. There were reformers across Europe and population growth on a scale that had not previously been witnessed. This was not just due to the fact that more food helped people to survive the harsh winters. Many girls benefited from an improved diet and consequently the age of menarche started to drop, allowing each female to have more children.[15] The populations of France, Italy, Spain, Portugal and Denmark rose by about a third; in Sweden and Norway they rose by two thirds and in Ireland by almost 90 per cent. The population of the continent as a whole increased by more than 50 per cent, from 125 to 195 million – a far higher total than ever before. This highlights the importance of a shared body of ideas and values that set landlords and tenants on a common path – towards wealth for the former and safety from starvation for the latter.

Enlightenment liberalism

Immanuel Kant described the Enlightenment as the ability to think for oneself, free from convention and dogma. Given such a broad definition, it is hardly surprising that it has been treated as an enormously elastic term. It is frequently taken to be a synonym for all the changes that distinguish the breezy, elegant world of Jane Austen's novels from the dark depths of the witch-burning seventeenth century. It is an intellectual bucket into which scientific concepts and rationalist theories are idly tossed, along with the rise of political economy and the decline of superstition. In that general sense, the Enlightenment started with Francis Bacon and Galileo in the early seventeenth century, incorporated the Scientific Revolution in its entirety, and did not come to an end until after the fall of Napoleon in 1815. This clearly is too vague a definition and too long a time span. For the purposes of this book, therefore, two intellectual changes frequently treated as elements of the Enlightenment are here dealt with separately: liberalism and economic theory.

At the heart of the Enlightenment was the 28-volume *Encyclopédie*, edited in Paris by Denis Diderot and Jean-Baptiste le Rond d'Alembert. This work, which was in the press for nearly twenty years (1752–71), was like an eternal flame around which the butterflies of genius fluttered – among them Montesquieu, Voltaire, Jean-Jacques Rousseau, Turgot and Louis de Jaucourt (who wrote a quarter of the *Encyclopédie* single-handedly). But the whole project was much more than the sum of its parts: it was an attempt to realign mankind's relationship with the natural world exclusively along the lines of reason, without recourse to magic, superstition or religion. Dividing all knowledge into three branches – memory, reason and the imagination – the editors created a taxonomy that left no room for matters such as divine will or spiritual intercession. Their purpose can be summed up in the single thematic heading to which they subjected the entire work: *understanding*.

Underpinning the ambitions of the editors and contributors was a self-perpetuating concept of social progress. Turgot explained this in

his *Philosophical Review of the Successive Advances of the Human Mind* in
1750. He started off with the premise of deism: that God was the 'prime-
mover' of the universe – a concept that Thomas Aquinas had originally
offered in the thirteenth century as part of his proof for God's existence.
In the language of the Enlightenment, God was the great clockmaker
who had simply set the world going and then left it alone. Slowly
humanity had emerged from its state of nature and passed through
three stages – hunter-gatherer, pastoral and agricultural – until finally
arriving at the fourth and final stage, the commercial. Along the way,
the ability to generate ever greater agricultural and manufacturing
surpluses had facilitated the transition from one stage to the next. For
Turgot, the evidence that this was indeed 'progress' lay in the fact that
mankind was constantly adding to the existing body of knowledge.
Thus humanity would continue to advance forever more, he reasoned,
on account of it being in our nature to enquire into things.

Progress could also be applied to political history. Montesquieu,
Voltaire and Rousseau were all heavily influenced by the constitutional
monarchy that had been established in England in 1688–9. Voltaire
spent three years as a political exile in England, during which time he
learnt English and developed a deep fondness for the country. 'How I
love English boldness!' he declared in a letter about Jonathan Swift's
Tale of a Tub (1704). 'How I love those who say what they think.'[16]
Unfortunately for him, he was so much in favour of the English theory
of government, and so keenly opposed to the French absolutist
monarchy that, after he returned to France, a copy of his *Lettres
Philosophiques* (1734) was burnt by the royal hangman. Voltaire took the
hint and left Paris for a second time. Thereafter he acquired a reputation as an intellectual maverick and a rebel – despite a stint as the
official royal historiographer in the 1740s. From 1760 he took up the
causes of various victims of state oppression, publishing essays and
tracts in defence of those unjustly tortured and killed by the state.
These acts of moral outrage, coupled with the phenomenal success of
his novella *Candide* (1759), in which he satirised the optimistic philosophy
of Leibniz's *Théodicée* and heavily criticised both the Church and the
government, made him a champion of liberty and a national celebrity.

The inequalities of society were even more sharply criticised by
Voltaire's contemporary, Jean-Jacques Rousseau. Like Hobbes and
Locke in the previous century, Rousseau's starting point in his *Discourse
on the Origin and Basis of Inequality among Men* (1754) was mankind in

a state of nature. Unlike Hobbes, who believed that natural Man was unable to comprehend morality and therefore had to have been wicked, Rousseau argued that in a state of nature, Man was neither moral nor immoral but essentially good because the evils that arose from society were not there to tempt him. Rousseau's natural Man did not have the language to express hatred. He was interested only in his self-preservation and in securing enough food, sleep and female companionship. He could not comprehend death. Rousseau's natural Man thus strikes the modern reader as a happy-go-lucky chap. But adversity forced him to guard against poor weather and the threats of wild beasts – not only for himself but on behalf of his fellow men. As Rousseau concluded:

> From the moment one man began to stand in need of the help of another; from the moment it appeared advantageous to any one man to have enough provisions for two, equality disappeared, property was introduced, work became indispensable, and vast forests became smiling fields, which man had to water with the sweat of his brow, and where slavery and misery were soon seen to germinate and grow up with the crops.[17]

In his most influential work, *The Social Contract* (1762), Rousseau set out to understand the limits of freedom within society. The book starts with the famous lines: 'Man is born free; and everywhere he is in chains. One thinks himself the master of others, and still remains a greater slave than they.' Rousseau goes on to argue that a state is unjust if it unduly represses the freedom of the individual. For a state to be legitimate it must be composed of two elements: a sovereign power, representing all the people to express the general will and devise the laws; and a separate agency, the government, which enforces those laws and the general will. For Rousseau it was important that people should participate in the government of the state, not merely be represented by it. Whatever might be required of an individual by the state should be performed immediately and without question, but nothing can justly be asked of a person unless it is in support of the general will. Ownership of property is granted back to its apparent owner by the state as a right. The book had a huge impact. Rousseau's and Voltaire's works provided intellectual arguments for liberalism and democracy and thus the strongest theoretical justification for the

French Revolution. Fittingly, the two authors died within a few weeks of each other in 1778, Voltaire on 30 May and Rousseau on 2 July.

Part and parcel of the social theory of the Enlightenment was the conclusion that society should be less intolerant of dissent. We need to remember that at the start of the century the persecution of minority religions was still increasing. The Edict of Nantes (1598), which had allowed French Protestants to worship freely, was revoked by Louis XIV in 1685. All the Huguenot churches were pulled down and their schools closed; hundreds of thousands of people were forced into exile. The Toleration Act of 1689, by which non-conforming English Protestants were allowed to worship in their chapels, was followed by a series of harsher measures against Catholics that same year, ensuring that they did not reside within ten miles of London. More English anti-Catholic measures were passed in 1700. But in the middle of the eighteenth century, with the arguments of Voltaire, Rousseau and Turgot enjoying popular support in France and further afield, the tide began to turn. Louis XVI finally granted freedom of religion to all Frenchmen in 1787. Four years later, Catholics were permitted to practise their religion in Great Britain. They still weren't allowed to hold offices or attend university, but it was a start.

Another indicator that liberalism was gradually permeating society was a changing attitude towards extramarital sex. Although the restoration of the monarchy in England in 1660 had seen the abolition of the legislation that required the death penalty for adulterers, prosecutions for illicit sexual acts continued well into the eighteenth century. Adulterers, fornicators and prostitutes were flogged publicly in London, carted around the city and shamed by being named on posters or 'black lists' put up in their parish of residence. On Sundays, clergymen would read out the names of offenders in church and force them publicly to confess their sexual transgressions. Some were sentenced by magistrates to do hard labour. In the first decade of the eighteenth century, over a thousand prosecutions a year were brought by societies established to police the morals of the city.[18] But gradually the fury abated. Not only could the London societies not keep up with the growth of the capital and its exponentially increasing appetite for sex but such prosecutions came to be regarded as unfair because action was only taken against the poor.[19] The idea of liberty added fuel to the debate. Could a prostitute be arrested for soliciting? No: as soliciting was not against the law, locking her up would be contrary to the terms of Magna Carta. And what about adultery? Was it against the natural law, and thus beyond the limits of tolerance? Or did it simply contravene the law of the Church, and was thus something for which the Enlightened response was *laissez-faire* – leave it be? Locke himself had been of the opinion that if a man had children with one or more women outside wedlock, it would not be contrary to natural law, but he was careful not to say so in public. The debate was most neatly resolved by the Scottish philosopher David Hume in his *Treatise of Human Nature* (1739–40), where he noted that lust was an appetite and 'confinement of the appetite was not natural'.[20] Indeed, procreation was the very basis of society.

By 1750, the idea that men and women were free to do what they wanted with their bodies in private was beginning to enjoy widespread support. The literary evocation of this principle can be seen in the novels that were published in England in 1748–9. These included Samuel Richardson's *Clarissa* and Henry Fielding's *Tom Jones*, both of which play with sexual themes outside wedlock. The same year saw the publication of John Cleland's *Fanny Hill*, an overtly pornographic novel in which almost all the non-bestial sexual vices appear. The equivalent visual stimulus was provided in France after 1740 by Boucher in his paintings, which were unashamedly erotic pictures of naked pretty young women in provocative poses. For those who wanted actually to partake of the pleasures of the flesh, prostitution became more obvious than it had been for the previous two centuries. From 1757 the names and services of all the good, bad and really bad prostitutes in the fashionable West End of London were published in a directory, *Harris's List of Covent Garden Ladies*. The royal tradition of taking mistresses, so dutifully observed by Charles II and Louis XIV, and shared by many members of the aristocracy on both sides of the Channel, was happily adopted by the emerging middle classes for whom sex was just another commodity. Southern European countries had always been somewhat more sophisticated when it came to sexual behaviour. The Catholic city state of Venice had always been comparatively tolerant of illicit love affairs but you could argue that it too became more liberal in the eighteenth century – it produced Giacomo Casanova.

Liberal ideas were also to be seen in the spread of humanitarianism. Elizabeth, empress of Russia, abolished the death sentence in 1744, much to the displeasure of most of her countrymen.[21] In Italy Cesare

Beccaria published *On Crimes and Punishment* (1764), in which he argued that there was never any justification for the state to take a life; life imprisonment was a far better deterrent, he argued, because of its long duration. Leopold II of Tuscany abolished the death penalty accordingly in 1786. Voltaire introduced the French edition of Beccaria's work in 1766 and it appeared in English in 1767. Even those countries that retained the death penalty reduced the frequency of their judicial killings. The execution rate in Amsterdam declined by a sixth; in London it went down by about a third.[22]

The effect of Enlightenment liberalism is even more noticeable when considering legal cruelty. Agonising punishments and officially sanctioned torture now seemed to say more about the tyranny of the state than the malice of the offender. In England the prohibition of 'cruel and unusual punishments' in the Bill of Rights (1689) meant that customary sentences such as loss of a hand or limb fell into disuse. The pillory was less frequently employed from 1775, the whipping of women was discontinued from about the same time, and branding was replaced with a fine in 1779. The last woman burnt in England for petty treason (the murder of a husband or employer) was incinerated in 1784, and the last woman burnt for high treason met a similar fate in 1789. Long before burning at the stake was finally abolished in 1791, humanitarian concerns led to an unofficial arrangement whereby condemned women were mercifully garrotted by the executioner as soon as the fire was lit. Some women were acquitted simply because juries felt that such a horrifying punishment was disproportionate to their crime. The same can be said for many men who would otherwise have been hanged. Englishmen were increasingly transported to America (until 1776) or Australia (from 1787) rather than being sent to the gallows. In the 1770s, John Howard campaigned for the reform of Britain's prisons, regarding imprisonment itself as a cruel and degrading treatment. In France too, the cruelty diminished. The last Frenchman to be burnt alive for consensual sodomy died in 1750 and the last for male rape in 1783. There was even a move to abolish the death penalty altogether in France in 1791. Unfortunately, any aspirations Revolutionary France had to create a more tolerant state went out of the window shortly afterwards. All things being considered, however, the cruelty of the French Revolution should be seen as an exception to the general trend of humanitarianism, not the end of it.

Economic theory

Until the late eighteenth century most European nations followed a series of economic principles that are collectively labelled mercantilism. The basic thinking was that there was only so much wealth in the world, and that the more of it you could amass, the less there was for your rivals. Governments therefore sought to limit the money available to foreign powers by preserving a positive balance of trade; at the same time, they looked to enrich themselves by profiting from their own citizens' trading activities. Ministers created monopolies and franchises – the monopoly on trade in the East Indies, for example – and then granted or sold those monopolies to companies, which sought to profit from the exclusive rights they had acquired. Domestic trade was exploited in a similar way, through the levying of tolls and customs. The system reached its apogee in France, where, prior to his death in 1683, Jean-Baptiste Colbert presided over a vast bureaucracy intent on exacting charges and fines, effectively milking every trade by means of regulation. Shortly afterwards, people started to voice criticism of such restrictive economic policies. In the 1690s the seigneur de Belesbat proposed that rather than spending valuable resources fighting the Dutch to capture their trade monopolies, the French should compete with them commercially – a radical new approach, in which liberty and private investment rather than state control were the platforms for success. Pierre le Pesant, sieur de Boisguilbert, similarly argued for free trade and the limitation of government interference. However, mercantilism remained firm. The idea of pursuing economic growth through a policy of encouraging free trade remained beyond the grasp of most political leaders.

Cracks finally began to appear in this economic stone wall in the early eighteenth century. One was the ability to increase the money supply through the creation of paper money. Another was inflationist theory – the belief that more money circulating in the economy was better for all. The combination of the two could have dramatic results. A Scottish inflationist, John Law, was appointed head of the French central bank in 1716, with responsibility for paying off the French national debt. Using his position as head of the new Mississippi Company in America, he issued banknotes guaranteed by the vast tracts of land waiting to be claimed in the New World. In this way he was able to flood the economy with cash, which in turn should have helped the government pay off the national debt. So huge were the sums held by investors as a result that a new word, 'millionaire', was coined to describe them.[23] However, schemes based on unrealisable assets are doomed as they depend on unshakeable confidence and infinite naivety. Law's system crashed in 1720, the same year as a similar share-based scheme in England, the South Sea Company, collapsed. The immediate consequence of these events was, of course, to scare off speculators, but others could see that economic theory should have had a part to play in limiting the damage. It became more important to understand what was going on in the economy.

This growing interest in economics was accompanied by the rise of statistics. By 1600 the English government had started collecting data on the number and causes of deaths in and around London, in order to quantify the effects of the outbreaks of plague. These figures were published annually, and in 1662 John Graunt used them to produce the first work of statistical analysis, *Natural and Political Observations Made upon the Bills of Mortality*. At the same time, a government minister, Sir William Petty, who had been a personal secretary to Thomas Hobbes, wrote several economic treatises in which he pioneered his 'political arithmetic', as he called it, or arguments based on 'number, weight and measure'. Not only did he start accounting for the national income, he also developed a primitive version of the quantity theory of money, which tries to explain the relationship between changes in the money supply and prices. Petty sought to establish the economic potential of a limited amount of cash and decided that its effectiveness depended on how quickly it changed hands. His statistical methods did not convince everyone: Jonathan Swift famously satirised the approach in his *A modest proposal* (1729), in which he calculated in Petty-like arithmetical language how the poor people of Ireland could make enough money to feed themselves by annually breeding and selling 100,000 surplus children to be eaten by the rich. Nevertheless, Petty indicated that by adopting a mathematical approach, an astute economist could calculate the path to national prosperity as surely as astronomers could compute the future position of the planets. In 1696 the statistician Gregory King went a step further when he drew up a detailed and surprisingly accurate compilation of the nation's wealth according to class and region. Parts of this appeared in print in Charles Davenant's *Essay upon the balance of trade* in 1699. It marks the first serious attempt to account for the wealth of a nation.

Into this arena stepped the first major economist of the Enlightenment. Richard Cantillon was an Irishman by birth and a maverick by nature. In Paris he participated in John Law's scheme, buying and selling the ludicrously overvalued Mississippi Company shares. But whereas Law blundered, Cantillon played a canny game. He could see where all this inflationism was heading, and secretly exchanged his own banknotes before the inevitable crash. Thus he was one of the very few 'millionaires' who did not lose his money. Later he lived in London, and prior to his murder in 1738, he wrote what is regarded as the first proper treatise on economics, *An Essay on the Nature of Commerce in General*. It circulated in manuscript form for many years until it was finally published in 1755. Cantillon adopted Boisguilbert's method of abstraction, in which the economist establishes a series of criteria for experimentation and maintains 'all other things being equal', thereby testing a single factor in a sort of theoretical laboratory. He developed a theory as to how the price of a commodity is determined, arguing that the key factor was not the cost of production but the demand for the item, the forerunner of our modern 'laws of supply and demand'. He established the importance of the entrepreneur in taking on the risk of the market, and theorised that interest was a reward for risk. He further developed the quantity theory of money that Petty had advanced. In the process, he started to cut the ropes of mercantilism that had bound European economies for so long.

When Cantillon's book was finally published, in France, it had a significant impact on a new generation of French thinkers. This was the first 'school' of economic theoreticians, known to history as the Physiocrats. Led by Dr Francois Quesnay, they fervently preached a gospel of free trade and laissez-faire – freedom from government involvement in business. Linking their economic theories with Locke's concept of natural rights, they suggested that there should be just one tax: a tax on land, which they believed was the source of all wealth. Their particular article of belief was a complex mathematical chart, the *Tableau Economique*, which Quesnay drew up in 1758 to show how the whole economy worked. The comte de Mirabeau, who was to become one of the leading figures of the French Revolution, declared it one of the three greatest achievements in the history of the world, alongside writing and money. Most people subsequently have found it completely unintelligible. At the time, however, it strengthened the idea that the economy was something that could be studied systematically. Among the rulers who adopted the principles of Physiocracy in governing their states were Carl Friedrich, margrave of the duchy of Baden, and Leopold II, grand duke of Tuscany. It marked a considerable turnaround from the days when government ministers dabbled in economics: now economic professionals were advising European governments.

Adam Smith was the man who wrote the bible of free-market economics and thereby hammered the nails into the coffin lid of mercantilism. Smith had read Cantillon and had met the Physiocrats. He was also a friend of David Hume, who had himself written on the quantity theory of money. Smith's magnum opus, *An inquiry into the nature and causes of the wealth of nations* (1776), was a magnificent encapsulation of a century of economic thinking. It discussed labour specialisation and its advantages, the uses of money, price levels, interest rates and the cost of labour, the nature of economic progress, the economic implications of the New World colonies, and the various systems of political economy. Crucially, Smith argued that the self-interest of merchants was not something that the state needed to guard against because the merchants, by becoming rich themselves, increased the wealth of the nation. He clearly laid out the argument in favour of free trade: high import tariffs encouraged smuggling; lower tariffs meant it was not worth smuggling tea and spirits into the country. He also showed that the old view of amassing wealth was simply wrong: countries did not gain by stacking up huge piles of bullion and doing nothing with it. His book was an immediate success. Most importantly, it came to the attention of the politicians. Lord North, the prime minister of the day, took on board Smith's arguments regarding taxation and free trade with Ireland. His successor, William Pitt the Younger, wholeheartedly embraced free trade, and in 1786 drew up an agreement with the French that made Smith's vision a reality.

The new economic thinking of the eighteenth century was not all about political economy, however; it was about private profit too. As we saw in the section on the Agricultural Revolution, people were beginning to invest in land so it would yield better returns. Crucial to such investment was the matter of capital. In order to buy the land or pay for its drainage and other improvements, entrepreneurs needed to borrow money. Thus the eighteenth century saw the rise of banking. There were about a dozen private banks in England in 1750. There

were 120 in 1784 and 280 in 1793.[24] By 1800 there were five banks in
Exeter alone, lending to people in Moretonhampstead and beyond – a
complete contrast from the start of this book, when cash was hardly
used in the south-west of England.[25] The loans these banks offered
had a far greater effect on the money supply than John Laws' inflationist issuing of banknotes. If a number of depositors placed £1,000
with a bank, and the bank operated a policy of retaining a reserve of
10 per cent of its deposits, then £900 could be loaned out. If that £900
was invested in a mill building, for instance, and the vendor placed
the sum he received with another bank, which also operated a 10 per
cent reserve policy, then that second bank could lend out £810. The
original £1,000 was now worth £2,710 on paper – after only two rounds
of deposit and loan. In this way banks could make capital go a long
way and a huge number of agricultural and industrial developments
could be paid for, greatly adding to the prosperity of the nation.

It would be inappropriate to end this section on economic theory
without mentioning one of the century's most important thinkers.
Thomas Robert Malthus was an English clergyman who had been
greatly affected by the work of Adam Smith and David Hume. Reacting
to the seemingly blind optimism of those Enlightenment writers such
as Turgot and the Englishman William Godwin who espoused the
belief that progress would never come to an end, Malthus applied the
principles of the new economics to the most fundamental question
at the heart of any society: whether all the people have enough to
eat. As he pointed out in the first edition of his seminal study, *An
Essay on the Principle of Population* (1798), throughout history a significant section of society had been unable to escape dire poverty, and
that was still the case – yet the Enlightenment optimists had failed to
explain why this was or how it could be alleviated. As Malthus wrote:

> I have read some of the speculations on the perfectibility of man and
> of society with great pleasure. I have been warmed and delighted with
> the enchanting picture which they hold forth. I ardently wish for such
> happy improvements. But I see great, and, to my understanding, uncon-
> querable difficulties in the way to them.

Malthus saw that Mankind's numbers increased in a geometric
progression, exponentially, while food supplies only increased arith-
metically. Thus it was not just in times of dearth that a growing

population could not be supported. If a country of seven million is easily able to feed itself, Malthus explained, its population will increase until the food that once fed seven million people has to feed seven and a half or eight million. Food prices will go up because of demand. However, poor labourers will find that because their number has increased, the value of their labour has decreased with oversupply. Thus a section of society is deprived of food through the natural process of multiplying. In reality, however, certain checks limit population growth. Looking back at history, Malthus observed that an over-large population is generally reduced by hunger, disease and violence. People might also take preventative measures to restrict population growth, including postponement of marriage, birth control, celibacy and abortion. Either way, the Enlightenment progressionists had been complacent. Far from society tending to a state of progress in which all its members experience ever-improving living standards, Malthus argued, quite the opposite is true.

Many people didn't like what Malthus said, then or subsequently: even today, sceptics react extremely negatively to the mere mention of his name. At the time, he was subjected to personal attacks and accused of heartlessness. Those who believed in progress wrongly saw him as an obstacle in their path, preaching a gospel of doom. They were shooting the messenger, of course: pessimistic economists are not to blame for a downturn in the economy; indeed, they do far less damage than the optimists. As for being heartless, Malthus was anything but. He was that rare thing, an economist genuinely interested in the plight of the poor, rather than being preoccupied with profit. He was quite right to say that the poverty trap had to be alleviated if progress was going to be for all and not just the few. The fact that his gloomy predictions were not fulfilled was not because they were wrong in themselves but because inventors and entrepreneurs were able to take advantage of fossil fuels and find better ways of fertilising soil and transporting food, thereby introducing new factors into the equation. As it happens, the factors that prevent his predictions coming true still depend on the ongoing supply of fossil fuels. Consequently, he remains one of the most important of all economic writers. His name and the concept of 'Malthusian checks' are frequently intoned by those trying to predict population growth and economic trends. But here, as the eighteenth-century economist least concerned with profit and most concerned with the poor, he stands for how far economic thinking had come from the days of mercantilism and royal monopolies.

The Industrial Revolution

Today it is widely accepted that the Industrial Revolution began in England in the eighteenth century. You might be surprised, therefore, to learn that the term was first coined in 1799 by a French diplomat in Berlin to describe what was happening in *France*. The explanation behind this apparent paradox is that the 'revolution' was a very gradual development. Initially it was imperceptible; it spread only very slowly from many disparate, localised beginnings in various parts of England to make its mark across Europe. Only at the end of the century was its revolutionary character recognised. As the historian Eric Hobsbawm noted: the Industrial Revolution 'was not an episode with a beginning or an end . . . for its essence was that henceforth revolutionary change became the norm'.[26]

Today we tend to associate the Industrial Revolution with steam power. At first, steam was just one aspect of the changes – and a relatively minor one at that. There were only about 1,200 steam engines employed in industry and mining in 1800; there were many more waterwheels creating much more power. Indeed, water still provided more than a third of the industrial power in Britain as late as 1838.[27] The real cause of industrialisation was commercial competition. If you were a mill owner and you wanted to prevail over your rivals, then you had to cut production costs and take full advantage of all the resources at your disposal. You had to change your working methods, adapt to new challenges and invest in people, machinery and buildings in order to maximise profit. In this respect, the Industrial Revolution and the Agricultural Revolution were two sides of the same coin: the desire to make money through greater efficiency in working practices.

In searching for the root causes of the Industrial Revolution we discover two distinct emerging markets. One of these was the demand for cotton and woollen cloth; the other was for coal and metalwork. The need to see these at the outset as two separate industrial revolutions that were eventually subsumed in a much larger one becomes apparent if we look at the sources of energy. The earliest cotton mills were built in the 1740s and were powered by animals or water; none used steam before 1780. Why, then, did coal production grow so much over the course of the century? Why were steam engines required to pump out the water from the deepest mines, which reached 300 feet in 1730 and 600 feet twenty years later?[28] Why was the world's first railway bridge (Causey Arch) constructed to transport coal from County Durham to Tyneside as early as 1726?

The answer to these questions is that there was demand for coal from industries other than textile manufacturing before 1780, albeit on a small scale. Blacksmiths and iron founders needed coal. So too did brewers of ale and beer, distillers of gin and whisky, manufacturers of salt, and makers of bricks, tiles and glass. All these crafts needed hotter temperatures than were achievable with wood. Traditionally they had made use of charcoal, but good-quality coal was much better. It was cheaper, too, as from the sixteenth century onwards, England was gripped by a firewood shortage. Timber was needed for house-building and shipping, furniture and most utensils, but a lot of woods had been cut down in the Middle Ages and more were cleared in the sixteenth century as land was required to house the rapidly increasing population. Lacking wood, people naturally turned to coal to satisfy their need for fuel. Houses with brick chimneys could burn coal instead of wood to heat water for laundry and general washing. Long before 1700 coal had become the staple fuel of London, where, over the course of the seventeenth century, the population had risen from about 200,000 to 700,000. Most of the capital's coal – 335,000 chaldrons (443,875 tons) per year – came from Newcastle and was shipped down the east coast of England, a well-established route that was able to supply London's seemingly insatiable demand for fuel. By 1770 that demand had doubled. Still the city and its suburbs carried on growing: by 1800 the population was over one million. At the same time, the rest of the country was undergoing a conversion to coal. Whereas the total output of England's coal mines in 1700 was about 2.6 million tons, by 1800 it was four times as much. Coal was becoming cheaper too, as the costs of extraction fell with the shift from many small, labour-intensive mines to fewer, larger and deeper enterprises. The proprietors of these large mines were able to organise efficient distribution systems: using the seas, rivers and canals, they managed to cut the transport costs to just a farthing per ton per mile.[29] The low costs further encouraged people to rely on coal, and so the cycle continued.

By 1850, Britain's national coal output had risen to more than 50 million tons per year.

The reason for the exponential increase in the demand for coal in the nineteenth century was that, through the widespread use of steam engines, it became essential to many different industries. But for most of the eighteenth century, the steam engine was nothing more than a machine for pumping water out of deep mines – a coal-burning machine for the production of more coal. The original idea is said to have been the brainchild of Thomas Savery from Modbury, in South Devon, who patented his 'invention for raising water and occasioning motion to all sorts of mill work by the impellent force of fire' in 1698. From that description, Savery clearly understood that the implications of his device went far further than mining, but in his book, *The Miner's Friend*, he emphasised that its principal use would be in draining mines of water. Having demonstrated his machine to the Royal Society in 1699 and obtained an extension of the protection of his idea by an Act of Parliament the same year, he had reason to hope his claims were about to make him rich. Unfortunately, his machine just wasn't efficient enough. But one developed by his fellow Devonian, Thomas Newcomen, was. The two men agreed a partnership and in 1712, at the Conygree Coalworks near Dudley, Newcomen installed the world's first working and commercially viable steam engine.

Newcomen's invention did not transform the world overnight. It required significant expenditure to install and used vast amounts of coal. But then the deep coal mines were exactly where the steam engine was needed most. The question was simply whether it was cheaper than teams of horses. To begin with, steam engines were just 11 per cent cheaper to run, but that was enough to persuade some owners to invest in them.[30] As mines grew deeper, the savings grew greater. Newcomen engines were installed at hundreds of mines in England. They were also exported: one was built in Sweden in 1727, and by 1740, there were engines at Vienna; Kassel in Germany; Schemnitz in Slovakia; Jemeppe-sur-Meuse, near Liège; and Passy, near Paris.[31] By 1750 the cost of steam power was down to 60 per cent of that of horses, and by 1770 it was just 40 per cent – a saving of 1½d per horsepower-hour. Steam engines now spread rapidly across Europe, all installed by British engineers. When the engineer John Smeaton toured the coalfields in the north-east of England in 1767, he found no fewer than 57 Newcomen steam engines at work. He recorded that they had a collective total power output of just 1,200 horsepower, so he set about redesigning them to be more efficient. Then in 1775 the partnership of Matthew Boulton and James Watt was established to exploit the potential of Watt's invention of a still more efficient machine. Through the use of a separate condenser, this used 75 per cent less fuel than a Newcomen engine and thus was attractive to those industrialists who did not have an unlimited supply of coal on their doorstep. John Wilkinson commissioned a Boulton & Watt steam engine in 1775 for his ironworks. Richard Arkwright had one installed at his cotton factory at Wirksworth, Derbyshire, in the early 1780s. It was from that point on that the revolution of steam power joined that of the factory system and the Industrial Revolution as we think of it was born.

The development of the cotton factory was the consequence of a series of technical innovations that allowed cloth to be made more evenly and cheaply. John Kay's flying shuttle, patented in 1733, was widely adopted by weavers in the 1740s and 1750s, doubling their output and creating a greater demand for spun thread. Lewis Paul and John Wyatt opened a factory using Paul's roller spinning machine in Birmingham in 1741. Although it closed four years later, Paul's concept of using rollers was adapted by Richard Arkwright for his water frame, patented in 1769. Arkwright, unlike Paul, was an astute businessman. Under his direction, the machine for spinning cotton became profitable, and when he died in 1792, his estate was worth about £500,000. Compare that to Mr Darcy's £10,000 per year and Mr Bingley's annual income of £5,000 in Jane Austen's *Pride and Prejudice* – not bad for a man who could not afford the fee for his first patent.[32] The money came from the large-scale mechanised systems employed in the factories he set up in Nottingham, Cromford, Bakewell, Masson, Wirksworth, Litton, Rocester, Manchester and elsewhere. These worked twenty-four hours a day, lighting up the night sky with the fire of industry. As has often been remarked, Richard Arkwright created the production-line factory long before Henry Ford. Another industrialist who accumulated a similar fortune was Josiah Wedgwood, the founder of the high-quality pottery factory that bears his name. A thorough and careful man, he laid out his estate at Etruria, in Staffordshire, to take advantage of a projected canal for the delivery of raw supplies to his factory and the distribution of the finished product. He insisted on the very best materials and the greatest cleanliness in the workplace, and his 278 staff were all specialists, organised in a strictly regimented fashion.[33] He provided housing on the estate for them, and developed an early form of sickness benefit to make sure they remained loyal. At the same time, he continually raised the standard of workmanship he expected. Accepting huge commissions from Queen Charlotte and Catherine the Great of Russia meant that *his* industrial revolution went beyond mere mass production and focused as much on the quality of the finished product.

From the 1780s industrialisation was well under way in Britain and beginning to catch the attention of businessmen across Europe. In the first half of the century the annual growth of the cotton industry can be measured at 1.37 per cent per year. From 1760–70 it was 4.59 per cent, the 1770s saw annual growth at 6.2 per cent, and in the 1780s it rose by 12.76 per cent. These significant shifts in industrial output were mirrored across almost all industries. The production of British pig iron increased from 30,000 tons per year in 1760 to 244,000 tons in 1806. The race was on for people to find the innovation that would make their fortune. In the years 1700–9, only 31 patent applications were granted in England; in 1800–9, no fewer than 924 were successful.[34] The next century saw even more industrial change, exponentially greater use of coal and increased innovation: in the 1890s, the British patent office received more than 238,000 applications. But the combined industrial revolutions of the eighteenth century had the 'Columbus effect' – of not only changing things for ever but showing people the path to the future.

Political revolution

As we have seen, the English Revolution of 1688 had a profound impact on political thinkers across Europe. The very idea that a parliament, representative of the people, could oust a king and select another, and impose on that new monarch a series of limitations on his power, shook the very concept of royal government. However, the English Revolution was primarily concerned with the relationships between the monarch and Parliament and between the government and the people. It was less interested in the way individual citizens stood in relation to each other. It was not until the political revolutions of the late eighteenth century that the idea of the equality of men was formulated and given political force.

The American Revolution began as an attempt by the colonists to resolve a lack of representation in government. It was a long-established principle in England that in return for granting extraordinary taxation, the elected Members of Parliament would have the ear of the king and be able to propose legislation. The American colonists had neither the ear of the king nor any influence over the government's legislative programme. They had no representatives at Westminster, even though they paid taxes to the British state. This was unconstitutional when viewed in the context of the 1689 Bill of Rights. The idea of the colonies electing MPs to sit in the British parliament had often been discussed but was always dismissed as being impractical. The Stamp Act of 1765, which laid a further tax burden exclusively on them, met with bitter opposition from colonists, who felt it was a violation of their rights as subjects of the English monarch. The Tea Act of 1773, which sought to charge tax on tea shipped to America even though it had been brought from the East India Company warehouses free of duty, met with violent resistance in the Boston Tea Party. An attempt at reconciliation failed, and the Thirteen Colonies (Virginia, Massachusetts Bay, Maryland, Pennsylvania, Delaware, South Carolina, North Carolina, Georgia, New Hampshire, New Jersey, New York, Connecticut and Rhode Island) set up their own governments. Each colony then declared itself a state, and together they formed the Continental Congress. Through this body they declared their independence from Great Britain on 4 July 1776. The text of the declaration began:

> When in the Course of human events, it becomes necessary for one people to dissolve the political bands which have connected them with another, and to assume among the powers of the earth, the separate and equal station to which the Laws of Nature and of Nature's God entitle them, a decent respect to the opinions of mankind requires that they should declare the causes which impel them to the separation.
>
> We hold these truths to be self-evident, that all men are created equal, that they are endowed by their Creator with certain unalienable Rights, that among these are Life, Liberty and the pursuit of Happiness. That to secure these rights, Governments are instituted among Men, deriving their just powers from the consent of the governed, – That whenever any Form of Government becomes destructive of these ends, it is the Right of the People to alter or to abolish it, and to institute

new Government, laying its foundation on such principles and organ-
izing its powers in such form, as to them shall seem most likely to
effect their Safety and Happiness.

The British government did not agree, and sent an army to make its
position clear. The war that followed lasted until the Treaty of Paris
was signed on 3 September 1783. By its terms, the British ceded to its
former colonies, which now called themselves the United States of
America, all the land east of the Mississippi and south of the Great
Lakes. By a separate agreement, East and West Florida were handed
over to Spain.

While these events were of huge importance for America and its
future development, it was the republican nature of the revolution
that had the greatest resonance elsewhere. There had been short-lived
republics in the past – for instance, the English Commonwealth (1649–
60) and the Republic of Corsica (1755–69) – but otherwise the only
established republics in Europe were small: the Italian city states and
the Swiss cantons. There was no precedent for the swearing-in of
George Washington as the president of more than five million people.
This had implications throughout the West – and nowhere more so
than in France, which had supported the Americans in their bid for
independence.

In France, the call for equality had a different context and meaning.
The authors of the American Declaration of Independence had used
the term 'equality' to express their belief that they had an equal right
to the liberties enjoyed by Englishmen. But it applied only to tax-
paying citizens; most of the Founding Fathers did not believe that all
Americans were equal with each other. Slaves were still regarded as
property, and in 1776 the liberty to enjoy your property still took
priority over a slave's right to equal standing. There were some early
calls for the emancipation of slaves. In 1780 the state of Pennsylvania
passed the Act for the Gradual Abolition of Slavery, by which all
children of slaves were declared free, so that slavery would gradually
die out over the course of a generation without anyone being forced
to give up their property. However, the principal slave states in the
American South did not follow suit. In France in 1789 there was much
less of a conflict between the desire for liberty and equality. The people
who demanded their liberty also wanted equality with those who
denied them their rights. It was thus a fundamentally different

revolution – against their own political structure rather than the imperial dictates of a distant nation.

It all began with a financial crisis. Hoping to get approval for his government's much-needed economic reforms, the king of France summoned a meeting of the Estates General – the first for 175 years. When the representatives of the people, the Third Estate, assembled, they declared themselves to be the members of a National Assembly and that they would proceed with the reform of the government with or without the representatives of the aristocracy and clergy. The king, in an attempt to stop them, closed the chamber. The 577 members of the National Assembly thus met instead in a tennis court on 30 June 1789. There all but one of them swore an oath promising that they would continue to meet until they had forced a constitution upon Louis XVI. Two weeks later, on 14 July 1789, the people of Paris stormed the Bastille – the Parisian fortress-prison that was the symbol of royal tyranny – and killed its governor. Leading noblemen started to flee the nation. Rioting broke out in the capital and spread to the countryside. What had begun as an attempt to impose a constitution on the king and his government turned into a full-scale revolution.

In August 1789 the comte de Mirabeau put forward the Declaration of the Rights of Man and of the Citizen. This document, which was approved by the National Assembly on 26 August, drew heavily from the social contract and the concept of natural rights as espoused by Rousseau and other political thinkers before him; it was also influenced by recent debates in America. It included 17 articles, beginning with 'Men are born and remain free and equal in rights.' It declared that 'the aim of all political association is the preservation of the natural and imprescriptible rights of man . . . liberty, property, security, and resistance to oppression', and that the limits of individual liberty could only be determined by law. It went on to state that the law 'must be the same for all, whether it protects or punishes', and that 'all citizens, being equal in the eyes of the law, are equally eligible to all dignities and to all public positions and occupations'. It provided for freedom from arrest except for breaches of the law, the prohibition of cruel punishments, the presumed innocence of suspects until proved guilty, freedom from religious oppression, freedom of the press, freedom of personal expression, the accountability of public officials, and a guarantee of the right of property.

All this might have been no more than the culmination of

Enlightenment political theory had it not been for what came next. The revolution became increasingly violent. From 5 October 1789, when the Parisian mob marched to Versailles with the city's cannon to force the royal family to return to the capital, it began to spiral out of control. In 1790 the peerage was abolished and the clergy was legally subjected to the authority of the secular government. Law and order broke down in many places, and massacres – both official and unofficial – took place throughout France. French royal authority, which not so long ago had been the most powerful in Europe, was abolished. The king was tried for treason and executed. Many aristocrats, including the queen, Marie Antoinette, followed him to the guillotine. Church lands were confiscated and the great cathedral of Notre-Dame was rededicated to the Cult of Reason. A new, revolutionary calendar was introduced. In the autumn of 1793 the Reign of Terror began. Many people were arrested under the wide-ranging provisions of the Law of Suspects, passed on 17 September. The anarchic horror of the revolution spread like fire. The state arrested and imprisoned hundreds of thousands of individuals and executed tens of thousands for fear that they might infringe the people's liberty.

The abuses that followed the Law of Suspects were a national tragedy but they should not distract us from the key point: the French Revolution was not simply *a* revolution, it was *the* revolution, the testing of one of the most far-reaching ideas of the whole millennium: namely that one man is worth the same as any other. This was a concept that had not existed in the ancient world, nor in the first millennium AD. While it was a Christian sentiment in origin, no Christian kingdom had ever tried to put it into practice. Yet Western society had been moving in the general direction of social equality for centuries. Changes from each century covered in this book are reflected in the 17 articles of the Declaration of the Rights of Man and the Citizen. The discontinuation of slavery in the eleventh century is highlighted in the first line, that 'men are born and remain free'. The legal changes of the twelfth century foreshadow the dictum that the sole restriction on liberty should be that of the law, and that the law is enacted to preserve the common good. The thirteenth century's desire for greater accountability is echoed in the principle that public servants should account for their actions, and that the government should not lock people up without reason. The active relationship between citizens and the state is adumbrated in fourteenth-century nationalism and parliamentary representation. The very concept of individualism, which we first encountered in the fifteenth century, resounds through the document. Sixteenth-century religious divisions are acknowledged in the clause promising freedom from religious persecution. The ideas of John Locke, the English Bill of Rights of 1689 and Jean-Jacques Rousseau's *Social Contract* all find their echo in the concept of natural rights, expressed in the second article. Of course it would be wrong to see all this as a continuous and uninterrupted march of society towards something measuring up to 'equality'. Equality itself is a nebulous concept – it lacks real meaning except when defined in relation to a specific scale of values. But if you could draw a graph of the rights of the ordinary man in relation to those of the rest of his society over the ages, the general trend line would resemble a stretched 's' shape: a curve starting with the discontinuation of slavery in the Middle Ages, rising very slowly from the Black Death to the early eighteenth century, and hitting the start of its maximum incline with the outbreak of the French Revolution at a gradient that it continued to follow for more than a century, only tapering off in the mid twentieth century as equality, or something close to it, was achieved.

The immediate consequences of the Revolution outside France were many and varied. To contain the revolutionary forces, Austria and Prussia declared war on France in April 1792; Great Britain was drawn into the conflict soon afterwards. The extremism, violence and injustice of the Terror also forced sympathetic reformers in these countries on to the back foot. Organisations such as the London Corresponding Society, whose members wanted to widen political representation to the working classes, had to accept that such a philanthropic vision was premature. The condemnation of the Revolution in the successful book *Reflections on the Revolution in France* (1790) by Edmund Burke, and the hard-hitting response by Thomas Paine in the even more successful *The Rights of Man* (1791) show how deeply opinion was divided. The failure of the French Revolution to address the question of the rights of women brought forth Mary Wollstonecraft's *Vindication of the Rights of Women* (1792) and rekindled the debate about women's position in society. Hence the French Revolution merely marked the *start* of the most radical gradient of our equality graph. Yet, as with many other developments discussed in this book, it showed the way forward and opened people's eyes to what was possible. Without the French Revolution, it is very unlikely that European thinkers in the next century would have approached the question of social reform on the basis of the equal value of individuals in society. Nor would political equality have become the moral default position in the Western world.

Conclusion

Had you enjoyed a bird's-eye view of Europe in 1800, your sharp vision would have noticed little change from what you saw in 1200. The cities would have been larger, and there were certainly more of them, but on the whole, the landscape was still predominantly rural. Even if you had focused on England, you would barely have noticed the gradual proliferation of mills and factories. Perhaps the silver lines of canals here and there would have caught your eye, or the odd mill building or mine workings. But by far the most obvious change would have been the enclosure of the vast majority of the country. No longer was the English landscape a patchwork of large fields consisting of separate strips of land farmed by different tenants. The modern pattern of small enclosures now came to predominate. But we shouldn't necessarily expect the greatest changes to have left the most obvious physical traces. Like eighteenth-century flying, some great developments did not have an impact on the land so much as on people's imaginations.

Having said that, it is worth bearing those fields in mind, for they represent the food supply and thus 'the greatest ongoing challenge facing our ancestors'. In going so far to meet that challenge, the eighteenth century transformed the entire relationship between mankind, our environment and God. Since medieval times, people had kept strict control of their neighbours' moral behaviour in the belief that immorality in the community would be communally punished by God, through a bad harvest, for example. Thus, if members of a community turned a blind eye to their neighbours' indiscretions, they too were guilty of a sin and deserving of punishment. However, as the food supply increased and fewer people went hungry after 1710, both these fears and the moralising that accompanied them diminished. At the same time, the enhanced understanding of the relationship between Man and his environment made people separate their belief in God from the causes of adversity. When in France the food supply dried up again in the 1780s, people blamed their fellow citizens rather than God. By this reckoning, the agricultural changes of the eighteenth century not only allowed the population to grow, providing the workforce for the Industrial Revolution; they also made society more tolerant, more permissive and less cruel.

The principal agent of change

None of the great changes of the eighteenth century was dominated by a single individual – at least not to the extent that, say, the advance of chemistry was dominated by Antoine Lavoisier. His name has not been mentioned until now – which seems to exclude him from consideration as the main agent of change – but it is probably fair to say that he, along with Isaac Newton, was one of the individuals who most affected our understanding of the natural world in this century. But Lavoisier's career curiously also reflected the undercurrents of the period. He led the way in identifying and systematically arranging the elements, in place of Aristotle's old system of 'earth, air, fire and water'. In breaking down compounds to find their constituent units and establishing the relationship of the particle to the whole, he resembled political thinkers like Rousseau, who was breaking down society to understand the relationship between the individual and the whole. Lavoisier's work on combustion and oxygen naturally led him to investigate respiration as a chemical process, and he demonstrated that breathing was indeed a gas exchange and thus a slow form of combustion. This went a long way to demystifying the processes of the body, secularising the understanding of life very much in the spirit of the editors of the *Encyclopédie*. He established the law of the conservation of mass, which states that for any reaction in any closed system the mass of the constituent elements must be equal to that of all the products at the end of the reaction. This quantitative approach is reminiscent of that of the economists who were at the same time trying to measure the wealth of nations. In this way, Lavoisier's work gives weight to the principle that the marquis de Condorcet outlined in his *Sketch for a Historical Picture of the Progress of the Human Spirit* in 1795: that progress in science will inevitably lead to progress in arts, politics and ethics. Unfortunately, neither Lavoisier's genius nor the parallels between his work and that of the social reformers was enough to save him. When the volatile substance of the French Revolution came into contact with the fact that he had once had a share in the collection of French taxes, there was an extreme reaction. He was sent to the guillotine on 8 May 1794, at the age of 50: a victim of Robespierre's Reign of Terror. If there was a point at which Kant's definition of the Enlightenment – thinking for oneself, free from dogma – ceased to apply, that was it.

So who was the principal agent of change? As this was the century that saw the first new form of motive power since windmills were invented, we ought to consider one of the engineers who made the Industrial Revolution possible. By this line of argument, it has to be the first great steam engineer, Thomas Newcomen. Although James Watt's steam engines were much more efficient, Watt only adapted about 500 machines that had been built to Newcomen's designs. Newcomen's impact was far greater, constructing 1,200 steam engines in Britain and Europe. Moreover, it was Newcomen who showed that steam power was commercially advantageous, and that was a huge achievement in itself. However, as noted above, most mills and factories in 1800 were still powered by waterwheels. So did steam engines really make that much difference to life in the eighteenth century? It seems to me that 1,200 steam engines replacing waterwheels is a change comparable to cannon replacing siege engines in the fourteenth century. The real impact of this technology lay in the future.

I would suggest that the man who most changed the lives of eighteenth-century people was Jean-Jacques Rousseau. His writings might be full of flaws but his ideas inspired the calls for tolerance, liberty and equality that collectively turned the events of 1789 from a financial crisis into a revolution. And there can be no denying that that revolution shocked everyone in the West – its kings and lords as well as its paupers.

1801–1900

The Nineteenth Century

Wherever you are reading this book, you are within touching distance of a nineteenth-century invention. If you are on a train or the Underground, your mode of transport dates from the nineteenth century. The same goes for buses: Paris, Berlin, New York, London and Manchester all had their first bus routes by 1830. If you are in a car, listening to an audio version of this book, bear in mind that both the internal combustion engine and the process of recording sound date from the late nineteenth century. If you are in bed or on a plane, you will be reading this with the help of electric light, which was pioneered in the 1870s. If you are relaxing with the book in a bath, note that the bath plug also has its origins in the nineteenth century. So does the loo, by the way: the first flushing water closets that could be mass-produced were exhibited in 1851 at the Great Exhibition. In fact, the very concept of a 'bathroom' is a nineteenth-century development: the earliest reference to one in an English house appears to be J. M. Barrie's line from 1888: 'What are politics when the pipes in the bathroom burst?' But I'm digressing. If you're reading a physical book, note that the paper is made of wood pulp, which was invented as a cheaper alternative to linen-based rag paper in the 1870s. If you're using a computer in the English-speaking world, you will see that the letters on the uppermost row of the keyboard are arranged QWERTYUIOP: that layout dates from a typewriter that went on the market in 1871. And if you're wearing a wristwatch, a pair of jeans or a bra – you're in direct contact with a nineteenth-century invention. You'd probably have to be naked in the middle of a jungle in order to escape the nineteenth century. But if you are, I'd be very surprised if you're reading this book.

The nineteenth century was the age of invention, even more so than the twentieth. As we saw near the end of the last chapter, applications for UK patents averaged 23,826 per year in the 1890s; by the 1990s, this had dropped by more than half, to an average of just 10,602 applications per year. Among the former were many innovations that we associate with the modern world – not least because we are constantly buying the latest versions of them. Electric toasters, fans, sewing machines and kettles were all invented in the nineteenth century. The first gas cookers went on sale in 1834, thanks to James Sharp of the Northampton Gas Company. Outside the home, gas street lighting was pioneered in London in 1807; by 1823, there were 40,000 gas lamps lighting the capital's streets, and most European cities were busy installing similar systems. Indoor electric and gas lighting was common by 1900: when the house in Moretonhampstead in which I began writing this book was rebuilt in about 1890 following a fire, it had gas lamps fitted to the walls of the entrance hallway and all the main rooms. It is very easy to forget the enormous difference that artificial lighting made to people's lives – not only in terms of safety when they went out of doors after dark, but also in the degree to which they could see colour and appreciate the things in their rooms, which had previously hardly been visible in the candlelight. And there were other crucial nineteenth-century inventions. Canned food was invented by a Frenchman in 1806 in order to supply Napoleon's army. The machine gun was developed at the start of the American Civil War (1861–5). Karl Benz patented his three-wheeled *Motorwagen*, the first commercially available car, in 1886. The Lumière brothers started making their remarkable short films on the streets of Lyons in 1895, and the following year, the world's first public cinemas opened in New York and New Orleans. Whereas Admiral Nelson died in 1805 on the deck of a leaky wooden ship that fired hot solid metal cannonballs at the enemy, by 1890 he could have travelled in an electric-powered submarine that fired explosive torpedoes.

As we have seen repeatedly in the earlier chapters, however, invention is not synonymous with change. Despite James Sharp's best efforts, most people did not cook with gas in the nineteenth century but continued to use solid fuel. It took time for cars and movies to become a part of our everyday lives. Even bras took a while to catch on, if you will pardon the pun. It is necessary to look beyond the mere novelty of something to assess the greater changes taking place in society, and to distinguish those inventions that made a profound difference to life in the West from those that just allowed us to do more easily what we had always done.

Population growth and urbanisation

You will be thinking that we are coming to the end of this book: we've dealt with eight centuries, so there are only two to go. You may be surprised to learn, therefore, that in historical terms we are not even halfway. The reason for this discrepancy is that history is not time, and time is not history. History is not the study of the past per se; it is about *people* in the past. Time, separated from humanity, is purely a matter for scientists and star-gazers. If a previously unknown uninhabited island were to be discovered it would have no history as such: its past would be studied by experts in natural history, botany and geology. We cannot write the history of the South Pole before mankind considered its significance and ventured to reach it. History is inextricably linked with what we have done, both as a species and as individuals. Thus a large country like Italy, with a population of 60 million and a huge cultural legacy from its past, has much more history than a small island with a tiny population. This is not being dismissive towards small islands; it simply reflects the fact that a country with 60 million people sees a million times more human experience every day than an island with just 60. There are a million more human exchanges, a million more social attitudes, and a million more diseases, aches and pains. We have to consider not just time passing but *human time* – that is, the volume of experience that a day or a year represents.

This volumetric approach to human time can be applied to compare centuries as well as different countries. If you add up all the days lived by people in Europe in the thousand years between 1001 and 2000, the relative proportions of the millennial population are shown in the pie chart on the right. If history were the same as time, the chart would show 10 equal divisions of 36 degrees. However, the differences in the chart are significant. We can see that the changes discussed in the first chapter of this book were actually experienced by about 3 per cent of the total European population over the millennium. All

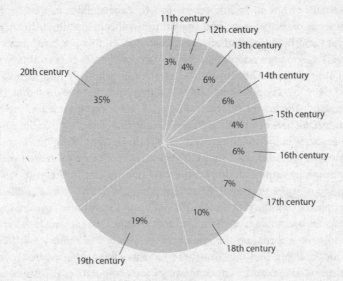

Proportions of European millennial population (person-days) per
century

those earth-shattering changes of the sixteenth century were dealt
with by less than 6 per cent. This is not to say that we should have
treated those centuries with greater brevity – they saw profound
changes that underpin everything that came later – but it does mean
that if we are going to focus on the changes that affected the greatest
number of people, then our judgement is going to be heavily skewed
towards the last two centuries. In fact, the above chart is a significant
underestimate of the modern bias in the West, for it does not include
the populations of the USA, Australia, Canada, South Africa or New
Zealand. Nor does it take account of the Westernised people in Latin
American, India and the Far East. Worldwide, only about a third of
history dates to the first eight centuries of the millennium. If the
significance of any given change is weighted according to the number
of people who experienced it, then we are now into the heavy
division.

Not every country saw population change at the same rate in the
nineteenth century. In France, the population increased from 28.7
million to 40.7 million (42 per cent), making this only its fourth
largest century of growth since 1001. The rise in Italy was far more

dramatic, from 18.09 million in 1800 to 32.97 million in 1900 – an increase of 82 per cent and the most significant rise in its history. Spain and Portugal similarly saw rises of 75 and 86 per cent respectively. Germany's population rose by 130 per cent, European Russia's by 181 per cent. The British Isles as a whole saw their total population rise by 153 per cent, but this figure conceals the fact that in England, the cradle of the Industrial Revolution, it increased by a staggering 246 per cent. Such a phenomenon was unthinkable just a century earlier. When the statistician Gregory King made his extensive calculations about Great Britain in 1695, he estimated that the population of London would pass one million in the year 1900 and that in the year 2300 the nation would be home to 10.8 million people. In fact, the population of London and its suburbs had already reached one million in 1800, and that of the United Kingdom as a whole hit 10.8 million the same year – a whole five centuries earlier than predicted by King. But these increases were dwarfed by those of the New World. The non-indigenous population of America (including slaves) increased by 1,335 per cent in the nineteenth century, that of Canada by 1,414 per cent, and newly settled Australia saw a rise of 72,200 per cent, from 5,200 to 3.76 million.

Population growth was driven for the most part by a combination of two factors: a larger supply of food and improvements in transport. The Agricultural Revolution did not stop in 1800: its legacy was continual innovation in farming methods and investment in new machinery. Crop surpluses and meat and milk yields increased. New, efficient means of storing and transporting these surpluses around Europe were developed. By 1860 Britain imported approximately 40 per cent of all its food. Famine still killed people by the thousand – in Madrid in 1811–12, in Ireland in 1845–9 and in Finland in 1866–8 – but after the arrival of long-distance train networks, food shortages in western Europe during peacetime became largely a thing of the past. Whereas before, people could only have as many children as they could feed, now they could have as many as they could breed. And that number was itself increasing with a gradual decline in maternal mortality. In 1700 the proportion of women who died in childbirth in London was 144 per 10,000, in 1800 it was 77 and by 1900 it had been reduced to 42.[1]

All these extra people had to find some way of making a living. In the past, many of them would have been employed on the land, maintaining the agricultural output on which their lives depended. But the rapid changes in agriculture meant that fewer and fewer people were required for farming. With better rotation systems, better ploughs and steam-powered machinery, farmers who were once barely able to produce enough food for their own household now could produce enough for many. In 1700 roughly 70 per cent of the population worked on the land throughout Europe. In 1801, according to the census of England and Wales, only 18 per cent were employed in agriculture; by 1901 that number had dropped to 3.65 per cent. People could no longer make a living in the countryside. All across Europe they left their villages and their rural roots and made their way towards the cities and towns where factories offered work. The same happened in America and Canada. Immigrants in the second half of the nineteenth century found most of the land already claimed. They too had to settle in towns.

The growing populations on both sides of the Atlantic needed to be housed. Long lines of urban terraces sprang up, especially in the cities of the highly industrialised and rapidly urbanising United Kingdom. Only the Netherlands had previously seen such a high proportion of town-dwelling, towards the end of its golden age, in

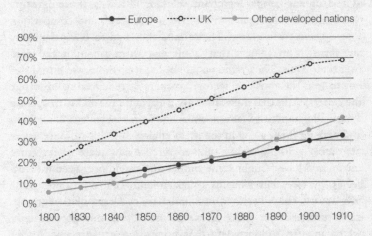

Percentage of population living in towns of more than 5,000 inhabitants in UK, Europe and other developed nations (America, Canada and Australia)[2]

1700. But the proportion of Dutch town-dwellers was now falling: the country had no coal and so it was slow to industrialise. By 1850 Britain had passed the Netherlands in becoming the most urbanised large nation in the world. Another tipping point was the year 1870 – when more people in the United Kingdom lived in towns than in the countryside. But even that does not reveal the full extent of the change. England went from being 80 per cent rural in 1800 to 70 per cent urban in 1900. Hot on its heels were the rapidly industrialising countries of Belgium and Germany (52.3 per cent and 47.8 per cent urbanised respectively in 1900), and not far behind them was the massively expanding United States of America (35.9 per cent).

Growing towns did not just consume food and drink but bricks and slates, wooden furniture and iron tools, coal and gas. The infrastructure that supplied them – the railways, pipelines, ships, wagons and horses – also required huge amounts of raw materials that had to be transported and manufactured. Cities and towns made things move, and needed entrepreneurs to borrow money from the banks and spend it in circles of increasing size and rapidity. But in order to be part of this economic cycle, the new arrivals from the countryside had to find work, and that either meant labouring (for which there was an overabundance of supply) or specialising in a trade. With so many people living and working in towns, those devoting their time to a particular trade had to compete, and competition forced them to innovate. They found they had to borrow money and invest in machinery that would give them a commercial edge over their rivals. Those who opted not to set up their own businesses but to join the ranks of factory workers were forced to exchange the general skills that their fathers had learnt for the ability to operate a particular, repetitive task on a machine, effectively de-skilling them and making them a cog in the great engine of industry. Whereas in 1800 most working people who required a new piece of furniture might well have bought some timber and carefully crafted it themselves, by 1900 they went to a department store to buy one made in a factory. This all created an inflationary cycle: population growth drove urbanisation, which drove industrial and transport growth, which in turn supported further population growth and further urbanisation, further specialisation, and so on. Population growth and urbanisation exponentially added to the changes wrought by the Industrial Revolution, ending the direct relationship between people and their land, which had existed since the beginning of farming in the Stone Age. To paraphrase the economist John Maynard Keynes, life in Europe for the vast majority of its burgeoning population was no longer a matter of how to survive; it was a question of deciding how to live.

Transport

Cities and towns could not have grown without a commensurate growth of the transport infrastructure that supplied them. Thus the nineteenth century built on the previous advances in road and canal transport that we discussed in the last chapter. It saw the construction of new canals, the extension of the turnpike trust network and, above all else, the advent of the railways.

Rails had been used to transport heavy goods for centuries – horse-drawn wagons on rails in particular had been employed for coal conveyance long before 1800. But in 1804 Richard Trevithick pioneered the combination of the power of a steam engine with railway transport in the mines of South Wales and the north-east of England. In 1812 the engine *Salamanca* was constructed for a colliery near Leeds, and in 1813 *Puffing Billy* was built by William Hedley and Thomas Hackworth for Wylam Colliery near Newcastle. Over the next decade George Stephenson, whose father worked at Wylam, produced a number of industrial steam engines for pulling coal. At the famous opening of the Stockton and Darlington Railway in 1825 Stephenson drove one of his engines, *Locomotion*, at 24 m.p.h., tugging a carriage full of passengers. His even more famous engine, *Rocket*, won the competition to power the 35-mile Liverpool to Manchester Railway in 1829. That line opened on 15 September 1830 with some drama, as the MP for Liverpool and lately resigned war minister William Huskisson inadvertently fell in front of *Rocket* and was run over, with the loss of a leg and much blood. Stephenson drove furiously – reaching speeds of 36 m.p.h. – trying to get the dying MP to Eccles to receive medical care but, alas! Mr Huskisson became a martyr to industry. His death ensured that every newspaper at home and many abroad carried the story in detail. This was another 'Columbus moment', when everyone could see the door to the future swing wide open.

Stephenson's *Rocket* set off a railway craze in England. Thousands of people flocked to buy shares in projects to bring the railway to their part of the world. It was a massively expensive business. A railway company had to obtain a parliamentary charter for setting up a new line – and that alone could cost thousands of pounds. It had to purchase the land over which the track was to be laid, sometimes from hundreds of separate owners. It had to build the rolling stock, engines, engine sheds and station buildings, and to employ suitable engineers and administrative staff to run the enterprise. Yet the idea inspired people. When railways were finally constructed they did not deliver the vast profits that had been expected but they did provide fast, long-distance, low-cost transportation. By 1840 there were 1,498 miles of railway in the British Isles; by 1850 there were 6,621 miles; by 1860 10,433 miles; and by 1900 21,863 miles. Or to put it in terms of usage, the railways carried 5.5 million passengers in 1838, 24.5 million in 1842, 30 million in 1845 – and then things really took off. The Railways Act of 1844 forced railway operators to make at least one train a day on every line available to third-class passengers at a price of no more than 1d per mile. In 1855 there were 111 million passengers every year; by 1900 the number had increased to 1.11 billion.[3]

It is no exaggeration to say that railways changed the world. The first passenger railway in America opened in 1830. Belgium and Germany saw their first lines open in 1835; Canada in 1836; Austria, Russia, France and Cuba in 1837; and Italy, the Netherlands and Poland in 1839. In 1841 the first international railway opened, linking Strasbourg and Basel. By 1850 Paris had six central stations: the Gare Saint-Lazare (1837), Gare Montparnasse (1840), Gare d'Austerlitz (1846), Gare du Nord (1846), Gare de l'Est (1849) and Gare de Lyon (1849). Those living in the vast expanses of North America had the most to gain from the new technology. By 1835 the Americans had laid down twice as much track as the British – the Pony Express was not long for this world – and the east and west coasts were finally linked in 1869. By 1900 more than 220,000 miles of track had been constructed in America: as much as in Great Britain, Germany, Russia, France and the Austro-Hungarian Empire put together. But Europe competed in style. From 1883 you could catch the Orient Express from Paris to Constantinople. Railways had joined up the world's dots.

	1840	1860	1880	1900
France	496	9,167	23,089	38,109
Germany	469	11,089	33,838	51,678
Belgium	334	1,730	4,112	4,591
Austria-Hungary	144	4,543	18,507	36,330
Russia	27	1,626	22,865	53,234
Italy	20	2,404	9,290	16,429
Netherlands	17	335	1,846	2,776
Spain	0	1,917	7,490	13,214
Sweden	0	527	5,876	11,303

Miles of European railway track in use 1840–1900 [4]

The consequences of this massive expansion went way beyond mere convenience. The railways brought a degree of homogeneity to society. Previously there had been no reason for all the clocks in a country to be set to the same time; it did not matter that 5 p.m. in Liverpool was not exactly the same as 5 p.m. in Manchester. But once trains connected the two towns and ran to a single timetable, the nation's clocks had to agree. Similarly, before the railways there had been no standardised way of spelling many place names: train stations' signs now established a de facto official spelling. Homogeneity extended to building too. Before the advent of railways, houses had been constructed with local materials. In Moreton and the Dartmoor area that was granite, in the Cotswolds limestone, in Sussex flint, and in Kent timber. The railways made it possible to transport cheap, durable brick to every town in the country, and the traditional building materials were increasingly set aside. So too were the local styles of building as standard 'modern' designs began to be employed. The railways destroyed localism in other ways too. Once upon a time the roads to a town were filled on market day with farmers driving their cattle and sheep on the hoof, ready to be sold to local slaughtermen. After the railway had reached Moretonhampstead in 1866 farmers sent their animals to a new livestock market, close to the station, for sale to middlemen who transported them by rail to the much larger town of Newton Abbot. Eventually the farmers cut out the middlemen and sent their animals direct to the slaughterhouse by rail; the local market closed down. The county of Devon, which had once had about 70 small market towns, now saw its trade concentrated in 20 large and medium-sized towns all linked by trains.

Today we view the coming of the railways in a very positive light – as a magnificent achievement – but we should also remember that 'modern life' was a traumatic experience for many hundreds of thousands of people at the time. Dislodged from the villages where they had been raised, many were unprepared for life in the cities because the culture they knew was heavily based on the complex set of reassuring relationships in a rural community. Thousands simply lost the ability to function as members of society. In England in 1845 every county was required to open a lunatic asylum where families could send those of their kin who were unable to cope. Reading through the admission registers of such asylums you come across hundreds of sad cases of women just staring blankly into the corners of rooms, or tearing off their clothes and ranting about their religious insights; and of men who had fantastic notions of how to make their fortunes in the cities, or who passionately wanted to have sex with Queen Victoria.[5] Even those who retained their sanity experienced the distress of seeing their communities dwindle and friends and family move away. Church attendances dropped after 1850 from about 40 per cent to 20 per cent by 1900. Some rural churches closed, and communities died with them. The cities and larger towns sucked the life blood out of rural England and the railways were the straws by which they did it.

For many, however, the trains opened up a world of opportunities. Young men and women could easily move all over the country, and thousands of hotels and boarding houses sprang up to cater for them. Before 1850 the vast majority of English people married someone from their own community or from the next parish. Afterwards, the numbers of people marrying someone from another county, or even another country, increased massively. Looking at my own family, my great-grandmother, Catherine Terry, is a good example of the new generation created by the railways. Her mother had been born in 1832 in Stowmarket, in Suffolk; she herself had been born in 1863 in her father's parish of Woodchurch in Kent. In 1883 she came to Plymouth with her father, met and married my great-grandfather, John Frank Mortimer, and thus came to live permanently in Devon. People of moderate means found that for the first time in their lives, they could travel long distances for business, for pleasure, for recreation and for love. Moreover, they could go back to their place of birth whenever they felt like it. That 215-mile journey from Plymouth to London – which had taken a week in the seventeenth century and still took 32 hours by stagecoach in 1822 – could be completed in just over six hours by rail in 1883. A brief note in my great-grandmother's hand relating to a trip not long after she was married reads: 'Travelled from Plymouth to Waterloo by train leaving 4.14 p.m. Arrived at Waterloo about 10.30 p.m.'

Railways were not the only way in which steam power improved travel. Although William Symington is widely credited with the first steamboat in 1803, and the American *Clermont* hauled people along the Hudson in the next decade, it was not until Isambard Kingdom Brunel built the SS *Great Western* in 1838 that a steamship was designed with regular ocean-going transport in mind. Up until then, ships crossing the Atlantic had had to rely on the wind. If there was no wind, or if it was against them, they could not move. If the wind was not quite in the right direction, the captain could tack but progress would be painfully slow. Steam engines changed all this. In 1843 Brunel's SS *Great Britain* was launched: the first iron-hulled, screw-propeller-driven steamship and the largest boat in the world at the time. Although the *Great Britain* herself eventually ended up on the Australia run, such innovations in ship design brought the time it took to cross the Atlantic down from the 14.5 days that the SS *Great Western* took in 1838 to 9.5 days in 1855 and 5.5 days by 1900. Millions of emigrants from Europe found their way to new continents and new lives, their journeys made affordable by the short duration of the voyage and the size of the ships that could transport them. In 1869 the passage from Europe to India and East Africa also became considerably quicker, safer and cheaper with the opening of the Suez Canal. Such advances in travel spurred the imagination of Jules Verne, the author of *Around the World in Eighty Days* (1873). The book's protagonist, Phileas Fogg, bets with friends at the Reform Club in London that he can circumnavigate the globe and return to the club within the said 80 days; he manages to do so by the skin of his teeth. But by the time the story was published such a journey was feasible in real life, due to the linking of the Indian railways and steamship lines. In 1889 an American journalist, Nellie Bly, performed the feat in just 72 days. It was a far cry from the three-year circumnavigations of previous eras.

The nineteenth century saw a revolution in road transport too. A steam car had been invented by a Frenchman, Nicholas-Joseph Cugnot, in 1769, but it had failed to catch on. In the early nineteenth century there were further attempts to produce steam-driven road vehicles to rival the railways, but there was always the dirty business of fuelling the machine: people wealthy enough to buy a steam carriage of their own did not want to be shovelling coal. From about 1860, however, a number of innovations put practical steam vehicles into the public eye. Steamrollers allowed better, smoother roads to be built. Traction engines could haul great weights along those roads, such as supplies of timber. And steam tractors and ploughing engines allowed farmers to dispense with horses.

The real revolution on the roads was not due to steam but human muscle. In the early 1860s the first bicycles – velocipedes – made their appearance. To start with, they were made entirely from wood, but soon they acquired metal frames and rubber tyres. By 1869 the penny-farthing was in production. Seen as rather dangerous, it was ousted from the market by the invention of the safety bicycle, pioneered by John Kemp Starley in 1885. By 1890 the geared chain drive had arrived, as had the chain guard, brake levers and pneumatic tyres. For hundreds of thousands of people, the bicycle meant freedom. Without the cost of maintaining a horse and stable, they could pedal 30 or 40 miles in a day, going to places that the train did not reach. Or they could take their bicycle on a train and ride from the nearest station to their destination. In this way a man or woman could set off alone and visit a place hundreds of miles away at no great expense. In terms of the freedom to travel, it wasn't the twentieth century that saw the greatest change but the nineteenth.

Communications

We have seen how closely communications and transport were linked in the previous century. The person sending a message might not have been galloping along to deliver it in person but *someone* had to carry it physically. In the nineteenth century that link was broken. But before we launch into the world of telecommunications it is necessary to look at an innovation in old-style message-sending.

In January 1837 Rowland Hill, a failed schoolmaster and low-ranking civil servant, submitted his pamphlet *Post Office Reform: Its Importance and Practicability* to the British government. In it he attacked the nation's inefficient and costly postal service. The Post Office charged double for two pieces of paper rather than one, increased the price for longer distances, and made the recipient liable for the payment. And as recipients frequently refused to pay, the cost of delivering a letter was forced on to other users. Hill proposed instead a uniform system: a 1d cost for letters of up to half an ounce sent anywhere in the kingdom, with the cost of the postage being paid by the sender. Proof of the payment was to be the display of a postage stamp. Great idea, you might have thought. Who could possibly object to such a system? The principal opponent was one William Leader Maberly, Secretary to the Post Office. He was more concerned that the Post Office should remain profitable than that people should send messages easily and cheaply. Hill won the first round of their battle, being given the job of introducing the Penny Post. He was vindicated when the Penny Black, the world's first postage stamp, went on sale in 1840 and hundreds of thousands of people used it. Whereas in 1839 three letters were sent every year per head of the population, in 1860 the figure had risen to nineteen.[6] But Maberly hit back. Despite the increase in the number of letters carried, the Post Office started to lose money, so in 1843, Hill lost his job. That same year, by a stroke of luck, Zurich and Brazil adopted his invention of adhesive postage stamps. Eventually, after a change of government in 1846, Hill was restored to his position – and became a national hero.

At the same time as Hill was working out how to communicate more efficiently by post, various inventors were developing a scheme to transmit long-distance messages instantly. The first successful electric-telegraph messaging system was invented by Francis Ronalds in 1816. In the back garden of his house in Hammersmith he strung eight miles of wire between two poles and successfully passed electrical impulses along it; these impulses turned the receiver's dial to a letter of the alphabet. Due to its effectiveness only in dry weather, Ronalds built an underground version encased in glass tubes. Realising the importance of instant messaging over long distances, he wrote to the Admiralty offering to demonstrate it. The reply, signed by the Secretary to the Admiralty, John Barrow, was possibly the biggest blunder in the history of technology. Barrow declared that 'telegraphs of any kind are now wholly unnecessary'.[7] Incredible though it seems, the

Admiralty believed that the semaphore system they had then recently adopted – that is, men waving flags at each other – was superior. Impressively, Ronalds rose above this setback. He published an account of his experiments in 1823 and went on to invent the means by which one could automatically record the readings of scientific instruments. Many years later, when the telegraph had become well established, he submitted a petition for recognition as its inventor. It fell on deaf ears until 1870, when the prime minister, William Gladstone, acknowledged the injustice and gave Ronalds a knighthood for his efforts.

The first practical applications of electric telegraphy did not take place until the late 1830s. In America, Samuel Morse and Alfred Vail developed a telegraphic system, sending their first telegram in 1838. By 1861 the US telegraphic system linked the east and west coasts. Independently, English inventors were bringing about the same revolution. Charles Wheatstone and William Fothergill Cooke patented a telegraph system and constructed it in tandem with a railway track owned by the Great Western Railway. In 1843 a telegraph wire connected Paddington and Slough. Two years later, Sarah Hart, a mother of two young children, was poisoned with prussic acid at Salt Hill, near Slough. The murderer calmly walked to the station and boarded a train to London. A neighbour who had heard screaming found Mrs Hart dying and called for help. The Revd E. T. Champnes followed the assailant to the station; he observed him boarding the train and promptly instructed the superintendent to send the following message to Paddington by way of the telegraph:

> A murder has just been committed at Salt Hill, and the suspected murderer was seen to take a first-class ticket for London by the train which left Slough at 7.42 p.m. He is in the garb of a Quaker, with a brown great coat on, which reaches nearly down to his feet; he is in the last compartment of the second first-class carriage.

The murderer was spotted at Paddington and followed on to a bus. He was arrested shortly afterwards. Eventually he was hanged for his crime. The newspapers gave extensive coverage to the event, highlighting the role played by the telegraph in the apprehension of the murderer. Suddenly people were aware that a new age had dawned. The military might have semaphore signalling but no waving of flags could compare with this revolutionary invention. For the first time in history, ordinary people could transmit messages over long distances faster than they themselves could travel. Murderers might well take a boat from Liverpool to New York to escape the long arm of the law, but from 1866, when a permanent submarine cable had been laid on the ocean floor, there was a good chance that the long arm of the law would be there to pick them up on arrival. From 1872 you could send a telegram to Australia, and from 1876 to New Zealand.

With the proliferation of telegraph lines, information could now be conveyed almost instantly to almost any city or town in the world. The last few miles from the telegraph office to someone's house might still require a post boy to make the journey in person, but that was a small delay in the scheme of things. The advantages to individuals, businesses and police forces of being able to send information so cheaply and quickly are obvious. It was even more important for governments. Think back to the previous century, when Turgot reduced the length of time it took to travel between Paris and Toulouse from fifteen days to eight. Now the transfer of information over that distance was instantaneous. If a crisis occurred in Toulouse, local administrators could refer the matter back to Paris and await the government's response. Moreover, they could now be expected to liaise with central government; if they did not, they could be held accountable. This was particularly important for the British, who controlled a worldwide empire: London could now give direct orders to the Viceroy of India, the High Commissioner of Canada, or the Agent-General in South Australia. As governments became more vulnerable to criticism from their parliamentary opponents, newspapers and the public, they saw the necessity of governing directly rather than trusting appointees to make decisions on their behalf.

All this went a step further in 1876 when Alexander Graham Bell obtained patents in Britain and America for 'talking by telegraph', as he referred to the telephone. In March that year, Bell uttered the first sentence over the phone to his technical assistant Thomas Watson, after he had just upset a chemical-filled glass jar: 'Mr Watson, come here, I want you!' In 1878 a telephone call was successfully made across 115 miles of wire from London to Norwich, and in that same year a telephone directory for New Haven was published. Two years later the Telephone Company produced the first London telephone directory, consisting of about 250 subscribers. Telephone numbers were developed shortly afterwards, as the number of subscribers grew unmanageable for operators. By 1886 telephone kiosks had been constructed, charging 2d for a three-minute call. By 1900 there were 17.6 telephones for every 1,000 people in the United States.[8] Communications had undergone a second revolution: first the telegraph had allowed one-way messages to be sent over long distances instantly; now the telephone made two-way exchanges possible. By 1900 Guglielmo Marconi was sending radio messages across the Channel from England to France, and by the end of 1901, he had transmitted a signal across the Atlantic. The first maritime SOS message was received in 1899 by the East Goodwin lightship, which resulted in the crew of the German ship *Elbe* being rescued by the Ramsgate lifeboat. It was all a very far cry from Lieutenant Lapenotière's 37-hour dash from Falmouth to London to deliver the news of the victory at Trafalgar.

Public health and sanitation

There are three things common to cities in all ages: stench, overcrowding and beggars. If a city lacks one of these, you can be sure it hasn't grown organically but has been laid out according to some philanthropist's or dictator's whim. At the start of the nineteenth century there were no such whimsically clean cities in the West. Every city was smelly and visibly populated by huge numbers of the needy poor. City streets were foul but the smells emanating from the cesspits beneath the houses in the slums were even worse. Neighbourhoods were rife with disease. Poverty bred illness and illness exacerbated poverty, dragging the urban poor down into a whirlpool of misery. The average age at death among labourers in Bethnal Green in 1842 was just 16; among the better-off in London it was 45.[9] But when it came to doing something about this situation, society was apparently unconcerned. The best prophylactic in most educated people's minds was to separate yourself and your household from the smells and the slums by buying a house in a leafy suburb. How and why the poor should have their way of life changed for them was not a matter of public interest. Finding out why they suffered more diseases was similarly not a medical priority. A French surgeon who wondered why the poor in one Parisian street had a death rate that was 50 per cent higher than that of the bourgeoisie in a neighbouring street concluded that immorality was to blame. If the poor could be taught to behave in a less dissolute fashion, they would live longer and so would their children.[10] Those who were sceptical about the efficaciousness of such moral improvements tended to fall back on the old idea that a miasma composed of gases from rotting matter created fetid air, which entered the body of a person and caused them to become sick. In their eyes, the poor were slovenly, hence they fell ill. Alternative explanations, such as that of Girolamo Fracastoro, who had suggested in his *De Contagione* (1546) that illnesses were spread by 'seeds of disease', had been forgotten.[11] There was a paralysis of prophylaxis.

The great exception to this paralysis was vaccination. In the 1790s the physician Edward Jenner had noted that dairymaids who had suffered a bout of cowpox developed immunity from smallpox. On 14 May 1796 Jenner deliberately inoculated an eight-year-old boy, James Phipps, with cowpox from a dairymaid who was then suffering from the disease. Six weeks later, he inoculated the lad with the dreaded *variola* virus. Phipps did not develop smallpox. Jenner was exultant, and tried to persuade the Royal Society to publish his findings; they were reluctant on the basis of so little evidence. Jenner accordingly carried out further experiments in 1798 and published the results in *An Inquiry into the Causes and Effects of the Variolæ Vaccinæ* that same year. His work immediately attracted a lot of attention. By 1803 there were editions in Latin, French, German, Italian, Dutch, Spanish, French and Portuguese. Charles IV of Spain had his own children vaccinated, and sent the royal physician Francisco de Balmis and 20 children from an orphanage to Colombia, where there had been a smallpox outbreak. The cowpox virus was kept alive during the Atlantic crossing by infecting one child from another. In this way de Balmis was able to vaccinate over 100,000 people in the Caribbean and Latin America. Gradually countries began to make smallpox vaccination compulsory soon after birth, although the United Kingdom itself did not do so until 1853. Other diseases continued to kill as they had always done, however, often finding an overflowing reservoir of victims in the increasingly crowded cities of Europe and the New World.

On to this murky medical stage entered the Hungarian physician Ignaz Semmelweis. Working in one of two free obstetric clinics in Vienna in 1846, he met pregnant women who would plead with him to be allowed to enter the other clinic. The reason was that his own clinic had death rates that fluctuated around 10 per cent whereas the likelihood of dying in the second clinic was much lower, about 2.5 per cent. No one could work out why this was. Everything about the two clinics was the same apart from the staff, and Semmelweis's clinic was staffed by highly qualified physicians whereas the other was staffed only by midwives. Then, in March 1847, one of Semmelweis's colleagues died after being accidentally stabbed with a scalpel by a student during an autopsy. Semmelweis noticed that his late colleague's corpse showed symptoms similar to those exhibited by the women who were dying in his clinic. He came up with the theory that 'cadaverous particles' had been transferred from the dead body to the student's scalpel in the course of the autopsy, and that these particles had killed his colleague. Semmelweis realised that this might explain why women were dying in greater numbers under the care of qualified medical practitioners, who conducted autopsies, than under the care of midwives. He therefore instigated among the medical staff of his clinic a regime of washing hands with chlorinated lime solution. Almost immediately, the death rate dropped to the same level as that of the other clinic.

Semmelweis's students were impressed and promptly set about spreading the news. Most initial reactions to his regime were negative, however. Physicians declared that it was reminiscent of outdated theories of contagion, like Fracastoro's 'seeds of disease'. Some feared that the theory of 'cadaverous particles' was too close to a magical understanding of the body, which they saw as the enemy of science. After Semmelweis's term of appointment came to an end, his contract was not renewed and the clinic returned to its old, unclean and deadly ways. In his next position, Semmelweis brought about a similar reduction in women's deaths through hand-washing and instrument disinfection, but still the medical establishment rejected his advice. Eventually, in 1861, he published his own account of his work, but poor reviews of his book only increased his frustration at not being heard. He became fixated on the rejection of his discovery by physicians more concerned with their own reputations than with helping their frightened patients. In the end he had a nervous breakdown and was committed to a lunatic asylum where he was beaten up by the guards and died of blood poisoning in 1865, at the age of 47. It was two decades before a paper by Louis Pasteur proved that Semmelweis's theory of 'cadaverous particles' was essentially correct.

While Semmelweiss was trying to save the lives of pregnant

Viennese paupers, the social reformer Edwin Chadwick committed himself to improving the lives of the London poor. He was instrumental in bringing about the Public Health Act of 1848, which encouraged towns to set up local Boards of Health to improve their slum dwellings, sewers, urban slaughterhouses and water supplies. Not everyone thought this was a good idea. One article in *The Times* declared that 'cholera is the best of all sanitary reformers'.[12] But Chadwick's cause received some unexpected support during the cholera outbreak of 1854, when John Snow, a London physician, charted the new cases of the disease and found that all the victims drew their water from a well in Broad Street. Snow had the handle removed to prevent any further infection, the cause of which turned out to be a cholera-infected cesspit a few feet from the well. Most importantly, he gave evidence to a House of Commons Select Committee that cholera was not directly contagious, nor caused by a miasma of fetid air, but transmitted by water. The solution, he declared, was to improve the drainage and the sewerage of the city. When the Great Stink descended on London in 1858 – a stench arising from untreated effluent so pungent that it forced Parliament to adjourn – the government commissioned Joseph Bazalgette to rebuild the entire sewerage system of the capital. The task occupied him until 1875. That same year, the second Public Health Act was passed, enforcing the provision of running water and sewerage to every new house and the compulsory employment of medical officers by every Board of Health. At the same time, Georges-Eugène Haussman was rebuilding Paris, complete with a new sewerage system under its boulevards. Modern cities were no longer allowed to decay in the mass of human detritus.

Despite his empirical research showing that cholera was waterborne, John Snow could not work out how the Broad Street well actually conveyed the disease. In 1861 Louis Pasteur stumbled on the path that would eventually lead to the answer. Experimenting with broths in Petri dishes, Pasteur showed that those left open to the air became mouldy in a short while, whereas broths that were closed to the air did not. He also noticed that broths remained clear of mould and fermentation when they were left open to the air but were protected by very fine dust filters. It followed that the broths were infected by some particle in the air rather than the mould being caused by a property of the broth itself. Pasteur's paper on the subject inspired another Frenchman, Casimir Davaine, who had previously found *bacilli anthraci* in the blood of anthrax-infected sheep. In 1863, Davaine published a paper that showed that anthrax was connected to the microorganism he had observed. In Scotland, the surgeon Joseph Lister also realised the potential in Pasteur's work, suspecting that particles in the air might be the cause of infection in his patients' wounds. In 1865 he started to apply carbolic acid to dressings and incisions in order to kill the microorganisms that caused gangrene, with positive results. In Germany, Robert Koch worked on the aetiology of anthrax, experimenting with the bacilli that Davaine had identified. In 1876 he established that the microbes created spores that were inhaled by the animal, or otherwise entered the bloodstream and multiplied in the blood, eventually killing the host. In 1878 Koch went on to identify the germ that caused blood poisoning, and in 1882 he found the microbe that was responsible for tuberculosis, one of the greatest killers in history. Pasteur himself experimented with inoculations against anthrax, chicken cholera and rabies. In a famous incident on 6 July 1885 he vaccinated a nine-year-old boy, Joseph Meister, who had been bitten by a rabid dog two days earlier. Joseph lived, as did a second boy three months later, who had been bitten trying to protect other children from a rabid dog. Germ theory, as Pasteur called it, had arrived.

All medical discoveries are, in one way or another, matters of public health, and many more advances could be discussed here. One great advance was the introduction of anaesthetics in the 1840s. Another was the successful use of Caesarean section. At the start of the century, this operation was a last resort, as it almost always resulted in traumatic loss of blood and the death of the mother. For the most part, early-nineteenth-century physicians preferred foetal craniotomy – the crushing of the unborn child's skull and the removal of the foetus in pieces through the vagina – to preserve the life of the mother. One of the very few known early cases in which both mother and baby survived was carried out around 1820 in South Africa by a British military physician, Dr James Barry, who was discovered at his death to have been a woman masquerading as a man all his/her career. From the 1880s, however, the operation was undertaken more regularly, and with increasingly positive outcomes for both mother and child. Over the course of the century, life expectancy at birth rose all across Europe, from approximately 30 years of age to roughly 50, and on that basis, the above changes amount to more than feathers in a few medical caps. The nineteenth century was when the West discovered what caused most illnesses, and worked out, in many cases, how to prevent them, how to cure them, and how to limit contagion.

Photography

Not long ago, I was interviewed about a medieval subject for a television programme. Shortly afterwards, a picture researcher phoned me to ask where she might find images of a certain character I had mentioned. When I replied that none exist, she said that in that case, they would drop all references to him. This little episode starkly reveals how our collective awareness of the past – and our knowledge generally – is shaped by visual sources.

There is a whole hierarchy to historical imagery: our ability to understand the past is closely related to the number, variety and range of images that survive. We find it much easier to imagine sixteenth-century individuals than medieval people because we can see their faces in portraits. The eighteenth century is even more recognisable, for we have street scenes and interiors as well as portraits. But of all the historical images available, it is photographs that have the most impact. One of the key reasons why the First World War is so much more meaningful to today's public than the Napoleonic Wars is because we can see photographs of the mud and the trenches, and the smiling troops on their way to the Front. We are familiar with the images of the corpse-strewn battlefields to which they were heading. When we see a colour photograph from the First World War, such as Paul Castelnau's autochrome of a French soldier in his blue uniform cautiously peeping across the top of a trench at the enemy lines, its realism strikes us with far greater force than the painted and engraved depictions of similar scenes in earlier battles.

Having said this, it is not on account of its future value to historians that nineteenth-century photography is discussed here as a significant change. It is rather because photography did more than any other form of illustration to reform society's image of itself. You could say it did for society what the mirror did for individuals in the fifteenth century. Think of it this way: if you saw a painting of a casualty from the Napoleonic Wars, you would have registered straight away that the artist had chosen to portray that particular man for a particular reason. The subject would almost certainly be an officer, and you could safely assume that the reason for the painting would have been to mark a conspicuous act of gallantry on his part. The painting itself would have taken considerable time to produce, and thus would only have been created after 'the moment' it depicts had passed: after our hero had ceased grimacing in pain and had had his wound dressed. Thus you would have understood that the artist had deliberately composed the picture, deciding how much of the wound to reveal and how much to conceal. In marked contrast, some of the photographs taken a hundred years later, during the First World War, show true horrors: the dismembered corpses of ordinary men and women among the wreckage of their houses, the earth spat up in the air in terrifying fury as a shell exploded, or a semi-naked woman blown in half with her baby when a mortar shell fell on a maternity hospital.[13] They show images of the instant of death as well as its immediate aftermath. They reveal the object in front of the lens whatever it happened to be, including many details that the photographer would not have noticed when he took the shot. Although there was of course still a great deal of intention in photographs, and many pictures were posed for propaganda and publicity purposes, people had come to believe that the camera captured the actual scene. Once the camera shutter had flashed open, the object itself told the story. It was no longer filtered through a genteel artist's imagination or memory.

Photography began with a series of pioneers working independently in the 1820s and 1830s: Joseph Nicéphore Niépce and Louis Daguerre in France, Henry Fox Talbot in England, and Hercules Florence in Brazil. From 1839, when the French government bought the rights to Daguerre's work as a gift to the world, daguerreotypes became the pre-eminent form of photography. They were hardly the sort of thing you'd slip in your wallet, being copper plates coated with silver nitrate and preserved beneath glass. Exposure times were long, too, so they were awkward to produce. Nevertheless, they proved immensely popular. For the first time, people who would never have dreamed of having their portrait painted could sit for a photograph. The daguerreotype was followed in the early 1850s by the glass-fronted ambrotype, and that in turn was succeeded from about 1860 by the tintype, on a metal backing, and the *carte de visite*, which allowed multiple copies of the same image to be produced on a card backing. This last image you *could* put in your wallet. Many middle-class men and women would hand out copies to family and friends in the same way that wealthy people in previous centuries had sat for miniature portraits to give to their loved ones.

The invention of photography would have amounted to little more than the puffing-up of middle-class pride if it had not been for the ability to publish photographic images. In this respect, the key invention was not Daguerre's method of photographing people in studios but Henry Fox Talbot's technique of creating negative images from which albumen prints could be made. The first photographic book, Fox Talbot's *The Pencil of Nature* (published in six parts, 1844–6), reproduced images of his family home, Laycock Abbey, as well as still-life images and landmarks such as the boulevards of Paris and the bridge at Orléans. Although it was not yet possible to mass-produce high-quality photographs, limited-edition technical books on botanical subjects could include albumen prints of photographs of specimens, not just engravings of drawings. In due course, the technology did allow for printed plates to be inserted into books, at which point the social sciences also benefited. John Thomson, who travelled extensively in the Far East in 1862–71, published views of China and Cambodia that showed English readers for the first time what a Chinese street scene looked like, or how the ruins of Angkor Wat, the great Cambodian temple, loomed out of the jungle. No amount of text could bring these things home to the reader so vividly. Although previously some travellers had produced incredibly detailed and beautiful steel engravings of the landscapes they had visited – the name of William Henry Bartlett leaps to mind – their images were still the result of artistic processes, not the action of the light from the object itself falling on a piece of silver-coated metal. They were not 'true' in the way that a photograph was true. While the public still had a huge appetite for steel-engraved contemporary images, as shown by the massive print runs of the *Illustrated London News* from the 1840s, this only indicates how much people wanted to see *any* image that related to a news story: if they could see a photograph, so much the better. Publications such as the *Illustrated London News* began to incorporate photographs – first in an engraved form, with the engraving trying to mimic the verisimilitude of the original image, and later, when the technology of the 1890s made it possible, in the form of a half-tone reproduction of the photograph itself. In the travel publications of the 1880s and

1890s, photographic plates depicting jungles, ruined temples and exotic people in faraway places became de rigueur. It was not enough to travel and tell the story of the expedition; you had to *show* the reader the marvels you had seen. Through such publications, European armchair travellers began to visualise the rest of the world.

By 1900 photography had become an essential component of publishing and journalism. The ability to create images that showed the actual scene in front of an observer's eyes increasingly conferred the obligation to do so. Images of the Crimean War taken by Roger Fenton were engraved for publication in the *Illustrated London News* in the 1850s. In the following decade, Mathew B. Brady employed a small cohort of photographers to document the American Civil War (1861–5) as it progressed; their work was published in engraved form in *Harper's Weekly*. With the images there to tell the story, reporting became more visceral too, describing in detail the scenes of battle. Photography also influenced public awareness of social changes in less violent contexts, such as the disappearing Native American way of life, or the housing conditions of people in slum tenements. Indeed, photography and textual journalism developed a common concern to depict the reality of social inequality and deprivation. Henry Mayhew's *London Labour and the London Poor* (1851) was a detailed textual description of poverty in London, pulling no punches in describing the unsanitary, hard conditions that people endured. John Thomson's *Street Life in London* (1878) revealed through images how the less fortunate eked out a living. Thomas Annan's *The Old Closes and Streets of Glasgow* (1872) recorded the character of the city's slums before they were pulled down by the city authorities. Jacob Riis's *How the Other Half Lives* (1890) combined text and images to show what life was like for the poorest inhabitants of New York's tenements, from hammocks in seven-cent boarding houses to tramps' sleeping places in basements.

Thus photography redefined our understanding of evidence and truth. It undermined the authority of the artist, whose storytelling images were appreciably more subjective than a camera. The witnessed moment could now be captured directly and shared amongst millions. Whether or not someone wanted to be in a picture became irrelevant. Crime scenes were photographed to preserve evidence of illegal deeds. Prisons kept images of all those who passed through their gates. 'WANTED' posters in towns on the American frontier carried mug shots of fugitives from justice. Police forces kept thousands of photographs of suspects. Before photography, criminals could only be identified by name, sex, eye colour and height, and there was no way of proving that one six-foot-tall middle-aged male with blue-grey eyes and receding brown hair was the same as another. From 1850 photography was also increasingly employed by scientists, especially in astronomy, with nebulae being photographed in 1880 and objects invisible to the naked eye in 1883. By 1900 the transformation was largely complete. A process of determining 'truth' that in 1800 had relied wholly on witnesses' perceptions and narrative skill had been displaced by a system that was predominantly based on objective evidence, thanks in no small part to photography.

Social reform

We have touched upon many different sorts of government in the course of this book but with the singular exception of Revolutionary France, they all had one thing in common: they saw it as their duty to protect their citizens from social change. They were conservative. In the years after the French Revolution, they became even more wary of political reform, tending only to allow it in order to preserve as much of the status quo as possible. In November 1830 the British prime minister, Lord Grey, justified presenting the Reform Bill to Parliament with the words: 'the principle of my reform is to prevent the necessity for revolution . . . The principle on which I mean to act is neither more nor less than that of reforming to preserve, and not to overthrow.'[14] When it was finally passed, the Great Reform Act of 1832 only slightly increased the franchise: from 516,000 men of property to 809,000 at a time when the population was 13.3 million.[15] It would take a while yet for the advocates of democracy to build up the pressure necessary to broaden access to the parliaments of Europe.

A major surge came at the end of the 1840s. A famine in 1846, which triggered an international financial crisis, led to widespread calls for reform. The French, who had briefly seen universal male suffrage introduced in the Revolution, now called for it to be reintroduced. When a reformist banquet in Paris was cancelled by the king in February 1848, thousands of protesters poured out on to the streets. The National Guard and the army joined them, and the king fled. A wave of revolutions followed across the whole continent: in Berlin,

Vienna, Budapest, Prague, Rome and many other cities. However, all
these uprisings were quashed: it seems that the restoration of universal
male suffrage in France was the revolutionaries' only lasting achievement. The reality was that the professional middle classes who were
the strongest advocates for change – lawyers, doctors and bankers –
were motivated by self-interest and wary of empowering the masses.
They certainly were not prepared to risk the anarchy of outright
revolution. Stable monarchies at least guaranteed their continued
enjoyment of their hard-won wealth and status.

In one important sense, however, the revolutions of 1848 did not
fail: they served as a powerful reminder to the forces of conservatism
throughout Europe that the events of 1789 could be repeated – and
not just in France. Every wave of revolution was like a high tide on
the beach – leaving a high-water mark to serve as a permanent
reminder of what might happen again. Even in Britain, which did not
see a revolution in 1848, there was increasing pressure for reform.
Foremost among the radical groups was the Chartists, a popular
movement that called for the adoption of a people's charter guaran-
teeing universal male suffrage. The year 1848 also saw the publication
of *The Communist Manifesto* by Karl Marx and Friedrich Engels, which
created a new intellectual framework for revolution. It set out in a
brief but influential form Marx's vision of the struggle between the
bourgeoisie and the proletariat over the course of history, and the
process by which a communist state could be established. It promulgated the idea that the worker – not land – was the primary source
of wealth, and that therefore the means of production should be
collectively owned by the industrial proletariat. After 1848, to many
workers, metaphorical Bastilles across Europe began to look
vulnerable.

Notwithstanding the impact of the events of 1848, the greatest
strides towards social reform in the first half of the century were not
made by revolutionaries but by single-issue campaigners. In England,
men like Edwin Chadwick gave themselves over to reforming the
living conditions of the poor. Anthony Ashley-Cooper, seventh earl
of Shaftesbury, devoted much of his career to improving the care of
the mentally ill and the conditions in which women and children were
employed in factories and mines. The name of the Irishman Daniel
O'Connell will forever be linked to the Roman Catholic Relief Act of
1829, after which Catholics were allowed to sit in Parliament.

Unfortunately, there isn't the space here to describe the many different initiatives to remedy cruelty, neglect and injustice. Therefore, we shall focus on four key aspects of nineteenth-century social reform – slavery, electoral representation, women's rights and education – which taken together should give an idea of how governments shifted from resisting social change to actively promoting it.

It is appropriate to deal first with the oldest and biggest social problem of them all: slavery. Enlightenment thinking had showed that slavery and natural rights were wholly at odds with one another. Yet Americans, who had declared in 1776 that they had a right to be freed from service to a foreign government, kept their own people in bondage. In 1780 blacks made up more than 20 per cent of the population of the newly independent United States, the majority of these being slaves. The awkwardness for the government of each state – and the reason for the hypocrisy – was that these slaves were private property, and private property was protected by the American constitution. It was the same contest between property and liberty that had led to slavery continuing in the Middle Ages. The state of Pennsylvania's proposal to abolish slavery gradually, by liberating the children of slaves, did not satisfy the enlightened thinkers and reformers who demanded an end to human bondage. In any case, only a few states adopted the same strategy; in the southern states of America and on other continents, men and women were only too keen to buy and sell humans. Feelings on the issue began to run high. In Britain, the Society for the Abolition of the Slave Trade was founded in 1787. Revolutionary France abolished slavery throughout its dominions in 1794, and despite it being reintroduced by Napoleon in 1802, all the signs were that it would soon be a thing of the past in the West. In 1807 Britain and America both outlawed the trade in slaves, thus bringing 300 years of the transatlantic slave trade to an end. In 1811 Spain abolished slavery throughout its territories. In 1833, after much pressure from William Wilberforce – who had first introduced a bill to abolish the slave trade into the British parliament in 1791 – slavery itself was abolished in all British-ruled lands, with the slaves immediately being set free. Parliament agreed to pay £20 million in compensation to slave owners. In America, the issue of slavery was a major cause of the Civil War in 1861. Following the victory of the North and the re-election of Abraham Lincoln, it was abolished throughout the USA in 1865. Four years later Portugal abolished slavery in all its colonies, thereby ending official endorsement of slavery in the West.

Electoral representation was the prime goal of middle-class campaigners in most European countries. In Britain, the Great Reform Act of 1832 ended the practice of a few landowners being able to place their men into the House of Commons without significant opposition. And, exactly as Lord Grey had intended, by relaxing the exclusivity of power just a little, the 'necessity for revolution' was averted. The Act bought three and a half decades' grace for the landed gentry, and the other two Acts that followed later in the century provided further extensions. You have to admit that the British landed interest did well in holding on to its political advantages as long as it did. Even after the third Reform Act of 1884, the electorate numbered only about five million men, out of a total population of 24.4 million. Other Western countries strode ahead of Britain in terms of electoral representativeness. By 1820 all white men could vote in America in every state except Rhode Island, Virginia and Louisiana. In 1870 all adult males in America were given the vote, whatever colour they were and whichever state they were from – although it should be noted that in some southern states, intimidation, beatings and other such measures were taken by whites to limit the ability of blacks to cast their votes. As we have already seen, all adult Frenchmen were given the vote again in 1848. By 1900 all adult males in Switzerland, Denmark, Australia, Greece, Spain, Germany, New Zealand and Norway were entitled to vote. Although it was not until the next century that the majority of Western nations adopted a policy of universal male franchise, the writing was on the wall.

What about votes for women? The first petition for women's suffrage in Britain was presented to Parliament in 1867; it was turned down. Calls for French women to have the vote were renewed at the same time – with similarly frustrating results. Some countries introduced limited voting for women: Sweden allowed tax-paying single women to vote in municipal elections from 1862. However, it was not until 1893 that a country extended the franchise to all its female nationals: this was New Zealand. South Australia followed in 1894, the rest of Australia in 1902, Finland in 1907 and Norway in 1913. Most European countries did not grant women the vote until after the First World War.

Why were women such a low priority in nineteenth-century

legislators' eyes? One reason was that governments felt that they had less to fear from women – they knew it was unlikely that hordes of females would man the barricades or that large numbers of men would take to the streets to demand the vote for their wives and daughters. But a much more important reason was the sexism that dominated Western society. Everywhere, women and men started from an unequal legal position. Therefore campaigners on behalf of women's rights had higher priorities than simply demanding the vote. Women were not allowed to attend university, practise a profession such as law or medicine, or hold public office. In English law – which was also the basis of law in America, Canada, Australia and New Zealand – the movable property of married women rightfully belonged to their husbands, including any money they earned. Women had to have their husband's permission to rent out or sell any houses or land they inherited. They were not allowed to make a will except with their husband's consent. A wife was not permitted to bring someone into the family home without her husband's agreement – and yet a husband was entitled to enter the houses of her kin in order to bring her back to the family home if she ran away. A man could legally beat his wife as much as he wanted, as long as he did not kill her. Wives were not allowed to give evidence for or against their husbands in court and could not legally leave their husbands. In remote parts of England, there was an unofficial form of divorce among the common folk that involved a man selling his wife to the highest bidder, normally just for a few pennies or shillings. Quite a few such wife sales are known to have been held in Devon.[16] When society believed a man could simply beat and even sell his wife, not having the vote seemed to many women an injustice of secondary importance.

Consider it from the point of view of Mrs Caroline Norton, née Sheridan, who declared, 'I do not ask for my rights. I have no rights. I have only wrongs.'[17] You might have considered her a lucky girl. She was beautiful, vivacious and had inherited the wit of her grandfather, the famous playwright Richard Brinsley Sheridan. But her father died in South Africa in 1817, when she was only nine, leaving the family penniless. For the next few years she lived with her mother and two sisters in a 'grace and favour' apartment at Hampton Court Palace. When the time came for her to be married, at the age of 21, her lack of dowry meant that her prospects on the marriage market were not good. Thus she accepted the proposal of the only suitor who paid her court: the Honourable George Norton, MP. The marriage was not just a failure, it was a tragedy. Soon after the honeymoon, her slow-witted husband started to hit her when they quarrelled. Their relationship was characterised by hatred and arguments about money. Her husband took their three young children away from her and refused to allow her to see them; one of them died not long afterwards. A small inheritance that came to her on her mother's death was taken by her husband, as was legally his right. She found refuge in her writing, from which she made a modest income, but this too disappeared into her husband's pocket. Beaten, publicly humiliated and ostracised, deprived of her children and forced to relinquish the fruits of her labours to the man who was the cause of her unhappiness, she decided to speak out against the system. She published a series of pamphlets on the injustice of removing young children from their mothers and succeeded in forcing Parliament to pass the Custody of Infants Act in 1839, by which mothers were given the right to look after their children until the age of seven. Having won success in this field, she then turned her attention to divorce.

At the time, the only way to obtain a divorce in England was to secure an annulment in an ecclesiastical court and then to obtain an Act of Parliament dissolving the marriage. It was a hugely expensive process that only the very rich could afford: on average just two divorces per year were granted in England between 1700 and 1857. Not only could the Nortons not afford to get divorced, it was entirely in her husband's interests not to grant Caroline her freedom. In 1857, partly due to her campaigning, Lord Palmerston pushed through the Matrimonial Causes Act, despite the opposition of the future prime minister, William Ewart Gladstone, and the bishop of Oxford, Samuel Wilberforce (son of the anti-slavery campaigner). Henceforth marriage became a secular contract that could be sundered in the divorce court, at reasonable expense.

The movement to alleviate the suffering of women, which Caroline had helped to start, soon gathered weight. In 1870 Parliament passed the Married Women's Property Act, which allowed wives to own property in their own right and to keep the wages they were paid. In 1878 Frances Power Cobbe published her seminal article 'Wife-Torture in England' in *Contemporary Review*, in which she related in distressing and graphic detail how working-class men regularly beat their wives, sometimes to death, for the most trivial reasons. It did the trick. The second Matrimonial Causes Act, which allowed women a divorce if their husbands were physically abusive, was passed that same year. By 1900 many of the legal injustices from which Caroline Norton and countless other women had suffered had been swept away.

It would be wrong to suggest that the British experience was typical of the Western world. While American states largely paralleled Britain in passing laws that allowed women to own their own property (in the 1840s), to manage it for their own benefit (in the 1870s), and to get divorced (most widely after the Civil War), most Catholic countries did not permit divorce until the twentieth century. Nonetheless, women everywhere campaigned for their rights. National Women's Rights conventions were held annually in America from 1850. The General German Women's Association was set up in 1865. The Society for the Demand for Women's Rights started in France the same year, and the following year, the American Equal Rights Association was established to campaign on behalf of all American citizens, regardless of race, creed or sex. In 1869 the National Women's Suffrage Association was established by Elizabeth Cady Stanton and Susan B. Anthony specifically to demand the vote for American women. Two years later a Women's Union started to agitate for equal rights in Paris. Anna Maria Mozoni founded the League for the Promotion of Women's Interests in Milan in 1881. In 1888, at a meeting of the National Women's Suffrage Association, the International Council of Women came into existence. It was about this time that the word 'feminism' started to be used. The ideas of liberty and equality, which for the last hundred years had inspired reformers keen to transform the relationships between men and their governments everywhere, were now claimed by women in their own campaigns for recognition.

Arguably the most significant breakthroughs made by women in the nineteenth century were those allowing them to attend universities and to gain professional qualifications. Although a few women had qualified as physicians and surgeons in the late sixteenth century, society had grown more intolerant of women in professional roles in the seventeenth.[18] No university, hospital or medical school anywhere in the world would admit a female student in 1800. This situation only slowly began to change. Oberlin College, a theological institution in Ohio, allowed women to attend classes from 1833 and awarded them degrees from 1837. In 1847 the wilful Elizabeth Blackwell managed to force her way into Geneva Medical College in New York, from which she graduated in 1849. Six years later the University of Iowa opened its doors as a co-educational university. In 1861 a 37-year-old journalist, Julie-Victoire Daubié, passed her baccalaureate at Lyons. In 1864 and 1865 two Russian women managed to pass the entrance examination to study medicine at the University of Zurich: one of them, Nadejda Souslova, successfully defended her doctoral thesis at examination in 1867. In Britain, Elizabeth Garrett Anderson educated herself privately and passed the examination to become a licentiate of the Society of Apothecaries in 1865. Subsequently she lectured at the London School of Medicine for Women, set up in 1874 by Sophie Jex-Blake. The first Frenchwoman to obtain a doctoral degree did so in medicine in 1875. Thus women gradually came to make the first significant breaches in the barricade protecting male-only higher education, mainly through the discipline of medicine. The first female colleges in Cambridge, Girton College and Newnham Hall, were founded in 1869 and 1871 respectively, although Cambridge University refused to confer degrees on women until 1948. University College London was the first British University to grant women degrees, in 1878, the same year that Lady Margaret Hall, the first Oxford women's college, opened. Swedish and Finnish universities admitted women from 1870, New Zealand universities from 1871, Danish ones in 1875, and Italian and Dutch ones in 1876. By 1900 women accounted for 16 per cent of all students at British universities and 20 per cent of those at Swiss ones, most of the latter coming from Russia.[19]

This catalogue of breakthroughs should not be taken to mean that women were accepted as equals in professional circles in 1900. The Society of Apothecaries that granted Elizabeth Garrett Anderson her licence amended its rules afterwards to prevent any more women gaining the qualification. The British Medical Association did likewise, barring women from joining its ranks for 19 years. Female physicians found many wards and jobs closed to them. Thus it was a moment of the greatest importance when in 1903, the Nobel committee, which had proposed granting a Nobel Prize to Pierre Curie alone for the work on radiation he had jointly undertaken with his wife, correctly added her name to the award. Marie Curie went on to receive a second Nobel Prize in 1911, thus becoming the first person to have been given two such awards. There could be no better advertisement for the general benefit to society of women's education. Ironically, the only reason Marie Curie had attended the University of Paris, and met her husband, was because the University of Krakow in her native Poland refused to accept female students.

While higher education was the essential proving ground for women, it has to be said that the women who made these break-throughs largely came from privileged backgrounds and had received a good basic education. For many people, social reform was not about getting degrees: it was a matter of having a clean water supply and enough food to eat, and learning how to read and write. Of these, education was the least important. This is why in 1800, when writing had existed for approximately 5,000 years, more than half the popula-tion of the developed world was illiterate.

The model for compulsory, free education was the Prussian system, which had been established in 1717 by Frederick William I and then developed further by his son Frederick II in 1763. The Austro-Hungarian Empire adopted the model in 1774. Horace Mann introduced the system to America in 1843, and Massachusetts (where Mann was the Secretary for Education) became the first state to make primary education free and compulsory in 1852. Spain followed the American example in 1857, and Italy did likewise in 1859. England and Wales saw school boards set up across the country as a result of the Education Act promoted by W. E. Forster in 1870, but not until 1880 was it made compulsory for children aged between five and ten to attend school. In France, Jules Ferry promoted the legisla-tion that made education compulsory in 1881. As a consequence, Europe and America both went from being largely illiterate to predominantly literate in the course of a generation. Some places such as Portugal remained a long way behind the pace, with 36.1 per cent male and 18.2 per cent female literacy in 1900, but France, whose male and female literacy rates had been about 47 per cent and 27 per cent in the late eighteenth century, reached 86.5 per cent and 80.6 per cent respectively in 1900. The comparable figures for the United Kingdom were a rise from 60 per cent in 1800 to 97.5 per cent in 1900 for men and from 45 per cent to 97.1 per cent for women. Literacy in the USA in 1900 stood at 89.3 per cent and 88.8 per cent respectively. This is a striking development – particularly as female literacy was almost as high as men's by the turn of the century. Indeed, in Canada, more women could read and write (89.6 per cent) than could men (88.4 per cent) in 1900.[20] There is no doubt that without the expansion of education, it would not have been possible even to entertain the notion of the equality of the sexes in legal, moral and financial contexts, or the equality of opportunity for all members of society, let alone set about trying to make these things a reality.

Conclusion

The nineteenth century presents us with an overwhelming tide of change. It contains within its temporal limits a whole series of astonishing transformations: from rural to urban living; from illiteracy to literacy; from agriculture to industry; from travelling on horseback to hurtling along a railway line at almost 100 m.p.h.; from sending a message from Britain to Australia in six weeks to sending a telegram in almost no time at all; from blind sexual prejudice to the ability for women to campaign for equality. So radical was the transformation of daily life that it is difficult to pick out any one major trend. If there is a single image that represents all the different factors, it is that of steam – steam trains and ships, mill and traction engines changed the world beyond recognition. But the most far-reaching change is that of social reform, or, as I described it earlier, the growing acceptance that one man is worth the same as any other.

One aspect of this shift that we have not yet discussed is the increase of leisure time. It is only when people stop struggling to feed and clothe their families that they have the time to take up games or hobbies. In the late nineteenth century, per capita incomes in Britain suddenly shifted upwards. Men and women increasingly took time off work to watch or play organised sports in the daytime and go to the theatre, music halls, classical concerts or the opera in the evening. They read novels, played the piano and even went on domestic and foreign holidays. If you ever wonder why many of the world's most popular team games – especially football, rugby and cricket – were invented in Britain, it is not just because so much of the world was ruled from London but also because British workers were the first to have sufficient leisure time to travel away from home to play regularly against other teams. By 1900 the working classes in many other parts of the Western world had enough time and money to down tools and kick a ball. If you cast your mind back upon the famine-struck centuries of the past, that is quite something in itself.

The principal agent of change

It is harder to select the principal agent of change for the nineteenth century than any other. Having discussed the matter regularly over the last few years, a shortlist of ten names has emerged: Alexander Graham Bell, Louis Daguerre, Charles Darwin, Thomas Edison, Michael Faraday, Sigmund Freud, Robert Koch, Karl Marx, James Clerk Maxwell and Louis Pasteur. Most people focus on the rivalries – between the inventors (Bell versus Edison), the medical researchers (Koch versus Pasteur) and the scientists (Maxwell versus Faraday). Mentioning Darwin's name often provokes the response that Alfred Russel Wallace had similar ideas about evolution and we only remember Darwin with greater clarity because his book *On the Origin of Species* (1859) became the focal point for the ensuing discussions about religion, science and evolution. Daguerre also has a strong rival in Fox Talbot, whose contributions to the development of photography were arguably more important. Over the years, the plethora of nine-teenth-century thinkers has led to animated discussions as to who was the most significant – to the point of table-thumping passion.

In my view, the two most deserving candidates are Charles Darwin and Karl Marx. My reason for considering Darwin relates to the matter of faith. As we observed in the chapter on the sixteenth century, whatever you believe, you believe in *something* – whether it's the creative power of God or the evolution of the species from a chemical accident in the primordial soup. However, there is a fundamental difference between believing that your existence is the consequence of God's will and believing that it is the result of a natural development unaffected by any spiritual force. You can pray to your Creator in good faith but not to an accidental conflux of chemicals in the aforesaid primordial soup. You can believe that a Creator requires obedience and worship; you cannot believe the same of the forces of evolution. If Darwin is to be credited with doing more than any other individual to destroy the belief that we can materially affect our circumstances on Earth through prayer to a spiritual power, then we should acknowledge his prime role in what is undoubtedly one of the biggest social changes of all time. But his agency in this respect is highly debatable. Church attendance in England had already dropped to 40 per cent by 1850 – nine years before *On the Origin of Species* was published. Moreover, readers were sufficiently sophisticated in their understanding of the Bible to realise that the inaccuracy of the Creation story in Genesis did not invalidate the other books, and especially not the New Testament. We should therefore see Darwin's impact on religion in a similar light to that of the physicians in the seventeenth century. Just because it was a medicine that made you better, and not a miracle, it did not mean that God had no hand in it. Whether or not you saw the hand of God in the medicine – or natural selection – was down to a personal belief system that was far more complex than the acceptance of a single scientific theory.

Karl Marx is therefore my choice as the principal agent of change of the nineteenth century. This is not because I see history as the struggle of class against class, or that I believe capitalism to be doomed to fail and the 'proletariat' bound to succeed – quite the opposite, as the last part of this book will show. Nevertheless, Marx conceptualised industrial labour as a historical force and helped create a mass movement for working-class emancipation that dominated politics from the third quarter of the century. His thinking was more than philosophical or economic commentary: it underpinned actual revolutions. Marx drove forward socialism, which, as George Orwell put it, has a 'mystique' of its own – the idea of a classless society – for which people are prepared to fight and die.[21] Marx's ideas led to the political organisation of labour, to workplace rebellions and industrial conflicts; they also triggered social welfare legislation that governments hoped would stem the tide of revolution. His vision of history as a grinding together of economic forces is persuasive, and while we might disagree with his predictions, on that particular point he was undoubtedly right. Socially and economically, we are bound by rules as old as society itself. And that understanding had a far greater bearing on the way people in 1900 looked at society and set about changing it than the important but abstract question of whether mankind was created or evolved.

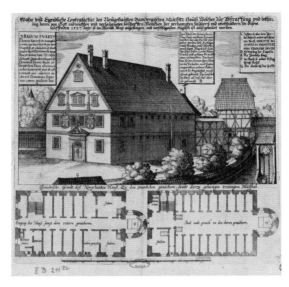

For all those who think of social history as a march of progress, the witchcraft terror of the late sixteenth and early seventeenth centuries carries a salutary message. The prince-bishop of Bamberg designed this building for the systematic imprisonment, torture and burning to death of witches.

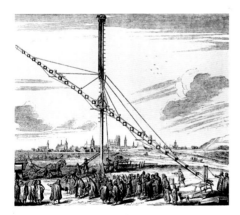

Initially the thinking was that the longer the telescope, the greater the magnification obtainable. This is Johannes Hevelius's 150-ft-long telescope that he built in Danzig (now Gdansk). It shows the lengths, literally, to which astronomers were prepared to go.

It was Isaac Newton who demonstrated to astronomers that, when it came to telescopes, size wasn't everything. Newton's reflecting telescope, although small, could magnify images 40 times.

A London opera rehearsal, painted by Marco Ricci in 1708. The gentlemen's wigs, the paintings on the walls, the assembly of instruments and the very performance of an opera reveal an elegant bourgeois event – something unthinkable in the London of a century earlier.

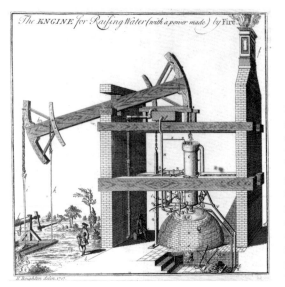

Dartmouth today is not what you would call an industrial hub but it was the birthplace of Thomas Newcomen, the man who invented and manufactured the world's first economically viable steam engine. This example, depicted in 1718, is one of 1,200 machines installed across Europe in the eighteenth century.

The Tennis Court Oath, after an uncompleted sketch by Jacques-Louis David. On 20 June 1789 all but one of the 577 members of the French National Assembly swore to continue to meet until the constitution of the kingdom was established. It was a seminal moment in the French Revolution – which became a testing ground for revolutionary ideas everywhere.

Thomas Allom's 1834 painting of power loom weaving. The Industrial Revolution developed from the need for individual businesses within the same industries to compete with one another. The power loom was invented by Edmund Cartwright in 1785; by the time of this painting there were 100,000 of them operational in Britain alone.

Plymouth to London in 32 Hours
WEAKLEY'S HOTEL, DOCK.
Extraordinary Cheapness combined with Safety
and Expedition.

Advertisement for the Plymouth to London stagecoach covering the 215 miles in 32 hours, from an 1822 local paper. Speeds of travel increased hugely even before the advent of the railways – this journey would have taken five days in 1700. The distribution of information also quickened: before 1700 there were no English newspapers; by 1800 several were being published daily in London, among them the *Morning Post*, the *Morning Chronicle*, the *Morning Herald* and *The Times*.

The rights to Louis Daguerre's method of photography were purchased by the French nation from him and given freely as a gift to the world. Hence 'daguerreotypes' became the most common form of photograph until the 1850s. This 1838 image of the Boulevard du Temple, Paris, is reputedly the first photograph of a person: the man having his shoes polished in the foreground and the shoe polisher were stationary for long enough to appear in the ten-minute-long exposure.

In England, the pre-eminent pioneer photographer was William Fox Talbot. This photograph shows Isambard Brunel's great ship, the *SS Great Britain*, being fitted out in Cumberland Basin in 1844. It was the first iron-hulled, screw-propeller-driven steamship, and the largest ship in the world when it was launched in 1843.

Orville and Wilbur Wright were determined to fly. First they experimented with gliders and then with engines. On 17 December 1902 this picture was taken by one of the witnesses of the first ever flight by a heavier-than-air machine, their *Flyer*, as it covered 120 feet in 12 seconds.

Autochrome photograph of a French soldier on lookout duty on the Upper Rhine, taken by Paul Castelnau on 23 June 1917. Photography in many ways undermined the authority of the artist: the realism of a photographic image of war or poverty is much harder to dismiss than a carefully composed painting.

Dr Nagai, medical instructor and x-ray specialist at Nagasaki Hospital, amid the ruins of the city after the nuclear bombing in August 1945. He died from radiation sickness shortly afterwards. In the twentieth century, war came to affect the whole of society, not just soldiers.

The Park Row Building, New York. Standing at 119 metres it was the tallest building in the world in 1900.

At 375 metres, the Petronas Towers – the tallest building in the world in 2000 – stood more than three times the height of the Park Row Building.

Earthrise, the first photograph of Earth from space and probably the most important photograph ever taken. It was snapped in a flurry of excitement on Christmas Eve 1968 by the crew of Apollo 8, as they were in orbit around the Moon. It provided an objective view of the Earth for the first time – and, disturbingly, it showed our world as a rather small, isolated planet.

The Twentieth Century

Early-twentieth-century photographs of my great-grandfather John Frank Mortimer show him wearing a business suit very similar to those worn today. Thinking about him, I realise that he had far more in common with me than he did with his own great-grandfather in 1800, of whom there are no pictures. He voted in national and local government elections. He spent evenings at the ice-skating rink as a young man, liked to bet on the horses, collected stamps as a hobby, married a woman from another part of the country, and his children (born 1904–8) played with teddy bears and dolls. He owned a bicycle and a gramophone, had a telephone, indoor lighting, running water and a range cooker in his home. Unusually, he also had a washing machine with a mechanically driven rotating tub, as the family business was dyeing and cleaning cloth. The streets in his home town of Plymouth were lit at night and patrolled by policemen. He and his wife Catherine took foreign holidays by railway and boat, and enjoyed excursions on Dartmoor at the weekends. They read a lot of novels between them, visited museums and attended public lectures by prominent and fashionable people. He lived to the age of 72; she to 82. Of course, there are differences between his life and mine. I can't skate, don't go to church, don't bet on the horses and have no interest in adding to his stamp collection. I haven't got a clue how to dye cloth, don't have any servants and my children weren't looked after by a nanny when they were younger; but otherwise the differences in our lives are only variations on similar themes. As a recreational activity I play guitar rather than collect stamps. I go to the cinema rather than the ice rink. I send emails instead of telegrams. The balances of work and recreation, need and desire, freedom and responsibility, solitude and sociability, education and experience were all much the same then as they are today.

Bearing this in mind, we might ask what really changed in the twentieth century – what changed *so much* that for many people it is inconceivable that any other century saw more change? It is a moot point. There is the story of a gathering of retired farmers in Somerset in the early 1960s debating the question of which invention in their lifetimes had made the most difference to working on the farm. The tractor, livestock trucks, the combine harvester, fertilisers, pesticides, the electric water pump, the electric fence and grain silos were all discussed. But everyone agreed that it was wellington boots that had had the biggest impact.[1] It is not always the most dramatic changes that make a difference to our lives, nor do they necessarily represent the greatest achievements. More than that, in the twentieth century, the things that we thought really marked a difference were those that involved comfort, efficiency, speed and luxury.

This should not surprise us. As we have seen, many of the most important and fundamental changes that ensured our survival occurred in earlier centuries. It was in the sixteenth and seventeenth centuries that violence in society most rapidly diminished, and we have been relatively safe since the mid eighteenth century, despite what stories of Jack the Ripper might lead you to believe. With regard to social reform, we saw at the end of the eighteenth-century chapter that a graph of the rights of the ordinary man in relation to his contemporaries in Western society would resemble an extended 's' shape. Many social changes in the West follow a similar pattern: a slow gradient at the start, a rapid escalation in the middle and a levelling off as the whole of society is affected and further change becomes harder or impossible. The development as a whole might be termed a 'civilisation curve'. We would see similar civilisa- tion curves if we were to chart the growing proportion of society able to eat a balanced diet, the increasing percentage of people who lived in a town, and the proportion of the adult population that had access to a car. Its shape is clearly shown on the right, in a graph illustrating the construction of the railways. As you can see, although peak mileage was not achieved until the 1920s, the changes in the twentieth century were minimal compared to those of the nine- teenth. Similarly, when we consider such phenomena as the food supply, urbanisation, literacy and homicide rates, while the zenith of the civilisation curve lies in the twentieth century, the steepest increments occur in earlier periods.

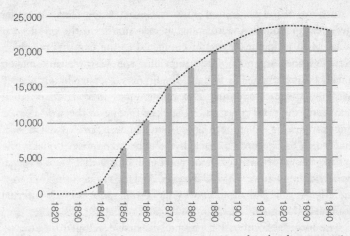

Miles of railway track in operation in UK and Ireland 1825–1940[2]

Having said this, feeding a European population of 729 million in the year 2000 was very different from feeding 422 million in 1900. Being separated from your family on the other side of the Atlantic was much more of a concern in 1900 than it was in 2000, when planes could cover the distance in a few hours. Racial discrimination in Britain was a less significant political issue in 1900 than it was in 2000 because there were very few people who directly suffered from it and even fewer who spoke out against it. Even more obviously, *not* having had an atomic bomb dropped on your country by 1940 (when such things didn't exist) was significantly different from not having experienced a nuclear conflict by the end of the Cuban Missile Crisis of October 1962, when it was a terrifyingly likely possibility. Context is everything when it comes to considering the twentieth century. It was not so much our lives that changed as the world in which we lived.

Transport

This book is concerned with 'the West' – and in the twentieth century, 'the West' spread like spilt ink across the world. Most countries came to embrace aspects of Western culture through huge advances in the nature, extent and integration of international transport links. Whether my great-grandfather wore a suit in 1900 similar to the one I wear today is an insignificant matter compared to the fact that people in New Zealand, Argentina, Japan and China were also wearing suits on a daily basis in 2000. The fact that the English language hardly altered over the twentieth century is a minor point in comparison to it becoming the third most widely spoken tongue in the world and a lingua franca for the globe. Markets that had been purely local in 1800, and which had become national by 1900, became international in the twentieth century. Forces of demand and supply that were international before 1900 were global by 2000. And the greatest single change underlying this globalisation was the proliferation of the internal combustion engine.

The internal combustion engine was first developed by Étienne Lenoir around 1860, and applied to a tricycle shortly afterwards. Its commercial potential was realised in 1886, when Karl Benz borrowed the money from his wealthy wife, Bertha, and patented his petrol-fuelled three-wheeled car. Karl might have been a great engineer, but he was a lousy salesman. In August 1888 Bertha decided to do something about her husband's shortcomings in the marketing department. She took one of his cars without telling him and set off from Mannheim with their two teenage children to visit her mother in Pforzheim, 65 miles away. The round trip of 130 miles was the first long-distance car journey ever made. People were astonished to see a horseless carriage trundling along the roads. When they saw it was driven by a woman, they were even more amazed. On the journey, Bertha acted as her own mechanic: clearing the fuel line with her hairpin and arranging for a shoemaker to improve the brakes by nailing leather on to the wooden brake blocks. But this trip was a great success: her expedition proved that the new invention was reliable. In 1894 Karl started to produce the four-wheeled Velo and sales took off. By 1900 his company was the largest manufacturer of motor vehicles in the world, turning out almost 500 cars per year.

The growth of the motor industry in these early years was astonishing. In 1904 there were already 8,000 cars, 5,000 buses and coaches and 4,000 goods vehicles on British roads alone. Motorcycles outnumbered cars in Britain from 1916 (153,000 motorcycles to 142,000 cars) until 1925, when the number of private cars caught up

(580,000 cars to 572,000 motorcycles). After the Second World War, the production of cars increased at an even greater rate. By 2000 there were 23.2 million cars and 825,000 motorcycles on British roads.[3]

Worldwide vehicle ownership was uneven, of course, especially in the early years of the century. But by 2000, numbers had increased universally. In 1960 there were 411 private vehicles per 1,000 people in the USA, 266 in Australia, 158 in France and 137 in Great Britain, but only 49 in Italy, 25 in Israel, 19 in Japan and 8 in Poland. Comparable figures in 2002 were 812 per 1,000 people in the USA; 632 in Australia, 576 in France and 515 in Great Britain. In the other countries, the imbalance had shifted: there were 656 cars per 1,000 people in Italy, 599 in Japan, 303 in Israel and 370 in Poland. Non-Western parts of the world still owned fewer cars: China had just 16 cars per 1,000 inhabitants in 2002, India 17 and Pakistan 12. Altogether the number of cars worldwide increased from 122 million in 1960 to 812 million in 2002.[4]

The other huge shift in twentieth-century transport was the advent of air travel. The first heavier-than-air flight – not using gas or hot air to produce lift – was the result of the Wright brothers' determined efforts. From 1899 to 1902 they experimented with the gliding aspects of flight before fitting a petrol engine and home-made propeller to their first *Flyer* in 1903. On 17 December 1903 Orville Wright took off – and covered 120 feet in 12 seconds. That is less time than it will have taken you to read from the start of this paragraph; nevertheless, those were 12 of the century's most important seconds. The Wright brothers remained in the vanguard of the development of the aeroplane for a decade, and frequently risked their necks in tests to improve the stability of their machines. By the end of 1905 they had made a flight of 24 miles. By 1908 they had even carried a passenger. In 1909 a series of contestants lined up in France to compete for a prize of £1,000 to be the first to fly across the English Channel: a race that Louis Blériot won on 25 July.

During the First World War many governments invested heavily in the development of aircraft when it emerged that they were by far the best means of quickly reconnoitring the enemy's advances and locating its artillery and naval vessels. But still aeroplanes were an unreliable mode of transport. Until the very end of the war they were too small and unsteady to be loaded with heavy explosives for effective bombing raids. After hostilities were over, the would-have-been bombers came into their own carrying long-distance post. And they finally breached the great barrier of the Atlantic. On 14–15 June 1919 John Alcock and Arthur Brown flew a modified Vickers Vimy non-stop from Newfoundland to Ireland. It was another Columbus moment that showed people the future: one day they too would fly between continents. Indeed, that same year, the first commercial airlines started business, flying passengers from London to Paris and back. Governments sponsored national airlines, especially when the threat of foreign companies dominating the domestic skies became clear. The first regular transatlantic passenger crossings were, however, by German Zeppelin balloons. Not until 1939 did an airline offer a regular aeroplane service between America and Europe.

Technical innovations during the Second World War in aircraft design, radio communications and radar swiftly added to the safety of flying. Rivalry between American airlines created pressure for planes capable of delivering profit. The result was the 21-seat DC-3, which first flew in 1935. It was now possible to cross the North American continent in 17 hours; crossing the Atlantic took 24 hours. Consequently, demand for air travel grew dramatically. The number of scheduled flights by British airlines between 1937 and 1967 increased from 87,000 to 349,000 per year. At the same time, the greater capacity of larger planes meant that the number of people carried rose proportionally far more: from 244,000 to 12.3 million.[5] In the 1970s 'jumbo jets' such as the Boeing 747, capable of carrying more than 320 passengers, further transformed the accessibility of flying. More and more people wanted to travel at the fastest speed possible, whether taking a domestic flight for work or going on holiday to an exotic island. And they also wanted to send goods and resources quickly around the world. Thus connections between different modes of transport were arranged for convenience: train lines were extended to sea ports and airports, airports were linked to major cities, and car parks were provided everywhere. The result was more or less the global transport network we have today.

The consequences of this transformation were huge – sometimes impressive, sometimes unsightly, and in some places traumatic. The world was commercialised, its political boundaries cut and re-cut by networks of trade and travel. Ancient balances of power were upset

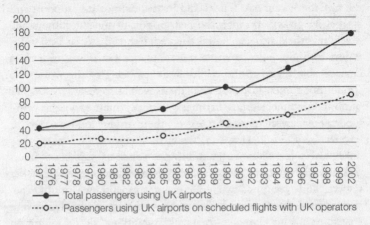

Number of passengers travelling through UK airports 1975–2000
(millions)[6]

by flows of Western capital and forced to readjust. Countries that previously had not realised they had resources of enormous value – for example, oil and uranium – were able to export them. Nations that dominated strategically significant transport routes were able to exploit their positions. The wealthy sectors of resource-laden and geographically well-positioned nations became rich on an international scale. The poorer sectors often grew more affluent too, as their social superiors spent their surpluses at home as well as abroad. Most countries grew more prosperous as trade unlocked the resources of the world.

All these transport links enabled machinery to be exported quickly and efficiently. They facilitated the distribution of artificial fertilisers and pesticides. They allowed agricultural surpluses to be shipped to markets in bulk. Any country that could afford to import produce was no longer vulnerable to harvest failure. The effect on the world's population was staggering. In the previous century it had grown by about 679 million – but 45 per cent of that growth had occurred in the developed world.[7] The population of the non-European world, which had remained between 307 and 356 million from 1200 to 1500 and then only grown at rates of 31 per cent in the sixteenth century, 18 per cent in the seventeenth, 37 per cent in the eighteenth and 60 per cent in the nineteenth, suddenly saw a 342 per cent increase in the twentieth century. Trucks and lorries did not just support this staggering growth through the distribution of food, fertilisers and machines; they also facilitated the rapid spread of medical aid – especially antibiotics – to more remote places. In this way, transport considerably added to the wealth and well-being of the world. We should not forget, however, that there were also losers as a result of this interconnectedness. In times of plenty, countries with insecure food supplies were tempted to export their surplus food cheaply, for low levels of profit. However, when they experienced crises, they did not have the wherewithal to buy the necessary supplies at international market rates, which reflected the scarcity of food in the region. Some countries where life was marginal therefore still experienced food shortages and famine. Nevertheless, many places were able to reduce or eliminate their vulnerability to harvest failure. In 2000 a much smaller proportion of the world's population was subject to periodic starvation than in 1900. The fact that in absolute terms there were more hungry mouths to feed was due to the fact that the world population increased dramatically – from 1.633 billion to 6.09 billion.

The transformation of the world's transport network, led by the West, resulted in the export of much more than just capitalism and business suits. Most of the Western world saw its own secular, democratic, materialistic, egalitarian, morally liberal values as the very peak of civilisation, and sought to confer these on the rest of the world. Businessmen and politicians in developing countries realised that it was sometimes to their advantage to adopt Western practices – or at least to accommodate them. Rapid modernisation forced the citizens of these nations to go through the same process of skills specialisation that had taken place in Europe and America in the nineteenth century. Many non-Western countries were therefore forced to come to terms with the Scientific, Medical, Agricultural and Industrial Revolutions in the course of just a few decades. It is no coincidence that as transport networks widened, food yields increased, population expanded, urbanisation increased and literacy rose. In 1900 only 13 per cent of the world's population lived in towns and approximately 20 per cent could read and write. In 2000 half the world lived in urban areas and over 70 per cent were literate. The whole world was forced to compete in a marketplace created by transport links and the movement of capital and goods. The only refuges from the march of capitalism were those nations where political barriers shut out economic competition, or the few countries where geographic barriers inhibited the advances of transport and global trade.

In an earlier chapter it was pointed out how transport was essential to the growth of a city. Obviously, food and other resources had to be transported into urban centres, and the larger the city, the more mechanised transport was necessary. Thus transport links were given priority over pedestrians, gardens, bicycle routes and individual living spaces, with dramatic consequences for our environment. By 2000 many people lived their whole lives in a metropolitan setting and rarely if ever saw what a rural region looked like, except in the cinema or on television. Half the world's population lived in a cityscape, coping with noise pollution, light pollution and air pollution, constantly harassed by the speeding cars and vans, buses and bicycles, motorcycles and lorries that thronged the multi-lane arteries of every city. Even at night transport dominated the landscape, through the stretches of barren tarmac that divided the blocks of human existence and the lights that flickered from red to green. Finding tranquillity in a city was difficult and frustrating. Many large conurbations required their citizens to travel for an hour or more before they could see something resembling the rural landscape of their forefathers. Perhaps as a result, many people started not to bother. Where once Edwin Chadwick had found slums and social deprivation, people in 2000 found everything they wanted – for both work and home life – and saw no reason to leave.

Transport systems, by extending the hinterlands of cities, focus attention on their centres. In the twentieth century property prices in the most desirable locations soared. In order to accommodate as many people as possible in such sought-after areas, architects began to build higher and higher, especially in commercial districts. The tallest building in the world in 1900 was the 986-foot Eiffel Tower in Paris – a non-residential showpiece. The tallest inhabitable structure was the Park Row Building in New York, at 391 feet. By 1931 the highest office in the Empire State Building towered 1,050 feet above the streets of New York, with its famous observation deck another 174 feet above that. Many buildings of a similar stature followed. In 2000 you could look down from the 1,230-foot-high top floor of the Petronas Towers over Kuala Lumpur – only a little higher than the observation deck of the Empire State Building but also only marginally taller than a mass of other skyscrapers. In the 1970s the skyline to the west of Paris was punctured by the construction of several dozen buildings over 300 feet in the business district of La Défense. The tallest building in Britain in 1900 was the Midland Grand Hotel (263 feet) at St Pancras Station in London; in 2000, it was the capital's somewhat taller One Canada Square (770 feet).

Transport transformed our social relationships as well as our physical surroundings. As explained in the last chapter, communities were progressively torn apart by the railways. The car inflicted further damage. The rise of cheap road transport between 1945 and 1960 killed off many local railway lines and left thousands of small towns with neither a market nor a rail link to the nearest commercial centre. Many people in rural areas were forced to live a more isolated existence or were increasingly dependent on their cars. It became difficult for older folk to continue to live in the homes they had known all their lives. They were forced into towns because they were too old to drive. At the same time, their grandchildren were forced to move to cities because they had to find work. By 2000 the majority of people in the West lived amongst strangers. Where once they would have known as many as 300 people in the three or four square miles around their home, now their kin and acquaintances were spread between many different towns and cities, often across the globe.

A similar process of estrangement took place between the providers of goods and services and their customers. In smaller communities, tradesmen were known to their potential clientele, and any failure to deliver was likely to result in damage to their reputation. In large towns and cities people could certainly draw on a wider range of suppliers but the weaker relationships between customers and the various businesses trying to undercut their rivals did not necessarily deliver the best service. When this point is applied to health matters, the difference is most telling. In a local community, where people had grown up together and lived among their own kin, the support networks for the ill and infirm tended to be much stronger than in cities, where community support was institutionally organised, expensive and impersonal.

Finally, transport not only delivered food worldwide and distributed the tools and equipment to farm the land more efficiently, it also led to agricultural specialisation. Global transport links forced farmers to compete internationally. Why should people in England buy wheat from hilly Devon if it could be shipped in more cheaply from Kansas? In order to maximise efficiency, farmers rapidly gave up the mixed farming they had practised since the Agricultural Revolution and concentrated instead on one form of production, such as growing wheat or rearing cattle. In America, intensive farming was practised after the mass production of the tractor between 1910 and 1920. In France, many regions were given over wholly to producing wine. In England, hill farms concentrated solely on meat production. By 2000 there was not one arable field nor a single dairy farm in the whole 12 square miles of the parish of Moretonhampstead, even though agriculture was still the second largest employer (after tourism). All the farms produced only beef and lamb. For the first time in history, most countries in the developed world gave up the possibility of being self-sufficient for their food and imported at least some of their requirements from cheaper overseas producers. In the 1950s Britain produced less than 40 per cent of its own food.[8] And this, of course, was the figure for the nation as a whole: urban areas were entirely incapable of self-sufficiency. Only when the country joined the European Economic Union in 1973 did its self-sufficiency recover, to over 70 per cent.

By 2000 the West was completely dependent on its transport infrastructure. Fossil fuels were required to operate the machinery that produced food for the nation and also to power the vehicles that distributed it. Every nation's fate rested on the steady production of petrol and diesel in a dozen or so refineries (just seven in the case of the UK). It is a salutary thought that if we were to lose them now, despite all the improvements in food production since the Agricultural Revolution, the majority of the population of the West would face starvation for the first time in 200 years.

War

It goes without saying that the conduct of war in the twentieth century altered beyond recognition. At the start it still lived up to its traditional definition of soldiers 'advancing in tidy ranks and dying in untidy heaps'. Tanks, trenches, poison gas and barbed wire soon changed the shapes and sizes of those heaps. The importance of fighter aircraft, bombers, submarines and the atomic bomb in the Second World War further changed them, often leaving no heap at all. In the second half of the century, the fear was that, at best, whole cities would be vaporised, or at worst, that everyone in the world would slowly die of starvation or radiation poisoning brought on by a nuclear holocaust. By 2000, however, it was clear that many traditional elements of war remained, and that in some civil wars, instant vaporisation might be a blessing. The atrocities perpetrated during the fighting in former Yugoslavia, in particular the rape and sexual torture of civilians and children, made it clear that the extreme brutality of previous centuries had not abated. As we saw when considering the decline of private violence in the sixteenth century: if you take away the potential violence of a higher power – the force that stops people attacking each other – they are liable to revert to a more aggressive, uncivilised state. Mankind remains potentially as cruel and inhumane as ever.

The point here, however, is not how warfare itself changed in the twentieth century but how war changed Western society. Here we need to look at many changes collectively, as we did in considering the impact of the Black Death. There were technological outcomes from warfare that had a significant bearing on civilian life. The Second World War in particular contributed directly to technologies as diverse as computing, the jet engine, cardboard milk containers and radar, and the development of penicillin and pesticides. The technology behind the German V-2 rockets led not only to modern military missiles but to the ability to fire a rocket into space and the advances of modern astronomy. It is arguable that any or all of these developments might have happened in the fullness of time, and that war was not so much a cause as an accelerator. Nevertheless, it hastened a great many developments simultaneously and contributed hugely to the transformation of life in the late twentieth century. There were of course social and economic consequences to the major wars, and global political repercussions too, including the institutionalisation of international relations. But the point at which to begin is the most important change of all, namely the fundamental transformation of the relationship between war and society.

Warfare before 1900 had normally only affected civilians where they either worked in munitions factories or lived in close proximity to the front line or in places through which armies and their supplies passed. The countries engaged in the First World War saw the advent of what historians have termed 'total war', in which the resources of an entire nation are devoted to the war effort. Human resources were deployed where needed for the pursuit of military ends. Social barriers were suspended to maximise the production of the war-orientated industries on the home front. Rationing was introduced, and transport systems were redirected to increase the efficiency of military supply lines. In reality, the 'totality' of warfare in the twentieth century went much further than these socio-economic arrangements suggest. Advances in aviation made everyone a potential target. Over a thousand Londoners were killed by aerial bombardment from Zeppelins in the First World War, and more than 28,000 died during air attacks on the city in the Second World War. Bombing raids over Warsaw, Rotterdam and many towns in Germany destroyed hundreds of thousands of lives and millions of livelihoods. They created rubble-strewn wastelands in the centres of towns and in some instances ignited wind-blown infernos from which there was no respite for any person, animal or national treasure. The RAF raids on Hamburg and Dresden were particularly horrific, killing nearly half as many civilians in those two cities as were wiped out in Hiroshima and Nagasaki in August 1945. The use of atomic bombs was itself another 'Columbus moment'. Two years later, atomic scientists in Chicago set the 'Doomsday Clock' at seven minutes to midnight in an attempt to draw wider public attention to the likelihood of mankind bringing about the extinction of human life on the planet through techno-logical means. In 1953 the clock was set at two minutes to midnight. By the time of the Cuban Missile Crisis in October 1962, world leaders were openly discussing the 'abyss' of nuclear warfare. In the event, the red button remained unpressed, but the potential violence was still there. Everyone on the planet was at risk of becoming a victim of war, even if he or she did not live in or near one of the belligerent nations.

The increasing deadliness of warfare is surely the greatest irony of human civilisation. Previous concepts of the end of the world had been inspired by Biblical stories of the Flood and the Last Judgement. It is hugely ironic that science – which had gradually replaced religious explanations of the world and was frequently used to undermine religious teaching – found practical ways to bring about the mass extermination and horrors that the Bible foretold. Even more ironic is the fact that scientists had devised this potential Armageddon deliberately, in response to the requirements of democratically elected leaders. Over the centuries the combination of absolute monarchical power, social hierarchy and religious doctrine had resulted in many wars and atrocities but it had never threatened to wipe out human life completely, as the alliance of democracy and science did in the late twentieth century. And yet the greatest irony of all is that most people in the West benefited from the escalating scale of warfare because it placed a new value on the individual. Indeed, total war – particularly in the first half of the century – brought with it many social and economic reforms that massively increased the political power, equality of opportunity and standard of living of the West's citizens.

Equality of opportunity was particularly pertinent at the start of the century for women, who still had some way to go to achieve social and economic parity with men. Labouring in munitions factories in the First World War permitted women to enjoy a great deal more freedom than they had previously known. Many found themselves employed for the first time, in sole charge of their household, and able to travel freely without a male companion. No doubt it led to a million and one private little battles when their husbands returned from the war – if they *did* return – but it was generally accepted that women had earned their greater freedom. The social pressure for the enfranchisement of women became overwhelming. In the aftermath of each world war there was a small rush of national Acts extending voting rights to women and any men who did not yet have them. The United Kingdom extended the franchise in 1918 to all men over 21 (19 if they had fought in the war) and women over 30 who were married, owners of property, or university graduates. Poland, Czechoslovakia, Austria and Hungary enfranchised their women the same year; the Netherlands followed in 1919, and the USA and Canada did likewise in 1920. The British government finally gave women the vote on the same basis as men in 1928. In 1944–5 the contribution made by women in the Second World War led to their enfranchisement in France, Bulgaria, Italy and Japan; Belgium followed in 1948. That greater political and economic independence of men and women heralded the decline of the servant class. Large houses that once had been staffed by cohorts of servants were closed up, often never to reopen. Just as the shortage of labour after the Black Death had accentuated the value of every labourer, so total war forcibly reminded society that every adult was useful, and led to the universal freedom to work and earn, vote, and (in the case of women) live more independent lives than before 1900.

Another social consequence of war was a turning of the tide against imperialism, monarchy and hereditary power generally. At the start of the century half a dozen empires ruled most of the globe. The largest was the British Empire, which included Canada, Australia, New Zealand, about two fifths of Africa, India, British Guiana and a number of Pacific islands. The French Empire consisted of huge swathes of North and West Africa, Vietnam and Cambodia, as well as colonies in India, China and the Pacific. The Russian Empire stretched from the Pacific Ocean to the Black Sea. Although far smaller in scale, the German Empire included territories in Africa and the Pacific. The Austro-Hungarian Empire encompassed not only the heartlands of Austria and Hungary but Bohemia, Slovenia, Bosnia and Herzegovina, Croatia, Slovakia and portions of Poland, the Ukraine, Romania and Serbia. The Ottoman Empire included Turkey, the Holy Land, Macedonia, northern Greece and Albania. One way or another, all these empires came to an end. The Russian Empire was overthrown by revolutionaries in 1917. The German, Austro-Hungarian and Ottoman empires were broken up in the aftermath of the First World War. Both the British and French empires underwent a process of decolonisation, hastened by the economic stresses suffered by each nation in the Second World War. As for the monarchies, at the start of the century, France, Switzerland and America were the only large Western countries that were not ruled over by a hereditary sovereign. Although many were constitutional monarchies, most kings and regnant queens still exercised a huge influence over their governments. By 2000 there were few kings left. The British, Belgian, Danish, Dutch, Norwegian and Swedish royal families still clung to their thrones, and the Spanish royal family had been restored after the demise of General Franco in 1975. But even in these nations they were subservient to democratically elected governments. Aristocratic power had similarly been almost entirely eradicated: even the British House of Lords was predominantly composed of appointed members after March 2000.[9]

The long process of holding hereditary rulers to account, which had begun back in the thirteenth century, had finally resulted in the near-extinction of the species.

The massive scale and horrific nature of modern warfare contributed directly to a series of attempts to develop international law and multinational organisations to limit the possibility of future conflicts. Even before the end of the First World War, philanthropists and politicians in England, France and the USA were proposing ways to limit conflict through international arbitration and the imposition of sanctions on aggressive states. The League of Nations was set up as part of the peace negotiations at Versailles in 1919. It proved a failure for a variety of reasons. It excluded the newly communist Russia, and failed to attract many other states, including the emerging economic superpower of the USA. It had no army and very little authority, as all countries on the council had a veto and were unwilling to take action against their potential allies. Its complete inability to do the most important thing it was set up to do – to prevent another world war – was demonstrated in 1939. However, its 20-year existence and modest successes along the way did create the sense that international relations could be institutionalised. The League's successor, the United Nations, was established in October 1945, also primarily in order to prevent a recurrence of international conflict. The actual activities of the UN have gone much further, of course, and it has involved itself in the social and economic well-being of people throughout the world. It also maintains the International Court of Justice in The Hague. Whereas the League of Nations never managed to bring together more than a quarter of the world's nations, the UN includes almost every sovereign state. War, you could say, brought the world closer together in the twentieth century. It also resulted in the revision of the Geneva Conventions in 1949 to protect civilians, medical staff and other non-combatants in war zones, as well as wounded and sick soldiers and shipwrecked mariners. These conventions, originally drawn up in 1864, represented the first attempts to limit the cruelty of war through an international moral code since the Peace of God and the Truce of God movements in the eleventh century. It is interesting to reflect on what this says about mankind's desire for peace and our seemingly uncontrollable propensity for violence.

Life expectancy

As we have just seen, the relationship between modern warfare and society is laden with irony. A particularly clear example is the fact that war had a positively beneficial impact on health. Obviously the people who were shot, starved, gassed, shelled, burnt or blown to smithereens would not see it quite like that, but the truth of the matter is that wars require healthy, fit populations to fight in the front line and to operate the munitions factories, trains and sources of food production back home. The First World War saw a rapid escalation in government care for the health of the workforce. Occupational health and safety became a significant issue, with lead poisoning, mercury poisoning and anthrax being the first ailments to be regulated, followed by illnesses such as silicosis and skin cancer. The same war saw the advancement of blood transfusions on account of the discovery of the first anti-coagulants, sodium citrate and heparin, and the mass production of chlorinating Lister packs with which to purify water. It saw the production of ammonia on an industrial scale for the first time, allowing for the manufacture of artificial fertilisers – originally to fuel the German war effort but later to feed the world. The psychological health of men suffering from shell shock forced the authorities to invest in research in mental health and care for sufferers. As for the Second World War, it saw the first antibiotic remedy, penicillin, mass-produced in readiness for the D-Day invasion of France in 1944. Today we take such medical innovations for granted but it is important to remember that before the introduction of antibiotics, simply grazing your elbow or knee could lead to blood poisoning. From 1944 a raft of diseases from meningitis to gonorrhoea were suddenly treatable. That little mould noticed by Alexander Fleming in September 1928, from which he developed penicillin, turned out to be one of the most important lifesavers of the modern world.

The medical discoveries made during the two world wars were hugely significant but so were the medical and social advances in peacetime. Countries in the West developed national healthcare systems with large public subsidies. National systems of pensions for the aged and infirm were organised, reducing the deleterious effect of advancing years on the poor in particular. Social reforms such as unemployment and disability benefits raised the standard of living for

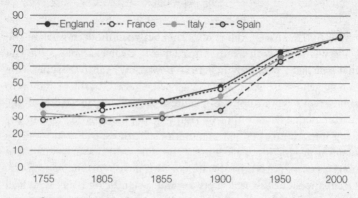

Life expectancy at birth in England, France, Italy and Spain

the needy far above their equivalents in the nineteenth century. Nursing, midwifery and obstetric surgery improved universally, resulting in a dramatic drop in infant mortality. In Britain, this amounted to a decline in stillbirths from 14 per cent of all confinements in 1900 to 6.3 per cent by 1930; and to 0.58 per cent in 1997. Neonatal mortality dropped from 3.2 per cent in 1931 to 0.39 per cent in 2000.[10] At the same time, maternal mortality plummeted. In 1900 about 42 mothers died per 10,000 confinements in Britain and about 80 per 10,000 died in the United States. By 2000 across the developed world the mother died in just two confinements per 10,000.

The above graph could be replicated for almost any country in the developed world. In 2000 male life expectancy at birth was 75 or more in Australia, Canada, France, Greece, Iceland, Italy, Japan, New Zealand, Norway, Singapore, Spain, Sweden and Switzerland. In the UK it was 74.8 and in the USA 73.9. Female life expectancy at birth was 80 or more in all the above countries as well as Austria, Belgium, Finland and Germany. In the UK it was 79.9 and in the USA 79.5.[11] Babies throughout the developed world could expect to live almost twice as long in 2000 as they had done in 1900. Of course, the lives cut short in infancy in the early part of the century are a distorting factor in gauging how much longer an adult might live, but even so, active life increased significantly. In 1900 the average American 20-year-old could expect to live another 42.8 years; in 2000 he or she could look forward to another 57.8 years. That meant 15 years more output: 15 years more experience for every scientist, physician, clergyman, politician and academic. This amounted to a massive expansion in the return on training. People lived into retirement, so that the less productive years – roughly the last 10 per cent of an individual's life – did not detract from a career.[12] A doctor who qualified at the age of 25 in 1900 who enjoyed the average life expectancy of 62.8 years and who did not work or only worked part time for the last 10 per cent of his life because of ill health would have had an active career of 31.5 years. In 2000 this man could have continued working full time for an extra 14 years, until he reached 70. But arguably, relieving people of the fear of dying young was an even more significant change. Who at the age of 56 would not welcome another 15 years of good health? You could quite reasonably argue that this prolongation of active life is one of the most significant changes we have covered.

The media

Pick up a copy of a newspaper such as *The Times* from 1901 and you will immediately notice the lack of a banner headline: the front page is given over almost entirely to small-print advertisements. This will probably strike you even more forcibly than the lack of photographs. A popular British newspaper like the *Daily Mail* had larger and more varied text, but its front page also lacked photographs and was dominated by advertisements. By 1914 the situation was changing. While the front of the *Daily Telegraph* was still riddled with small ads, other papers were prioritising the big news. To our eyes, the *Glasgow Evening Times* for 2 August 1914 looks almost modern, with its headline: THE WAR CLOUD BURSTS. The header text beneath reads: THE DECLARATION OF WAR. GERMANY'S GRAVE RESPONSIBILITY. WILL BRITAIN FIGHT? FATEFUL CONFERENCE TODAY. American newspapers by this time had not only banner headlines but half-tone photographs. The shift to these more eye-catching and opinion-forming models forced politicians during the First World War to pay attention to the newspapers' messages and their ways of selecting and presenting the news. Journalists increasingly had access to political leaders, who not only needed the support of the newspapers to get themselves re-elected, but also wanted their policies and decisions to be reported

in a certain way. It was the beginning of an intimate yet uneasy relationship between power and the press.

Alongside the growth of mass-market newspapers, the film
industry provided both entertainment and news. The first films had
been shown in Britain in the final years of the nineteenth century
and their popularity had spread rapidly as they were screened at
fairs and in music halls. In 1906 the daily viewing figures were about
4,000 in London alone. The first purpose-built cinemas in Britain
opened in 1907, a year in which 467 British films were released. As
a result of the 1909 Cinematograph Act, cinemas had to be licensed
by local authorities, but this did not lessen the rapidity with which
venues opened. By the end of 1911 the town of Plymouth (population: 112,030) had at least a dozen cinemas, most of which seated
more than 300 people.[13] In 1910 the showing of the first newsreels
brought images of current events to towns up and down the country.
In the 1930s people were exposed to a welter of current political,
social and moral messages, as voiceovers could deliver information
far faster than captions. By 1939 the UK saw 19 million weekly visits
to the cinema. Thirty-one per cent of cinema-goers went every week,
13 per cent twice a week, 3 per cent three times a week and 2 per
cent four times; only 12 per cent of the population never set foot in
a cinema.[14] If that seems impressive, consider this: attendance
increased by two thirds over the course of the Second World War,
reaching a peak of 31.5 million weekly cinema visits in 1946. That is
the equivalent of every adult in the country going to the cinema at
least once a week. With newsreels shown before many feature films,
people from one end of the country to the other were simultaneously fed with news. But the cinema was not just a means of
spreading information more widely and rapidly, it also created international stars whose faces were instantly recognisable by millions
and whose opinions were taken seriously by their admirers. Together,
mass-circulation newspapers, magazines and films focused national
attention on key moral and political themes and bound the country
together in a series of national debates. If nations were more thoroughly integrated in the twentieth century than ever before, the
media were an important part of this process.

The inventions that took that integration a stage further were radio
and television. Starting in the USA with local broadcasts relaying the
results of elections in 1920, radio stations began to deliver the news

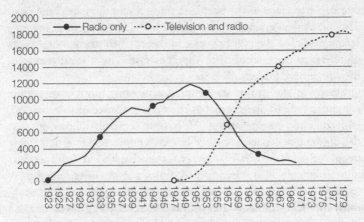

UK radio and television licences purchased annually (thousands)[15]

faster than ever before. The BBC, the world's oldest national broadcasting organisation, was founded in 1922. In 1934 it added regular television broadcasts to its radio output – although television was suspended during the war (as it was in France and Russia). In 1946 there were more than 10 million households in the UK with a radio licence. In 2000 23.3 million households were licensed to receive TV and radio programmes.

By 1960 every country in western Europe had regular television broadcasts and many Latin American and eastern European countries were starting their own services. Although the delivery of news was not in real time or continuous, the ability to interrupt programmes with news flashes meant that important information could be broadcast to the nation almost on the spur of the moment. National events were sometimes broadcast on several channels at once. Millions watched the same soap operas and discussed the moral dilemmas arising from the storylines. TV programmes raised issues that the entire viewing public felt obliged to consider. Strikes, marches and social protests came to national attention through regular bulletins. Ethical matters such as bullying in schools, homophobia and the inequality of pay between the sexes became matters of national debate through drama as well as reporting. Even if you happened to grow up in a remote, all-white community, the message was forcefully circulated that racism was deeply damaging, divisive and morally wrong. Trends set in the capital one day were seen as the next best thing across the nation the next. The media thus gradually wove people's awareness together in every country. In 2000 the rich financier in the city had many more points of reference in common with the poor farmer in a rural area than their respective forebears had had in 1900.

Finally, at the end of the century, a growing number of people started to look to the Internet as their prime source of information, education and entertainment. We have to remember, however, that the digital age arrived very late in our scheme – so late that many people in 2000 had barely been touched by it. The Internet was only formed in 1969 when four American universities linked their specially designed computers. Even though dozens of research establishments soon joined them it was not until the growth of Tim Berners-Lee's World Wide Web, which went live in August 1991, that its potential to become a public media channel was recognised. The fact that it was a royalty-free system led to its extraordinary growth. Just as with the railways in the nineteenth century, there was an immediate rush of excitement that failed to yield profits, yet by the end of 1995, over a million websites were active worldwide. By December 2000, 361 million people were using the Internet: 5.8 per cent of the world's population. In the United Kingdom, 28 per cent of adults had access to the Internet at home.[16] In its first nine years the World Wide Web had a significant impact, drawing the world together in communication just as newspapers, films, radio and television had done on a nation-by-nation basis earlier in the century. By 2000 it had not yet resulted in sufficient online shopping for people to discuss the demise of the high street; nor had it had the political networking impact that would result in the Arab Spring of 2010–11. Nevertheless, people could see how it would change the world in the not-too-distant future.

Electrical and electronic appliances

While I was writing this book, my family and I stayed briefly in an old cottage in Suffolk. One night, storms brought the power cables down and over the next few days the electricity companies struggled in vain to restore the supply. It provided a salutary reminder of how dependent we are on electricity. The cooker in the house was electric, so there was no means of heating anything up – not even the water for a cup of tea. The kitchen utensils were rendered useless. We were deprived of all forms of entertainment and communication as the television and radio fell silent, and after a short while, our laptops lost their charge. The vacuum cleaner became a glorified dust box. The fridge freezer lost its cool – in both senses of the word. We could not take a shower or a bath. The dishwasher, washing machine and tumble dryer were all out of service. Most distressingly, so was the coffee machine. I don't use an electric shaver, and I sadly have no need for a hairdryer, but these too would have been denied me. Of course, there were no lights either. As I sat jotting down ideas for this book by candlelight, I reflected on the whole electrification of our lives.

At the start of the twentieth century, there was only one common domestic electrical appliance – the light bulb. But even that was only to be found in a minority of homes, as many people still used gas lights. After the First World War, however, companies started to advertise a rapidly growing stream of electrical gadgets for the home. The electric kettle, which was first produced by Compton & Co. in 1891, came into its own in 1922, when the Swan Company manufactured a device with an internal heating element.[17] As mentioned in the previous chapter, gas cookers had existed since the nineteenth century but had not sold in large numbers due to the need to be connected to a supply. However, as the electrical grid spread through towns in the early twentieth century, the occupants of new flats and houses switched to electric cookers. Housing estates were built in the 1930s with electricity already connected and a pre-installed electric cooker for the proud new homeowner. The first commercially successful fridge went on the market in 1927, allowing fresh food to be preserved much longer than previously. By this time, the marketing of electrical appliances was firmly directed at women who were expected to do the cooking, cleaning and housework. A sales catalogue printed in January 1935 shows a young woman on the front with the caption 'Every housewife wants Magnet household labour-saving electrical appliances'. The inside of the catalogue offers two types of kettle, a toaster, an iron, a hairdryer, an upright vacuum cleaner, an upright floor polisher, six types of electric fire, a cooker, a 'wash boiler', a 'stimulator' (a sort of exercise machine) and a car engine radiator (for putting under the bonnet in cold weather, to prevent damage).[18] A flood of other electrical appliances followed. By 1970, almost every home was packed with such gadgets and many more besides: audio and television equipment, electric drills and other power tools, electric blankets, juicers, alarm clocks, Teasmades, lawnmowers and so on.

Most of the appliances in daily use had been refined into a reliable form by 1970, only changing gradually thereafter. However, the 1970s saw a shift in the sort of gadgets that you might buy as the microchip crept into our consumer goods. The first ones I encountered as a boy were in pocket calculators. Less than ten years after seeing my first computer in the late 1970s, their use in schools was obligatory: all my undergraduate essays from 1986 had to be word-processed. By 2000 microchips were in everything from car dashboards to children's toys. They also hugely increased the reliance on electricity in the workplace. In the office of the 1960s you might have found a telex machine and banks of electric typewriters. In the 1970s the photocopier became common, as did cassette-storage dictation machines, the fax machine, shredders, pocket calculators and, at the end of the decade, computers. By 2000 desktops, printers and scanners were de rigueur. Governments and businesses had mostly ditched their antiquated paper-based filing systems. And with the advent of the Internet, there arrived new systems for the storage, manipulation and distribution of data.

You could say that all this makes relatively little difference – that electrical gadgets, computers and fax machines didn't change the nature of what we do, they just allowed us to do the same thing more quickly. Switching on an electric heater was considerably faster than lighting a coal fire in a hearth every morning but not very different in its effect. Sending an email was much the same as writing a letter and having it delivered immediately, rather than the next day. Where was the change in that? But speed mattered. Labour-saving devices in the home and the office allowed more time to be spent working or producing. Information could be transferred almost instantly; it did not have to be copied out prior to being sent. Large datasets could be consulted in a fraction of the time it took to peruse a card index, especially if you had to read the awkward handwriting of a predecessor. The twentieth century saw a significant change in how much we could accomplish and the types of intellectual tasks we could tackle, largely due to electronic devices.

Sitting in that cottage in Suffolk in the glow of a candle made me think of other aspects of our increasing dependence on electricity in

the twentieth century. Domestically we underwent a similar process
to the de-skilling that the working class experienced in factories in
the nineteenth century. Before the Industrial Revolution anyone
working in a workshop made their own tools as well as their own
products – it formed part of a young man's apprenticeship. A wheel-
wright would know how to instruct the blacksmith to produce everything from the type of plane he needed to the iron tyre to fit the rims
of his wheels. The majority of men learnt the carpentry skills they
needed to mend the doors and shutters of their houses, or to make
new furniture for their growing families. But when factory owners
introduced production lines, their workers were required to fulfil only
the function of the machine he or she operated. The use of the
machine required no experience in making tools, and the ability to
use the machine was not a transferable skill. Machine work therefore
had the effect of de-skilling the workers and keeping them in an
unskilled state. A similar process affected all of us in our homes in
the twentieth century. Any housewife who could manage a kitchen
in 1900 knew how to bake bread in an oven heated with hot coals or
burning furze or faggots. She also knew how to sieve, mix and blend
ingredients. Have you ever tried making a consommé without electricity? Or a jelly, starting with the fruit and the shavings of the soft
antlers of deer? Over the twentieth century we lost vast amounts of
domestic knowledge, much of it practical and basic – such as how to
build a fireplace for boiling a large amount of water quickly (which
is different from a fireplace for cooking), how to iron clothes cleanly
without an electric iron, and how to store food for months without
a fridge. A major consequence of our growing dependence on
electricity was our shrinking ability to get by without it.

It goes without saying that much the same thing applied to our
workplaces. The shift from card indexes to data storage seems, on the
face of it, a process that could easily be reversed. Card indexes are
hardly the world's most difficult technology, after all. But the change
was more complex than that. As the year 2000 drew near, professional
advisers warned that many computer systems would not survive the
transition from the two-digit date '99' to '00'. People began to become
aware of just how vulnerable their electronic systems might be. It was
then that the complexity of computerisation became apparent: not
only was the data now stored in potentially a less robust system, it
would not be possible to return to a non-electronic system if computers

did indeed prove vulnerable. To do so, you'd have to begin writing out your card index all over again. Computerisation was a one-way street.

It is difficult to appreciate the significance of a change that is ubiquitous. As we saw with regard to clocks in the fifteenth century, we take inventions for granted very soon after they come into our lives. However, one way to assess the significance of a particular change is to ask yourself how easy it would be to undo it. After several days without power in that Suffolk cottage, I ended up thinking that it would probably be easier to reverse all the nineteenth-century changes mentioned in the previous chapter – to dig up the railways, reintroduce slavery and the subjugation of women, and take away the vote from all but the rich – than it would be to give up our dependency on electricity. All our record-keeping relies on it. We need it for the systems that allow society to operate, from our bank accounts and payments by debit and credit card to the records kept by doctors, dentists and the police. Without electricity, modern trains would not run – because of the signalling as well as their own power – and planes would collide. Stock markets would cease to operate. The logistics of our food supply would collapse. The gadgets with which we entertain and amuse ourselves would no longer work, and nor would many essential domestic tools. And yet the entire electrical and electronic system has an inbuilt vulnerability. If we were to experience a solar storm as powerful as the Carrington Event of 1859 – which knocked out the nascent telegraph system and wrapped the world in an aura akin to the Northern Lights – it might well destroy the functionality of all the satellites, communications systems, computers, hairdryers and coffee machines in its path. Then the significance of the twentieth-century shift to electrical dependency would be fully appreciated.[19]

The invention of the future

You might recall the opening line of the fourteenth-century chapter, where I explained that medieval people did not understand social history. It goes without saying that they had even less idea of the future. The astounding Roger Bacon might have deduced in his thirteenth-century friary that it was possible to build cars, flying machines, suspension bridges and diving suits, but he did not have any vision of the future as such. His reasoning was simply that these engineering projects were not beyond the bounds of possibility. The future and the past did not occur to the medieval mind, which was utterly consumed in the ever-continuing present. Gradually, however, the sixteenth century ushered in an awareness of the past. By the eighteenth century, the sense of Western society constantly changing had developed into a concept of progress as outlined by Turgot and Condorcet, and that led to people imagining the future. Hegel theorised that liberal values would continue to prevail, leading to an 'end of history' as everyone in the world adopted the same, most beneficial form of government. For Karl Marx, of course, this was socialism, and he was by no means alone in thinking that a socialist state was the desired end product of mankind's development. At the end of the twentieth century the historian Francis Fukuyama looked at the trajectory of the West up to the fall of the Berlin Wall and saw the rest of the world gradually buying into the values of liberal democracy.

The future was not found only in the pages of political analysis and utopian ideology. Science fiction introduced it to people who had no interest in Marx or Hegel. In the 1880s several novels dealt with what the future might hold through the conceit of a central character falling asleep and waking up in the future. The best known were Edward Bellamy's *Looking Backward 2000–1887* (1888), in which America is envisaged as a socialist state in 2000, and William Morris's *News from Nowhere* (1890), which presents the author's own socialist vision of a future society. Such works were set in the real world; they were idealistic expressions of what their authors *hoped* would come true in their own societies. Many commentators drawing on the 'progress' of the West also saw the future through rose-tinted spectacles. One John Elfreth Watkins Jr, writing in *The Ladies' Home Journal* in 1900, made a number of predictions about life in the year 2000. He declared that trains would travel at 150 m.p.h.; automobiles 'would be cheaper than horses'; farmers would own 'automobile hay-wagons'; photographs would be 'telegraphed around the world'; a university education would be free to every man and woman; there would be 'aerial warships and forts on wheels'; people would buy 'ready-cooked meals' from stores in the same way they bought bread from bakeries; and food would not be sold exposed to the air. Less successfully, he predicted that hydroelectric power would have replaced coal in the home; mosquitoes and flies would have been eradicated; there would be no wild animals; medicinal drugs would no longer be swallowed; and strawberries would be grown as large as apples. Politicians who entered the uncertain business of predicting the future also tended to be optimistic. In 1930 the earl of Birkenhead wrote that over the course of the next hundred years, 'warfare will not increase in savagery. The civilised world is rapidly becoming a single economic unit . . . the disaster of one nation involves all nations.'[20]

While at the start of the twentieth century various ideas of *a* future – what *might* happen – existed among the literati, historians, philosophers, politicians and their many readers, the understanding of how to predict *the* future – what would *actually* happen – was still a long way off. The idea that society would progress to a certain point and cease to change when it had reached a state of happiness was a common trope. Indeed, most visions of the future were happy. But then the First World War broke out. It shocked many believers in progress out of their complacency. How could so many enlightened nations and empires wreak such terrible destruction on one another? After the First World War it was disconcerting to read the grand narratives of older historians who praised every revolution that had been instrumental in bringing about the present world order – only for the supposedly superior modern age to prove itself more destructive to human life than all the superstitious, hierarchy-riddled monstrous regimes of the previous five hundred years. At the same time people were coming to terms with the fact that socialist revolutions did not necessarily result in the socialist paradises that Edward Bellamy and William Morris had been looking forward to, not to mention Karl Marx's vision of a communist society. A series of dystopian visions followed, most famously Aldous Huxley's *Brave New World* (1932), H. G. Wells's *The Shape of Things to Come* (1933), and George Orwell's *1984* (1949). Arguably most prescient of all – especially because it was written before the First World War – was E. M. Forster's short story 'The Machine Stops'(1909). This imagines Earth in such a polluted state that life on the surface is unsustainable; people are forced to live in cells beneath the ground, their lives controlled by an Internet-like machine, which allows them to communicate with one another through video links and delivers all their needs. Gradually they become so dependent on the machine that they lose touch with the natural world and all sense of meaning in life. When the machine breaks down, they have no resources or skills on which to fall back, and they perish.

The imagining of futures good and bad received a rocket boost in 1957 as the first satellite, Sputnik 1, went into orbit around the Earth, triggering the Space Race. The landing of two of Apollo 11's crew on the Moon in July 1969 only encouraged space-related science fiction to look far beyond the reachable horizon. By the time of that famously small but giant step, however, a much more important consideration of the future was under way. In 1956 the geologist M. King Hubbert had predicted that the production of oil would follow a mathematical curve, reflecting a gradual increase in production at first, followed by a steep increase for a period, rising to a peak, and then a sudden decrease as oil reserves are used up, tailing off as the last resources are exploited over a longer period of time. The whole graph resembles the shape of a bell. Using the formula for the curve, Hubbert predicted the exhaustion of the world's then-known oil reserves in 1970. Fortunately for us, more reserves were subsequently found. But the point was easily transferable to other resources, such as natural gas, coal and copper. The patterns of usage could be predicted, and the reserves available could be estimated, so that precautions could be taken against the exhaustion of supplies – in theory, at least. It has to be said that in the late twentieth century governments showed no determination to limit the exploitation of the world's mineral resources; they seemed to believe that any shortages would result in higher prices that would, in turn, encourage the development of alternatives. Nevertheless, awareness that the world's resources were limited made many ordinary people worry about the future. The publication of *Earthrise* – the first photograph of Earth taken from space, snapped by an astronaut aboard Apollo 8 in orbit around the Moon on 24 December 1968 – stunned many people. Looking at that image of the Earth it became abundantly clear that no matter what religion you believed in, or how brilliant mankind's technological capabilities, the resources of that small planet were all we had. Earlier that year, Paul Ehrlich's book *The Population Bomb* had predicted that hundreds of millions of people would starve to death in the 1970s due to a series of Malthusian checks on the exponential population growth over the previous decades. More accurately, the report of the Club of Rome, a group of about eighty economists and other intellectuals, entitled *The Limits of Growth* (1972), demonstrated how calculations of remaining resources based on current usage were inappropriate when the use of those resources was still increasing every year. The

Earth's resources were thus being consumed much more rapidly than people realised. By this time UNESCO had started regularly to calculate long-term population figures. In 1968 its statisticians estimated that the population of the world would reach a peak and stabilise at 12.2 billion in 2075; in 1990, they anticipated that the stable population would be 11.6 billion and that it would not be achieved until 2200. At the same time, the authors recognised that there was considerable room for error: they admitted that the world's population could be as high as 28 billion in 2150, if high fertility was maintained, or as low as 4.3 billion, if low survival rates were to apply.

At the end of the twentieth century predicting the future was part of usual practice in many professions. Besides UNESCO's mapping of population, urbanisation, ageing, poverty and education, economic forecasters were trying to predict economic trends for the next few months or years, with mixed results. Meteorologists tried to predict the weather, with varying accuracy. Organisations monitoring public opinion and market researchers tried to predict everything from the outcome of forthcoming elections to the likelihood of success for certain products on the supermarket shelves. Demographers calculated the increase and ageing of the population in specific cities and planned for the accommodation, education and transport needs of the future. Local authorities looked ahead for the location of new housing estates and sites for long-term rubbish disposal and mineral extraction. National authorities developed strategies for future infrastructure and defence. And at international level, scientists increasingly monitored the Earth's frozen regions for signs of global warming. By 1988 this had become a serious issue, with forecasts of rising water levels as the polar ice caps melted and the destruction of many of the world's coastal towns, not to mention the extinction of a large number of species. The century that had begun with dreams of socialist utopias and felicitous predictions of human progress ended with millions of people looking anxiously into the darkness of the unknown.

Conclusion

Deciding what to include in this chapter has not been easy. I have selected the above six changes to represent aspects of daily life as well as some more disturbing underlying themes. There will be those who protest that I should have included whole sections on space flight and mobile phones; others will be vexed that I have not focused on the Russian Revolution and the Great Depression; still others will be most disappointed not to find even a mention of Elvis Presley and Marilyn Monroe. Indeed, I am sure there will be men who claim that Marilyn Monroe's curves had a bigger effect than all the civilisation curves mentioned above. But as I stated in the introduction to the chapter, there was a need to balance the changes in the way we live with the changing context of our lives – and the impact of space travel on most of us does not begin to compare with the impact of war or the petrol engine. Incredible though the Moon landing was, if it hadn't happened, life would not be so very different today whereas, if the two world wars had not been fought or the petrol engine not become popular, life would barely be recognisable today.

As I see it, there were three huge changes to the context of life in the West in the twentieth century: globalisation, the threat of mass destruction, and the unsustainability of our standards of living. In dealing with globalisation under transport, mass destruction under war, and unsustainability in the last two sections, I hope I have drawn sufficient attention to these issues. Can we say which amounts to the biggest change of the century? Only if we draw the line at the year 2000 and exclude all events after that as irrelevant to the discussion. The world was not destroyed in a nuclear holocaust by 2000; electrical systems had not burst into flames as a result of a solar storm, nor had global warming or population growth led to pandemonium and mass deaths. Therefore I cannot help but conclude that the most important difference between the world in 1900 and in 2000 was transport and its consequences.

In reality, however, we cannot draw the line at the end of 2000 – not if we wish this study to have relevance today. As I stated in the chapter on the nineteenth century, history is not about the past but about people, and the most important reason for studying society in different times is to understand ourselves – how we react in a range of different situations, why we behave the way we do, and what may happen to us in the future. Those other two contexts still matter. We cannot disregard the threat of nuclear war because it did not break out in October 1962 or afterwards: we still live with the risks. Nor can we set aside population growth just because it had not resulted by 2000 in the famines predicted in the 1960s and 1970s. Indeed, we are more aware than ever before that our way of life is unsustainable. I am conscious therefore that my selection of transport as the major change of the twentieth century is a hollow one. It might be accurate as far as a study of the past goes, but the importance of twentieth-century changes in historical terms is still open to revision. As we have seen, if a cataclysmic solar storm were to strike the Earth tomorrow and destroy the world's economy and transport infrastructure, we would have a very different view of our growing dependence on electricity in the last century. And that is something we need to bear in mind when we turn to the conclusion of this book and try to establish what all of these changes mean for us in the twenty-first century, and in the centuries to come.

The principal agent of change

With one exception, the foremost candidates for the twentieth century are obvious. The Wright brothers, through their persistent attempts to power their gliders, did not just show the world that flying was possible, they advanced aviation so rapidly that they led the world in discovering how to fly safely. Albert Einstein's work on relativity marks him out as not only one of the most recognisable icons of the century but a key figure in those sciences that made warfare so dangerous and radioactive metals so valuable. He played a role in persuading President Roosevelt to authorise the Manhattan Project, which led to the development of the atomic bomb. A third candidate we need to consider is Joseph Stalin. He persecuted and terrorised millions of his own people, established the vast empire that was the Soviet Union, industrialised his nation, equipped his empire with nuclear warheads and played major roles in both the defeat of Hitler and the start of the Cold War.

The least obvious candidate, and therefore one who deserves a word of explanation, is Fritz Haber. A German Jewish scientist, his invention with his brother-in-law Carl Bosch of the Haber-Bosch process, to make ammonia fertiliser, enriched the whole world. Estimates of the number of people who are alive today on account of his invention are in the hundreds of millions, and even in the billions. What a great benevolence to mankind, you might say, what a saviour of lives! But the same man was also responsible for the invention of chemical warfare. He not only invented chlorine gas but personally supervised its use against English and French troops at Ypres in 1915. His creation of nitric acid for explosives and his ammonia fertilisers were said by Max Planck to have prolonged the First World War by a full year. His life story is thus one of the most conflicted you could ever possibly come across. He hoped by his contributions to the war effort to prove himself a German patriot despite being a Jew; but his wife, who was also a scientist, was so distressed by his work on chemical warfare (not to mention his disregard of her career) that she shot herself on the day he was promoted to the rank of captain. Worse was to follow. After the First World War, Haber led the team that invented the cyanide-based insecticide Zyklon B. It was this chemical that was used to murder vast numbers of Jews in the Nazi death camps during the Second World War. It seems fitting that as we draw to the end of this chapter, we face a final war-related irony: that the man who saved more lives than anyone else was also responsible for millions of deaths. Just as Lavoisier seems to epitomise the eighteenth century, so Haber seems to epitomise the twentieth, with all its contradictions and tragedies. Ultimately, however, we have to recognise that he himself was not responsible for the application of his inventions for all their destructive purposes. He was simply a scientist who tried to please his political masters. The real agency was that of the politicians who opened the gates of genocide and war.

As a result, the principal agent of change in the twentieth century has to be Adolf Hitler. He was responsible for the outbreak of the Second World War. His aggressive national supremacism and its violent consequences significantly damaged nationalism as a political force, even though it had prevailed in Europe for centuries. He imagined and brought about the Jewish Holocaust, and directly caused massive loss of life on the fields of battle, as well as colossal destruction across Europe, Africa, Russia, the Middle East and the Far East. His threat to create an atomic bomb led Einstein to push the American govern-ment towards the Manhattan Project. And as there must be a silver lining to every cloud, the war he started resulted in a huge number of technological and medical advances that had a positive benefit in the second half of the century, from the exploration of space to the use of penicillin. There is no doubt that the world would be a very different place today if he had never lived.

CONCLUSION

Which century saw the most change?

> There is, I feel, an age at which the individual man would wish
> to stop: you are about to inquire about the age at which you
> would have liked your whole species to stand still.
>
> Jean-Jacques Rousseau, *Discourse upon Inequality* (1754)

The last ten centuries have been paraded before you like a row of
particularly ugly beauty queens, each one smiling for the audience
despite her missing teeth, plague sores, famine, wartime grief and
uncomfortable revolutions. As with so many beauty pageants, it is
possible to make a case for any of the candidates to be the winner. It
is tempting to rank them in chronological order, starting with the
eleventh century, for without the changes of the eleventh, those of
the twelfth would have been impossible, and without the twelfth
century, the thirteenth would have been very different, and so on.
However, it is necessary to resist this temptation for the simple reason
that the achievements of one century, even if they were fundamental
to those of later centuries, do not necessarily represent a greater
degree of change. For the same reason we must resist the illusion of
modernity – the sense that our most recent achievements, being the
most sophisticated and most dazzling to our eyes, embody the most
change. This book is not about achievement per se. Mankind's existence is not a race to the stars; it is not even a race to the truth. It is
rather a balancing act – a constant shuffling along a tightrope in the
hope of reaching a better place while constantly risking disaster. And
regularly looking backwards while doing it.

For what it's worth, my own impression is that the sixteenth and
nineteenth centuries were the ones that saw the most change for my
predecessors in the house in which I now live. However, my feelings
on the subject are irrelevant here. It is important for me to set aside

my own thoughts and possible prejudices in order to develop criteria on which to base a final, objective decision. Those criteria will provide not only the context to the subsequent consideration of the question but also the framework for explaining why the question *matters*.

Deciding these criteria, however, has its own problems. In the course of writing this book, I met an investment banker at a reception in London who assured me that the most important development of the last thousand years was the telegraphic transfer of money. The reason, he explained, was because without it 'I could not take advantage of business opportunities quickly enough, and thus I would not be able to do what I do.' Even when I suggested that Columbus, Luther, Galileo, Marx or Hitler might have had a more significant impact on the world, he did not budge. It reminded me of something I had heard from an Iraqi ship's carpenter in a cockroach-infested back-street bar in Singapore in August 1990. This man told me that the officers of his ship had purposely run the vessel aground because Saddam Hussein had just invaded Kuwait and they would have had to join the army on their return. The carpenter himself was more than happy to be marooned in the Far East and paid in US dollars: he had already spent several years fighting for Saddam against Iran and had vowed never to do so again. If he weren't in Singapore, I asked him, where in the world would he most like to be? 'London,' he said without a moment's hesitation. 'Why?' 'Because you can buy medicines there twenty-four hours a day.' The investment banker and the ship's carpenter obviously had very different priorities from each other, but they both illustrate how we naturally judge what is important in life according to our own experiences.

Stability and change

Given the widespread assumption that society is prone to ever greater changes at an ever faster rate, it is fascinating to reflect that in many respects the reverse is true: things have a tendency to become more and more permanent. To illustrate this, imagine that you find yourself at a spring in the primeval forest and that you can see ahead of you a high place where you might find shelter. If no one has ever passed from where you are now to that high point before, any route through the forest is possible. The pioneer will probably take the easiest path, negotiating the waterlogged ground and the fallen trees. If in due course another way turns out to be faster, the previous route will be abandoned. Soon a preferred path will become established. After centuries of use, it might become a road. If so, eventually someone will take possession of the land on either side, allowing it to be cleared and farmed or used for building. Then all the alternative paths will become impassable: everyone will follow the one route prescribed. Further change will be very difficult.

This simple model accounts for many aspects of our society. The 's'-shaped 'civilisation curves' we encountered in the eighteenth, nineteenth and twentieth centuries, reflecting the slow start, rapid acceleration and eventual levelling-off of so many changes discussed in this book, reach that final stage because the new behaviour becomes universal. When 100 per cent of the adult population has the vote, no further increase is possible. And when something has become firmly established, it is very difficult to change it. Every newly elected politician must wonder at how little power he or she has in reality, being restricted by so many conventions. You can see this crystallising force in everything from the adoption of units of measurement to laws and professional standards. Over the course of time, certain patterns of behaviour become enshrined in tradition so that alternatives become less familiar, less attractive and even threatening to the established order. A tribe of hunter-gatherers may well have to face disruption and upheaval when the herd of wild animals on which they rely moves to a new grazing ground thirty miles away but their lack of permanent structures allows the people to adapt relatively easily. If the herd moves, the tribe moves too. In a modern town, however, a lack of food in all the shops within thirty miles would be a far more serious problem. The most significant changes are experienced when society is forced to deviate from its entrenched patterns of behaviour. When New Holland was renamed Australia and New Amsterdam became New York, the process in each case was straightforward. Imagine trying to rename Australia and New York today: you'd face a logistical nightmare, political upheaval and communications mayhem. The more firmly established our patterns of behaviour, the more difficult it is to give them up. The lighter our footfall on the planet, the less significant the shifts in our behaviour and the smaller the change.

Why, then, has change not ground to a halt over the centuries? It stands to reason that if there is a tendency for our patterns of behaviour to crystallise, then generally things should be changing less and less. The explanation of this paradox is another paradox: the more things are set in stone, the more things change. Stability itself is a destabilising factor. In economic terms, as Hyman Minsky pointed out, stability leads to complacency, over-lending and a boom-and-bust cycle of prosperity and depression. And as regards population, as Malthus explained two hundred years ago, stability leads to population growth, which in turn puts pressure on food supplies. In addition, the systematic exploitation of any finite asset or resource is liable to lead to depletion, thus eventually forcing change. Traditional fishing grounds become overfished. Continual farming of the same piece of ground removes the nitrogen from the soil, making it infertile. When the rich seams of ore run out, a mine becomes redundant. On top of these factors, many people earn their salaries froom effecting change. Builders, architects and town-planners alter the landscape as part of their jobs. Scientists, inventors and entrepreneurs similarly are required to develop the ways in which we live. Then you need to consider cultural clashes. A steady influx of immigrants to a small island may be initially welcome but attitudes are likely to change when their overwhelming numbers start to erode the island's culture. Even deliberate attempts to resist change tend to result in new patterns of behaviour. Whereas old buildings were once regularly pulled down to make way for new ones, now the old structures might be preserved and new systems put in place to make sure they are not altered. The only way for a community *not* to experience constant social change is for it to be isolated and self-sufficient, with enough resources to satisfy all its requirements, no risk of it exhausting them, no need to defend them, no need to avail itself of technological advancement, and a death rate that corresponds with the birth rate. It is doubtful that any such community exists today – although it is possible that some tribes in the Amazonian rainforest still live according to ancient patterns.

Having said this, the fact that we can hypothesise what it would take for a society to undergo little or no change means we can also hypothesise the criteria that *do* affect change. The key word is 'need'. If a society has no need to do something it is not currently doing, then the chances of it changing are greatly reduced. If we concentrate on that essential point we can measure change over a series of centuries by its consequences – that is, by the extent to which it met society's most important needs. We therefore have to ask what those needs might be.

A scale of needs

What causes a significant social development? It is not that someone has a great idea and everyone else follows suit; it is *never* as straightforward as that. The social context has to be right for a good idea to take root. The compass was known for centuries before it was regularly used for crossing the world's oceans; many people had questioned the practices of the Roman Catholic Church long before Martin Luther; Francis Ronalds's telegraphic system was rejected by the Admiralty – and so on. As we have seen so often in this book, it is not the invention that results in a major change so much as the adoption of that invention by a significant proportion of the population. There has to be sufficient demand for the change in question in order for the invention to take off. That said, the 'demand' is not always consciously expressed. Few people *demanded* to fly long distances at high speeds in 1900. However, the advantages of airborne transportation were immediately obvious. Military commanders, for example, could attack an enemy's capital city without a full-scale invasion. People could travel around the world for business or pleasure. The potential was always there for a series of rapid developments to follow the invention of a suitable engine to propel an aircraft. Had the internal combustion engine been around just 60 years earlier, in 1800, the passenger railway might never have been invented: there would have been no demand for it.

What creates a level of demand sufficient for a single invention to change the world? Looking back over the last thousand years there seems to be a fundamental shift in the thirteenth century. The Four Horsemen of the Apocalypse – Conquest, War, Famine and Disease – have caused changes throughout human history but society was particularly vulnerable to these threats in the first two centuries we examined. In the eleventh century the development of castles, the resistance to Viking invasions, and the spread of the influence of the Church were all closely connected with the threat of conquest and war. In the twelfth century, the population expansion was associated with the provision of food, and the changes in medicine and the rule of law sought to address disease and 'war' (in the sense of social disorder). But in the thirteenth century, money entered the picture. People now did all they could to avoid financial disadvantage (unless they were friars). Some strove to make themselves wealthy, the most successful city merchants rivalling the power and status of the old aristocracy. People started to reject the old adage that God had created three estates ('those who fought', 'those who prayed' and 'those who worked') as Europe shifted significantly to a form of international dialogue that was not solely driven by the dictates of kings or noblemen but took account of merchants and markets. The emerging desire for personal enrichment has been an underlying factor for change ever since. Sixteenth-century explorers, seventeenth-century bourgeois, eighteenth-century agrarian reformers and nineteenth-century industrialists were all motivated by dreams of riches. The twentieth century saw businessmen and women turn self-enrichment into an art form, as they played 'real-life Monopoly' with the world's assets. As a result, I would argue that the primary forces underlying change over the last millennium were: the weather and its effect on the food supply; the need for security; the fear of ill-health; and the desire for personal enrichment.

This set of four primary forces doesn't guide us directly to the century that saw the most change but it does give us some starting points. All four loosely correlate with the hierarchy of needs drawn up by the American psychologist Abraham Maslow in 1943.[1] He defined these needs as follows: the physiological necessities for life (i.e. food, water, air, warmth); safety, including health; love; personal esteem; and self-actualisation. The order is important: if a man has insufficient food, it doesn't matter whether his contemporaries are producing great art or travelling by train. As Maslow states,

> For our chronically and extremely hungry man, Utopia can be defined very simply as a place where there is plenty of food. He tends to think that, if only he is guaranteed food for the rest of his life, he will be perfectly happy and will never want anything more. Life itself tends to be defined in terms of eating. Anything else will be defined as unimportant. Freedom, love, community feeling, respect, philosophy, may all be waved aside as fripperies which are useless since they fail to fill the stomach.

If a man is fed and watered he will be most concerned with security next; only if he is safe and healthy will his mind turn to love, emotional support and personal esteem. Finally, if every other need is met, his preoccupation will be 'self-actualisation'. This Maslow explained in a number of ways – as the pursuit of truth, beauty, satisfaction and meaningfulness, among other things – but for our purposes it can be summed up in his phrase 'a musician must make music'.

Maslow's hierarchy was very much a product of its time and it doesn't wholly correlate with what we find when we look at earlier centuries. Many of our forebears placed religious matters before security or food – those sixteenth-century people who chose to be burnt to death for their beliefs rather than recant, for example, or the medieval lord who opted to fight on a crusade rather than live in peace on his estates. In their cases, self-actualisation took priority over everything else. And while Maslow regards 'freedom from prejudice' as an aspect of self-actualisation, before the rise of liberalism in the seventeenth century people believed that there was virtue in enacting their prejudices, and so self-actualisation was quite different. Having said all this, Maslow's work clearly shows that certain needs take priority over others. It doesn't matter much whether or not you have the latest mobile phone if you are suffering from plague. We have to give greater weight to the need to eat and drink, and to be warm, safe and healthy, than to changes of luxury and convenience. The relative importance of ideological factors is more difficult to assess. For those on hunger strike for their political beliefs, ideology matters more than the need for food; for those who stand up against racial prejudice, their belief can be more important than their personal safety. With this caveat about the variable position of ideology in the hierarchy, we can determine a more historically representative scale of needs by which to evaluate changes in society:

1. Physiological needs: whether the members of a community had enough food, heat and shelter to sustain life, or not;
2. Security: whether the community was free from war, or not;
3. Law and order: whether members of the community were safe in peacetime, or not;
4. Health: whether they were free from debilitating illnesses, or not;
5. Ideology: whether the members of a community were free from moral requirements and social or religious prejudices that prevented them from satisfying any of the needs below, or made them forgo any of the needs above, or not;

6. Community support: whether they had sufficient companionship within the community in which they lived, including emotional fulfilment, or not;

7. Personal enrichment: whether they were personally enriched and able to realise their ambitions, or otherwise personally fulfilled, or not;

8. Community enrichment: whether they were able to help other members of the community with regard to any of the above.

Generally speaking, if the answer to one of the above is 'no' for an individual or a section of the community, then the progression stops there (bearing in mind the caveat regarding the variable position of ideology). If the answer is 'yes', the next criterion is the one that defines their need. Obviously not every individual in society found themselves facing the same needs at the same moment in time. In the Middle Ages, if a nobleman was healthy and his country was at peace, he might see all eight of his needs met while not even the first would have been satisfied for the peasants who tilled his soil. Nevertheless, the whole scale applied to everyone, wherever they found themselves on it. It thus defines the collective needs of a society and allows us simultaneously to evaluate a large number of significant changes that otherwise would be impossible to quantify collectively. Measuring the ability to meet physiological needs, for example, allows us to measure the effects of agricultural change and transport at the same time, as well as an element of social reform. Changes in law and order permit us to gauge developments in morality as well as the efficacy of justice. If a change does not correlate with one of these needs, then it is, to use Maslow's word, a 'frippery', and can be disregarded.

Social change in relation to the scale of needs

PHYSIOLOGICAL NEEDS

The best test of whether the members of a community had enough food, heat and shelter to sustain life (or not) is to examine whether the population was expanding. Very simply, if the population was growing, they did. If it was contracting, that does not necessarily mean they did not have enough food – contraception, emigration, disease or war might have been the cause – but a population facing significant food shortages for long periods of time could not expand. Increases in the food supply should thus be relatively easy to quantify.[2]

The data in the Appendix (page 347) point firmly to the nineteenth century as seeing the greatest change in Europe (116 per cent), with the twentieth century being in second place (73 per cent), followed by the eighteenth (56 per cent), the twelfth (49 per cent) and the thirteenth (48 per cent). Not every European country followed the same pattern. In England, the nineteenth century saw by far the greatest population growth (247 per cent), followed by the sixteenth century (89 per cent) and then the twelfth (83 per cent). In France, the thirteenth century was the period of greatest growth (71 per cent), followed by the twelfth (48 per cent). But in terms of changes in the overall access to food across the continent, the nineteenth century was pre-eminent.[3]

What about those periods when there were downturns in the food supply? Famines have occurred in all centuries: even the plentiful nineteenth century saw millions of Irish people starve during the potato blight of 1848. But dearth was more commonly experienced in the early centuries, when communication links were worse. While we cannot quantify the seriousness of the problem before about 1200, after that date the worst food shortages were experienced during the multiple famines of 1290–1322 and 1590–1710. However, because famine reduces interest in other aspects of life – as Maslow says, to the starving man, Utopia is defined by food – it limits changes in society. You don't start painting like Leonardo to while away the hours while you are starving. The periods of famine were tragic but they were of short duration and of minimal long-term social impact. The alleviation of hunger was the principal shift in respect of physiological needs, and thus the nineteenth century saw the most significant changes.

SECURITY

A comparison of the military dangers that European communities faced is more problematic. We could simply total up the number of years that each country was at war but this would not provide an

accurate picture for the earlier centuries, when war consisted of a
series of short, bloody campaigns and a wary peace thereafter. In 1001,
fighting was endemic in many regions. Later on, long wars such as
the Hundred Years War and the Eighty Years War were more clearly
defined, in that declarations of war remained in place for decades,
but these conflicts only saw intermittent fighting. The length of these
wars reflects the lack of a permanent peace, not the lack of a ceasefire.
Alternatively, we could limit our test to those wars fought on home
soil but that would rule out both world wars as far as the British were
concerned – apart from southern England being the target of bombing
raids – which would result in an even more unrealistic view of the
impact of war.

What we really want to measure is the change in the sense of
security – the vulnerability to military force as well as the length of
a war. To this end it is instructive to draw on the work of the sociolo-
gist Pitirim Sorokin.[4] In 1943 he tried to measure the relative impact
of war in several ways. In one exercise he calculated the number of
casualties in all the wars he could identify in a sample of four countries
and related them to the total population for those countries, arriving
at the following comparison:

Century	Population (millions)	Military casualties	Military casualties relative to millions of population	Percentage change
12th	13	29,940	2,303	–
13th	18	68,440	3,802	65%
14th	25	166,729	6,669	75%
15th	35	285,000	8,143	22%
16th	45	573,020	12,734	56%
17th	55	2,497,170	45,403	257%
18th	90	3,622,140	40,246	-11%
19th	171	2,912,771	17,034	-58%
1901–25	305	16,147,500	52,943	211%

Pitirim Sorokin's estimate of military casualties for England, France,
Russia and Austria-Hungary

In using Sorokin's figures, however, we need to be aware of a few problems. His casualty estimates for the earlier centuries are based on chronicles, which are very patchy, and his population figures for these early centuries are certainly too low. His figure for the twentieth century only includes military casualties for the first 25 years – he was writing in 1943 – and did not take into consideration the catastrophic loss of life in the Second World War. Another issue for us to consider is that his figures relate only to soldiers, not civilians. However, despite these problems, Sorokin's quantitative assessment is a good place to start to think through the issues.

To begin with Sorokin's low population estimates for the early centuries: if he had had access to more accurate population data, the ratio of military casualties to population would have been even lower than the figures in the above table. With regard to the partial coverage of the twentieth century, we can make allowances for that – in respect of both the very high figures in the first half of the century and the smaller number of deaths in the second half. As for civilian casualties, it is reasonable to suppose that high military casualties reflect many civilian deaths: with a few exceptions (such as the Napoleonic Wars), military engagements before 1950 did not normally try to spare civilians, and effective methods of killing large numbers of soldiers were equally effective at killing large numbers of non-combatants. If we were to add the deaths of the Second World War to Sorokin's figures, then there is no doubt that the twentieth century saw the greatest impact of war on society. In terms of increased vulnerability – the very opposite of meeting our need for security – the introduction of total war accentuates this conclusion. Second place goes to the seventeenth century, not only on the strength of the above figures but also because the total number of deaths, including civilians, resulting from the Thirty Years War was in the order of 7.5 to 8 million. Most German states saw more than 20 per cent of their populations wiped out, and several saw more than 50 per cent mortality in this hitherto unprecedented conflict.[5]

The ways in which war affects a community depend on many factors but clearly the deployment of killing weapons is a key element. Although we think of the Middle Ages as a particularly bloody time, their wars were hugely inefficient as they were conducted by expensively equipped soldiers travelling on poor-quality roads or across dangerous seas. Those soldiers had to do their killing by hand. Only civilians finding themselves in the immediate vicinity of an army died from violence or war-related famine or disease. Commanders generally were wary of battles as they could not afford to risk losing troops on foreign soil. Small wonder that the numbers killed were relatively low. Sorokin measured the casualties of war slowly rising from 2.5 per cent of twelfth-century armies to 5.9 per cent in the sixteenth century. He noted a sudden jump in the seventeenth century to 15.7 per cent as the firearms revolution of the previous century was applied to deadly effect. The mortality level stayed roughly the same until the twentieth century, when it leaped again, to 38.9 per cent.[6] There is a strong correlation between the expanding sizes of armies and the efficiency of the weapons of the time to wipe them out. Add to that the increased ability in the twentieth century to transport large armies, and by 1945 to deliver weapons of mass destruction by air to practically anywhere in the world, and there can be no doubt that the twentieth century was the one that saw the greatest change. The first two places in this category are thus relatively easy to assign.

Using Sorokin's estimates, it was the fourteenth century that saw the third highest increase in the deadliness of war. This was the century of Edward III's longbows and English nationalism, not to mention the wars in Italy (which are not included in Sorokin's figures). However, Sorokin's four-nation sample is skewed towards England and France and thus to the Hundred Years War. Against the fourteenth century we might set the sixteenth: the reason why the *seventeenth* century was so deadly was because of the massive changes in armies and weaponry in the years just before 1600. The unmeasurable eleventh century should also be considered: the plethora of castles, the defensive force of feudal lords and the stabilising role of the Church across the whole continent added hugely to the safety of communities that were at risk of war with their neighbours, while Viking attacks also dropped off. Whereas the people in 1001 had had no option but to flee from an attacker, who might arrive without warning, in 1100 at least they had a refuge and a protector. That change is reflected in the significant population rise of the twelfth century. As for the nineteenth century, there is no doubt that the technology of weaponry changed hugely; however, most European nations' wars were fought outside Europe and the continent itself enjoyed long periods of peace. Thus I would suggest that the third, fourth and fifth centuries of change, in order of meeting or threatening the security of society, were the eleventh, sixteenth and fourteenth.

Generally speaking, civil wars are not included in the above reckoning. Sorokin did attempt to measure political unrest: he devised an index method whereby a disturbance was rated as to whether it was merely local or national, and how long it lasted, and then he plotted all the disturbances to see whether changes increased or decreased. However, his results are completely useless to us as they are heavily based on the availability of sources. Unsurprisingly, he found that the countries with the best documentation (England and France) were the most prone to disturbances and those with the poorest records (Ancient Greece and Rome) the most tranquil. However, it is worth noting that civil unrest of a class nature was rare before the Black Death. Civil wars in that early period were largely conducted for control of the throne. They were fought on home soil, by definition, and so they entailed fewer transport difficulties and less innovation. They also tended to be bitter, as there was no glory in being on the losing side of a civil war: it was not unusual for victorious commanders to massacre their defeated adversaries. Perhaps for that reason there have been relatively few full-scale civil wars in Europe since the seventeenth century, with the exception of separatist wars such as the Irish struggle for independence in the early twentieth century and the Yugoslav and Georgian wars in the 1990s. The most important exceptions for the last century were the Russian Revolution and the civil war that followed, and the Spanish Civil War. While riots and civil disturbances are still common today, significant numbers of fatalities (in excess of 1 per cent of the population) are exceedingly rare in the West and do not come close to rivalling the horrors of international conflict. And when a civil war involved the whole of a nation, as happened in seventeenth-century England and twentieth-century Spain, the factors that made the seventeenth and twentieth centuries stand out for their international deadliness also applied to these conflicts.

LAW AND ORDER

Most nations have kept records of reported crime only since the late nineteenth century. In addition, these are unsuitable for our purposes as they vary in accordance to what is considered illegal. Some things that now are considered criminal were permitted at an earlier date and some things that were previously crimes (such as homosexual acts) are now legal. Crimes tend to be defined in relation to the

changing values of society. However, there is one area of crime that
is not relatively defined and for which we have some good statistics,
namely homicide. And although we have no data for the early centuries, we can be reasonably confident that those years were not significantly less deadly than the fourteenth century.

As mentioned in the sixteenth-century chapter, the homicide rate
in the largely Protestant north of Europe declined by roughly 50 per
cent every hundred years from the end of the fifteenth century until
1900. It remained low through the first half of the twentieth century
and started to rise again in the 1960s. The sixteenth century saw the
largest decrease in respect of the murder rate, followed by the seventeenth. The fifteenth saw both the third biggest decrease in the north
and, at the same time, a significant increase in the homicide rate in
Italy, a dichotomy that confuses the averages but gives rise to two
changes at the same time. Fourth most significant on statistical grounds
is the eighteenth century. We need to bear in mind the introduction
of the systematic application of law in the twelfth century, for it must
have had an effect on the homicide rate if only by removing some of
the more ruthless killers from society. On qualitative grounds I suggest
that it should be prioritised over the tiny fluctuations of the nineteenth
and twentieth centuries, and placed fifth.

HEALTH

In measuring changes in health, we are assessing two things: the relative propensity to fall sick and the relative ability to aid recovery. There
is no doubt which century saw the most change with regard to the
former: the Black Death takes all the grim prizes. Medical practitioners
were incapable of preventing deaths; they could only advise people
to flee from infection. But while the Black Death had the single biggest
impact, life expectancy at birth was much the same before and after
the event. In thirteenth-century France it fluctuated between 23 and
27 years, just as it did in the seventeenth century.[7] In England life
expectancy in the late thirteenth century was just over 25; in the late
sixteenth, it fluctuated around 40 and then fell back to around 35 until
the end of the eighteenth century.[8] Although there is every reason to
suppose that the cohort of people alive in 1348 saw their own life
expectancy cut by up to a half, the outlook for people born before

Country	1750	1800	1850	1900	1950	2000
Sweden	37.3	36.5	43.3	54.0	70.3	79.75
Italy	32	30	32	42.8	66.0	79.2
France	27.9	33.9	39.8	47.4	66.5	79.15
England	36.9	37.3	40.0	48.2	69.2	77.35
Spain	(28)	28	29.8	34.8	63.9	78.85
Average (unweighted mean)	32.42	33.14	36.98	45.44	67.18	78.86
Changes in the mean	–	0 years 9 months	3 years 10 months	8 years 6 months	21 years 9 months	11 years 8 months

Life expectancy at birth in five European countries[9]

and after the Black Death was not so very different. Despite the plague recurring every eight or so years, and wiping out 10–20 per cent of a town or city each time, the higher standard of living and better nutrition enjoyed by the survivors compensated in part for this. As a consequence, health as measured by life expectancy at birth saw most change in the modern period.

Clearly, the increase in life expectancy of 33 years over the course of the twentieth century was far in excess of anything that had been seen before. The nineteenth century is in second place on quantitative grounds. Changes in life expectancy were relatively small before 1800; thus the plague-ridden fourteenth century seems to merit third place in the centuries that saw changes in health. On qualitative grounds, fourth and fifth places should go to the seventeenth century, which saw the near-universal adoption of medical strategies and the end of Galenic medicine, and the sixteenth, which experienced the rediscovery of anatomy, the introduction of chemical remedies, and considerable advances in professional medical knowledge.[10]

FREEDOM FROM IDEOLOGICAL PREJUDICE

The four preceding needs, all of which are matters of life and death, can in some way or other be quantified. With the remainder, little or no quantification is possible. Ideological prejudices are especially hard to rate, even qualitatively, due to their number and variety. We have to bear in mind trends such as the eleventh-century discontinuation of slavery, the formal abolition of slavery in the nineteenth century, and the passing of laws against religious minorities in the interim. We need to consider the humanitarianism of the eighteenth and nineteenth centuries, and the gradual introduction of legislation to protect women and children. Specific minorities experienced periods of intolerance at different times. The Jews were thrown out of England in the thirteenth century, expelled from Spain in the fifteenth, and persecuted by Hitler with horrific vindictiveness in the twentieth. The persecution of 'Egyptians' (Gypsies) started with their expulsion from many European cities and states in the fifteenth century. In England, the Egyptians Act of 1530, which required them to be expelled from the country, was replaced by a second such Act in 1554, which required them to be hanged. Lastly we can hardly ignore the common prejudices against the poor, children and women.

In trying to determine the greatest change in respect of all these things, let us consider discrimination on the grounds of race, religion, sex and class.

- Racism in the medieval West was largely confined to intolerance towards those on the periphery of Christendom and Jews; there were periods of greater bitterness as a result of the Crusades, and more violent anti-Semitism, but on the whole these waxed and waned over the first five centuries. With the exploration of Africa, racism acquired a new dimension, which quickly gave rise to a fear of black men in the late sixteenth century. It was amplified by the reintroduction of slavery for sub-Saharan Africans. This racial prejudice does not seem significantly to have abated until the eighteenth century. It lessened further in the twentieth.
- Religious discrimination rose and fell periodically in the Middle Ages – for example, with the Albigensian Crusade in the thirteenth century, Lollardy in the fourteenth and the Hussite Wars of the fifteenth. But it reached its apogee in the sixteenth and seventeenth centuries, with the torture and burning of heretics, and the Wars of Religion. It abated gradually in the eighteenth century, lessening further in the nineteenth, until it remained only a localised problem in the secularised West in the late twentieth century.
- In the Middle Ages, sexual prejudices were largely secondary to those of class. A noblewoman, for example, was next in command

in her household after her husband and thus far more important than all the other men. The ritualised humiliation of females, such as the gang-rapes of young women by up to half the town's young men that regularly took place in cities such as Dijon in the fifteenth century, was directed by class prejudices as much as by misogynistic attitudes. However, while status might have distorted the sexism of the time, to the benefit of those of high rank and the detriment of those at the bottom of society, sexual prejudice still prevailed on a daily basis in most walks of life. As the Garden of Eden story demonstrates, it was implicit in the entire Christian belief system. In the sixteenth and seventeenth centuries, the subjugation of women started to be challenged by the growth of widespread female literacy and the publication of some proto-feminist tracts. At the same time, however, the witch craze increased the persecution of women, especially those of low status. The seventeenth century also saw the disempowerment of women due to the growing rigidity of gender roles: women were required to maintain a house while their husbands went to work. In northern Europe and America, puritanical views on sex led to women being subject to moral judgements and punishments for sexual transgression, including the death penalty in extreme cases. Their situation changed for the better in the eighteenth century, and hugely improved in the nineteenth as married women were allowed to own property and, in some countries, divorce their abusive husbands. But it was the twentieth century that saw the most significant improvements in the status of women.

- Class prejudice fluctuated considerably in the early centuries, as countries first ceased to recognise slavery and then altered the conditions of serfdom. The rise of the towns in the thirteenth century gave many unfree peasants the opportunity to escape the bonds of their servitude. The depopulation caused by the Black Death placed a far greater value on the worth of labourers. The sixteenth century saw the impoverishment of the lower classes accentuate the differences between the haves and the have-nots. The real wealth of the ordinary working man in England greatly diminished at that time (as shown in the table below), and did not start to recover until the eighteenth century. In the late nineteenth century, attempts were made across the whole of the West to assert the equality of men of all classes. The drive to break down class barriers continued into the twentieth century

	1271–1300	1371–1400	1471–1500	1571–1600	1671–1700	1771–1800	1871–1900
Average index value	51.5	74.7	98.5	51.2	49.9	56.1	113.1
% change	–	45%	32%	-48%	-2%	12%	102%

Real wages index for building craftsmen in southern England
$(1451–75 = 100)$[11]

If we take all these shifts into consideration, there seems to have been a basic upward trend in the status of those at the very bottom of society throughout the millennium – from slavery to serfdom, then villeinage, and eventually free labouring – with the lower echelons of society benefiting from higher remuneration and political power in the twentieth century. This can be seen as a long, albeit uneven, decline in class prejudice. However, against this decline we have to set the significantly greater prejudice with regard to race, religion and gender noticeable from about 1500. It assumes the form of a bell-shaped graph – an arc of intolerance – rising steeply in the sixteenth century and reaching a shallow peak in the seventeenth before declining gradually in the eighteenth and rapidly in the nineteenth and twentieth centuries. The financial hardship of the working class also follows this arc. Therefore, in terms of the impact of ideological prejudice, or the freedom from it, I rank the sixteenth century first (the steepest rise of the arc of intolerance), the nineteenth second (the steepest fall of the arc), and the eighteenth third, followed by the twentieth and seventeenth centuries.

COMMUNITY SUPPORT

When considering love, Maslow's third requirement, we have to suppose that there has not been an enormous change through the centuries. Boy meets girl is one of the few constants over the millennium. There were some changes: in the Middle Ages, for example, the lack of money to keep a family prevented many men from marrying. At the same time, feudalism imposed restrictions on whom an unfree peasant could marry. Thus we might look at the decline of feudalism in the fourteenth century and the rise of peasant incomes in the fourteenth and fifteenth (as shown in the table in the previous section) as signs that men and women were able to find love more easily than their predecessors. However, it was not in a feudal lord's interest to stop his unfree peasants from marrying and breeding, so we should not overstate the significance of this factor. While a few people at the bottom of society were unable to marry the person of their choice until about 1400, the same can be said for those of higher status too, as the marriages of wealthy people were generally arranged by their families. In terms of love, these were not necessarily bad marriages, or at least no worse than the marriages of those who married for love and found themselves falling out of love. The real problem lay in being shackled for life to a hateful or uncaring partner. Therefore, the most significant change in finding emotional fulfilment was undoubtedly the ability to divorce, which was a development of the nineteenth and twentieth centuries. At least that way, if you made a mistake, your life was not completely blighted. As for same-sex romantic love, this was a capital offence throughout most of Christendom from the Middle Ages until the nineteenth century. The last men executed for sodomy in England, James Pratt and John Smith, were hanged in 1835; as with divorce, the centuries of greatest change were the nineteenth and twentieth.

Romance is not the only form of love we have to consider: there is also the affection and support of neighbours and friends. In this respect, it is community integration that has altered most over the centuries. To begin with, most people lived in the countryside in small, self-sufficient groups; they had to be supportive of their members in order to function. Town-dwellers similarly depended on being recognised and respected in their communities. Many urban codes included the ultimate sanction that townsmen who repeatedly broke a by-law were to be banished. This was a far more serious punishment than it would be in modern times, as it entailed a man losing his friends and supporters – all the people who would vouch for him in court, protect him on the streets and lend him food or money. Such a dependency on the community prevailed even when cities began to grow larger in the late Middle Ages; where you were from was an important part of your identity, whether that place was a huge city or a small village. In the sixteenth century, with the Reformation and the growth of travel, it began to break down in the cities but remained strong in the smaller towns and the country. However, the advent of the railways in the nineteenth century dealt a huge blow to community integration everywhere. Large towns and cities could not provide the same bonds of support and collective reassurance that small towns and villages once offered their inhabitants. More often than not, the residents of a street in a large town or city had not grown up together, and recent proximity could not replace the trust and fellow feeling of lifelong familiarity. People began to live further and further apart from their families and friends. In the mid nineteenth century, they began to emigrate in ever larger numbers. Thus the nineteenth and twentieth centuries again feature as the ones that saw the most change, due to the destruction of supportive communities and the creation of alienating conurbations. A deciding factor in respect of which of the two saw the most change is the degree of urbanisation. In most Western countries, more people lived in the country than in a town or city before 1900 – only England and Holland saw more than 50 per cent urbanisation at an earlier date. Therefore with regard to the Western world as a whole, it seems that the twentieth century saw more change than the nineteenth.

As with previous elements in our scale of need, it is necessary to select the third, fourth and fifth centuries of greatest change. I would suggest that the sixteenth century should be placed third, because of the growth of travel and the religious divisions that split communities. The Black Death puts the fourteenth century in fourth place, on account of the deaths of so many people and the disintegration of many settlements, which proved unsustainable after depopulation. The twelfth century should be fifth, I think, because of the growth in numbers, community security and collaborative efforts to clear the land.

PERSONAL ENRICHMENT

Obviously there were wealthy people in every century in the past, to the extent that most centuries saw a greater disparity of wealth than the twentieth; but there can be no doubt that the twentieth century witnessed the greatest change in the disposable income of the population as a whole. Angus Maddison's study of the world economy estimated per capita GDP for western Europe as follows:

Country	1500	1600	1700	1820	1913	1998
UK	714	974	1,250	1,707	4,921	18,714
France	727	841	986	1,230	3,485	19,558
Italy	1,100	1,100	1,100	1,117	2,564	17,759
Germany	676	777	894	1,058	3,648	17,799
All western Europe	774	894	1,024	1,232	3,473	17,921

GDP per capita (1990 international dollars)[12]

As the table shows, people's purchasing power in the twentieth century increased by more than 400 per cent. The second largest increase in general wealth occurred in the nineteenth century. For the centuries before 1800 it is less easy to discern, as Maddison simply assumed standard levels of increase (15 per cent per century in many cases). However, more recent studies by economic historians give an indication of the changes in the period 1300–1800, as shown below. The fact that the increase in per capita GDP in the first half of the nineteenth century exceeded any previous century's growth confirms that it should take second place. Third on this scale would be the eighteenth or the fourteenth century. However, in the sixteenth century, the average per capita GDP across three of the six countries

Country	1300	1400	1500	1600	1700	1800	1850
England (GB after 1700)	727	1,096	1,153	1,077	1,509	2,125	2,718
Holland (Netherlands 1850)		1,195	1,454	2,662	2,105	2,408	2,371
Belgium			929	1,073	1,264	1,497	1,841
Italy	1,644	1,726	1,644	1,302	1,398	1,333	1,350
Spain			1,295	1,382	1,230	1,205	1,487
Germany			1,332	894	1,068	1,140	1,428
Average (mean)	1,186	1,339	1,301	1,398	1,429	1,618	1,866
Change		13%	-3%	7%	2%	13%	15%

GDP per capita (1990 international dollars)[13]

– England, Italy and Germany – fell by 20 per cent – more than it rose in either the fourteenth or the eighteenth; the figures for Spain and especially for Holland skew the picture. The thirteenth-century shift to a market economy is also important, even though it is unmeasurable. Given the number of markets and fairs that were founded, and the fact that this development underpins the entire shift to a money-based economy, it seems that the thirteenth century should be placed third on qualitative grounds. The sixteenth then would be fourth due to the massive decrease in the per capita wealth of several countries, with the principal exception of Holland. The fourteenth century should be prioritised over the eighteenth because the changes outside England in the latter were small or negligible, whereas the per capita wealth of the peasantry after the Black Death increased across the whole of Europe.

Not all forms of enrichment take the form of money. The beauty of Italian Renaissance art and early-nineteenth-century music were responses to a demand for fine art and romantic orchestration; both therefore can be said to fulfil a need. However, we cannot assume that one century *needed* another century's cultural values. Besides, as Maslow's hierarchy makes clear, such elevated needs only have significance for people whose other, more pressing requirements have been met. Our modern desire for cultural enrichment is greater than ever because fewer of us are hungry, cold, in danger or seriously ill than ever before. But it should go without saying that that greater need is only satisfied by more of us having more disposable income to pay artists, writers, musicians and film-makers. If society did not have the surpluses to feed its artists, there would not be any art. Thus the quantitative assessment of changes in real income is the best way of measuring all forms of enrichment – cultural as well as financial – across the centuries, without having to make subjective judgements about the aesthetic value of, say, Donatello compared to Dali.

COMMUNITY ENRICHMENT

People's ability to enrich their communities underwent a transformation over the millennium. In the eleventh century only aristocrats could afford to give to their community as only they had disposable assets, such as manors and mills, to donate to the Church or to hospitals who looked after the poor. Likewise, they alone had the wealth and land to build bridges or to grant their tenants the right to gather firewood freely from their woodlands. By the thirteenth century merchants were also among the benefactors of society; by the sixteenth it was predominantly the taxpayer who supported both the community and the nation state. And so it has been ever since. In the modern period a massive amount of money was given to the community through income tax, indirect taxes such as value added tax and capital gains tax, inheritance tax, and local taxes. There can be no doubt that the twentieth and nineteenth centuries saw the greatest changes in the ability to enrich the community as benefits were given to the needy that had hardly existed previously, such as unemployment benefit, old-age pensions and disability support. The levels of tax paid today dwarf those demanded in the Middle Ages and early-modern period. Thus the ability to enrich the community is largely dependent on the growth in GDP, and the changes to be noted in this section are more or less the same as those we encountered in the context of personal enrichment. It seems unnecessary to discuss them again here.

SUMMING UP

As far as society's most essential needs go, and using the closest quantitative assessment available, the results are undoubtedly in favour of the modern world.

Need	Measurement	1st	2nd	3rd	4th	5th
Physiological needs	Population growth	19th	20th	18th	12th	13th
War	A measure of casualties relative to population, supplemented with qualitative assessment	20th	17th	11th	16th	14th
Law and order	Homicide rate, supplemented with qualitative assessment	16th	17th	15th	18th	12th
Health	Life expectancy at birth, supplemented with qualitative assessment	20th	19th	14th	17th	16th

Need	Measurement	1st	2nd	3rd	4th	5th
Ideology	Qualitative	16th	19th	18th	20th	17th
Community support	Qualitative	20th	19th	16th	14th	12th
Personal enrichment	Partly qualitative assessment, supplemented with per capita GDP & number of markets founded	20th	19th	13th	16th	14th
Community enrichment	As above	20th	19th	13th	16th	14th

By all the needs that I have defined as important, the twentieth century emerges first in five out of the eight categories. In fact, if you chart these changes with a points system – five points for a prime position, four for a second, three for a third – there is an unmistakable pattern. According to our scale of needs, the twentieth century saw the most change. Although I have no doubt that the Black Death was by far the most traumatic single event that humanity has ever experienced, our adaptability meant that we managed to restore most practical

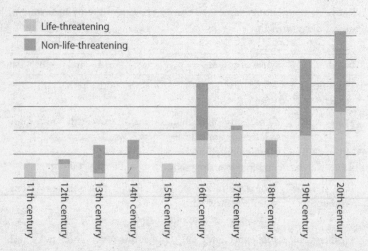

Aspects of change described in this book related to the scale of needs

aspects of our lives within a relatively short time. In the twentieth century that same adaptability moved us further and further from our ancestors as we willingly embraced alternative patterns of behaviour. Therefore it looks as though I have to eat humble pie and admit that the TV presenter in December 1999 was right and I was wrong. Having said this, I maintain that I was not wrong to doubt her, for her opinion was based on an unqualified assumption about the relationship between technology and social change. Moreover, as I hope most readers will realise by now, it is not the answer itself that matters. It is what we have found out in the consideration of the question that is important. Breaking down the overarching concept of change into smaller facets has allowed us to glimpse the dynamics of long-term human development. We can see that not all change is technological: it includes language, individualism, philosophy, religious division, secularisation, geographical discovery, social reform and the weather. In fact, the fundamental innovations before 1800 were based on very few technological innovations. But since the mid nineteenth century we have been practically living on another planet. Our lives and livelihoods now depend on the economy, not on the land, and therein lies a world of difference.

The end of history?

People living in the West today would appear to have met practically all their needs. While the least wealthy 10 per cent will no doubt argue that this is not true for them, it goes without saying that there will always be a poorest tenth of the population, who will feel disadvantaged. However, their relative poverty today looks hugely privileged next to that of the poorest 10 per cent in 1900. The injustices and inequalities that remain today are the by-products of the systems by which we have satisfied the requirements of the majority of the population. But what next? If so many social factors followed these civilisation curves to an apogee by 2000, will the twentieth century *remain* the one that saw the greatest change?

This question has much in common with that posed by the historian Francis Fukuyama in his book *The End of History and the Last Man* (1992). Ever since the Enlightenment, various historians, economists and philosophers have postulated that one day society will have progressed to a point from which it can go no further. Eventually everyone will accept the best all-round form of society, whether that be a liberal democracy or a socialist state, and the political development of the world will slow down and stop. That progression – from hunter-gatherer to the final state of society, which Fukuyama believes will be liberal democracy – has been called Universal History. As Fukuyama says, there will still be 'history' in the form of events when Universal History reaches its final conclusion. Wars will still break out. Diseases and inventions will continue to plague and benefit mankind. But these will be no more than ripples in a calm sea. Politically the world will have arrived at the ideal and unchanging state. Everyone will be fed, educated and looked after with regard to their health. Ideologically there will be no need for anyone to oppose the government under which they thrive. Fukuyama thought that the fall of the Berlin Wall on 9 November 1989 indicated that Western liberalism was the political paradigm that would prove enduring, and wrote *The End of History* to support that thesis.

With so many civilisation curves indicating the culmination of so many beneficial changes, it is reasonable to conclude that the latter part of this book (if not the whole of it) supports the concept of the 'end of history'. We have charted the path to an egalitarian, liberal democracy that more or less corresponds with the political model that Fukuyama thought would become accepted everywhere. How could any future century see more change than that? Once you're there, you're there. But while such a conclusion is reasonable, it is incorrect. Like Fukuyama and every renowned political economist before 1945 – with the notable exception of Malthus – we have considered just one side of the vast economic exchange that underpins humanity's existence: we have only considered the demand side. That is to say we have examined what we want: what our needs are, how we might make ourselves and our nations wealthy, how we might distribute our riches, and how we might satisfy ourselves. But every economic exchange also consists of a supply side. Fukuyama – like Hegel, Marx and other, less important protagonists of Universal History – neglects this side of the exchange between mankind and our environment.

The 'supply side' of this relationship is the availability of resources, from such basic things as water, land, air and sunlight to timber, coal, metal ores, oil and natural gas. In the past it was taken for granted that sufficient land and natural resources existed, and the only question to be debated was who controlled them. However, when the photograph *Earthrise* was published in 1968, it revealed in a beautifully simple way how small the Earth is and how limited our resources are. As it happened, there was no immediate threat to the high living of the sixties, and the pessimism did not take hold. The fleeting attention of the world moved on to other things. Only a few earnest souls tried to alert political leaders to the impending over-exploitation of the world's resources. Most people in positions of responsibility decided that worrying about such things was premature, distracting and a low priority while there was the important task of encouraging business, international competition and, above all, economic growth.

It should be obvious to all that endless growth of manufacturing and food production on a planet of limited size is impossible. Some optimistic economists argue, however, that unending *economic* growth is possible, despite our limited resources. This is because economic growth is measured by Gross Domestic Product: everlasting increases in GDP are theoretically possible due to the potentially endless recycling of resources, with value added at each stage. These optimistic economists often cite copper as an example. If the copper in old electrical goods is recycled into the newest technological devices, its value is increased and it adds to the growth of the economy. When the new electrics become old and fail, they too are recycled into a better, value-added product and the growth cycle continues, without the need for more copper. However, most of the resources on which the world depends do not fit this model. As shown in the volumetric approach to history discussed at the start of the chapter on the nineteenth century, in Europe, over half of the last thousand years of human life was experienced in the last two centuries. This means that per capita consumption of mineral resources has been far higher over that time than before 1800. If we were to draw a pie chart showing the last thousand years of metal consumption, it would indicate that virtually all of it was consumed after that date. And with regard to the everlasting-growth optimists' favourite example, more than 95 per cent of all the copper ever mined – since the start of the Bronze Age – was used in the twentieth century.[14] Almost all the oil produced over the last millennium was drilled, pumped and burnt in the twentieth century. As for coal, the amounts consumed in the twentieth century far outweigh the quantities burnt in the nineteenth century, and

consumption levels before 1800 were negligible by comparison. Even
the use of iron is largely a modern phenomenon. Currently annual
steel consumption in Europe is about 400–450 kilograms per capita,
and almost double that in some highly industrialised manufacturing
countries. Before 1800, it probably would not have exceeded 10 kilograms. On this basis 95 per cent of iron has been used since the
Industrial Revolution. The world's iron supplies are plentiful, but to
produce steel you need coal – between 0.15 and 0.77 tons of it per ton
of steel, depending on your method.[15] And although steel and copper
can be recycled, coal cannot. Nor can natural gas. Nor can oil. Thus,
with an even bigger population emerging from all this consumption
of the world's resources, the argument that we can achieve unending
economic growth by recycling copper and steel is wholly
unrealistic.

The demand side of mankind's exchange with the Earth and
the supply side are thus in agreement. The twentieth century was
not only when we satisfied more of our needs than ever before but
also when we exploited Earth's non-renewable resources at an
unprecedented rate. It was therefore unlike any previous period. In
socio-economic terms, we were living on a newly discovered planet.

The problem is, of course, that we only have one planet, and using
up a large proportion of its most useful resources in one century was
not particularly clever if we wanted to satisfy our needs and live
comfortably for ever more. The political thinkers of the past –
Hegelians, Marxists and nineteenth-century liberals – never saw the
importance of the supply side of the mankind–Earth exchange. They
were only interested in what mankind wanted, or, to be more precise,
what they wanted for mankind. For Marx, socialism was a matter of
who controlled the resources, the means of production and the
markets: if these were all controlled by the proletariat, then the proletariat would benefit. However, if the number of people on the planet
were to double, sharing all those resources would halve those available
to the initial proletarian cohort, and thus population expansion would
gradually impoverish the proletariat, whether or not they owned the
means of production. Furthermore, even if the population remained
stable, not all those proletariat-owned resources would last an equally
long time. Some oil-producing countries would run out before others,
destroying the economic and social well-being of those who depended
on them. Eventually a few countries would be left with the only

profitable oil reserves remaining in the world, placing them in an economically dominant position over those whose resources had run dry. Marx's vision was, like all utopias, a midnight point beyond which, even if the hour hand had reached it, it would have remorselessly ticked past.

Some people still believe that we will never exhaust our resources. When the Reconquista was completed in 1492 and Christendom reached the end of its possible expansion, Columbus promptly crossed the ocean and found Hispaniola. Five years later, Cabot reached Newfoundland. Such a spirit of adventure is not dead, these people say; it will lead us to the stars. Unfortunately, the twentieth century also ended that dream. As this book has shown, the 'spirit of adventure' is really a euphemism for fortune-hunting, or the pursuit of profit. Columbus and Cabot were inspired by dreams of wealth. So were the governments that backed them. The exploration of the African coast only proceeded past Cape Bojador because Gil Eanes found gold and slaves there. In the eighteenth century, people did not start improvising with new agricultural techniques in the hope of feeding the world; they did so for profit. But in the twentieth century we came to understand the limits to our expansion: we discovered that it would never be profitable to leave the solar system. It might one day be worthwhile mining the ore of Mars for metals that are rare on Earth. However, I strongly suspect that the multi-billion-dollar price tag on sending missions to barren, freezing, airless Mars will mean it is always cheaper for governments to form an alliance with a resource-rich state, or attack one that is economically or militarily weaker. Beyond Mars there is no hope of expanding in a commercially viable way. The remaining planets in this solar system are not suitable bases for humans to settle and start mining. The next solar system, Epsilon Eridani, is 10.5 light years away, and its planets are not in a habitable zone. The nearest planet after Mars on which we could possibly settle is Gliese 667Cc, 22 light years away. Just getting there would be a huge problem. The fastest we have propelled a manned spacecraft is about 25,000 m.p.h.: at this speed it would take us 589,248 years to make the journey. And then we would need to get back again – a round trip of over a million years. That length of time is never going to excite investors, whatever the promised return. And there is no guarantee there will *be* a return, in either sense of the word.

To go beyond this is to venture into the realms of science theory, if not science fiction. In April 2010 Stephen Hawking postulated that a gigantic spacecraft containing enough fuel to burn solidly for two years could approach half the speed of light (334,800,000 m.p.h.).[16] If it had enough fuel for four years, it could reach 90 per cent of the speed of light (602,640,000 m.p.h.). In this way you could get the return trip to Gliese 667Cc down to about 58.6 years.[17] I am quite happy to accept Professor Hawking's assurances that, due to the properties of space–time, the people on board the craft would only experience half the time elapsing while they are travelling at 90 per cent of the speed of light: 37.3 years instead of the full 58.6. However, I find myself wondering where you would put a fuel tank containing over half a billion tonnes of liquid oxygen and liquid hydrogen (for the journey there and the journey back again). Indeed, would a craft that heavy be able to take off? Could it be fuelled in space? Clearly, this is why I'm not a rocket scientist. And come to think of it, nor is Stephen Hawking. What I do know is that no profit is going to be gleaned from a voyage to another solar system. It is not the technological limitation that persuades me this will always be true, it is a combination of the sheer distance, the fact that we cannot travel as fast as light, and the cost implications. It will always be commercially more expedient to barter, negotiate or otherwise fight for Earth's resources than spend trillions of dollars on a desperate attempt to send a few people on a very, very long quest to another solar system.

So space, our so-called 'final frontier', does not offer us a solution to the problem. It does, however, focus our attention on the forces that are likely to act upon our nature in the future. Many of the changes discussed in this book have something in common: they are about a breaking of boundaries. Geographical boundaries were smashed by Columbus, Cabot and other early explorers. Boundaries of perception were shattered by the supernova of 1572, and by the microscope and the telescope. Social boundaries were dismantled in the French Revolution and, in the nineteenth century, by reformers throughout the West. The Earth's atmosphere was breached in the twentieth. Many of these boundary crossings can be understood in terms of the 'Go West, young man' paradigm. You go westwards, you find the boundary, you cross it, you discover, you acquire, and you become rich. This paradigm characterised the expansion of the Vikings, Normans, Crusaders and explorers of the New World. It underpinned scientific discoveries, world exploration and economic growth. But with the recognition of the approaching exhaustion of our fossil resources on Earth, this boundary-breaking mentality is out of date. The challenge now is not one of expansion but self-containment: a series of problems with which the all-conquering male is ill-equipped to deal. We, *Homo sapiens*, have never before had to face the problem of our own instincts threatening our continued existence; they have always been for our benefit, the survival of our genes. The frontiers we face now lie not on the horizon – or even in space – but inside our own minds.

The principal agent of change

The ten chapters of this book have put forward ten very different individuals as the principal agents of change. Indeed, it would make an extraordinary guest list for a dinner party: Pope Gregory VII, Peter Abelard, Pope Innocent III, King Edward III, Christopher Columbus, Martin Luther, Galileo Galilei, Jean-Jacques Rousseau, Karl Marx and Adolf Hitler. Four Italians, three Germans (one Austrian-born), two Frenchmen and one king of England whose ancestry was entirely Continental. Which one of these was the major agent of change of the entire millennium? Or was it someone else – someone whose influence affected people over several centuries? Aristotle, perhaps? Or Isaac Newton?

There is no doubt who was the principal agent of change of the millennium. It was God. Personally, I don't believe in God. However, my personal beliefs are irrelevant here. Even though He does not exist (in my opinion), He had more influence on the Western world than anybody that did. Such is the magnitude of that irony that I have no qualms about preserving the old-fashioned capitalisation for references to Him in this book. It was the Catholic Church's perception of God's will that was behind the Peace of God and the Truce of God movements and the discontinuation of slavery in the eleventh and twelfth centuries. God was the sole international agency acting for peace throughout the Middle Ages. It was the Christian community's worship of God that made the West accept the authority of the papacy. It was Christian monasticism that led to the twelfth-century renaissance and the beginnings of learning and science in the West. Before the thirteenth century, religious men were almost the only guardians of literacy. After printing was invented, it was the study of God in the Bible that taught common men and women how to read, and thus gave women the chance to express themselves to significant numbers of other women for the first time. Widespread literacy led to better government administration and bureaucracy, which in turn caused the decline in personal violence. It was their understanding that they were exploring God's Creation that led so many scientists to devote their lives to uncovering the mysteries of the universe and the properties of botanical specimens from around the world. It was the belief that God's healing power worked through them that gave so many seventeenth-century physicians the confidence to try to help the diseased and infirm. In the nineteenth century it was the understanding that God had made everyone equal that persuaded many people that arguing for the equal rights of men and women, black and white, rich and poor was the only defensible moral standpoint. Only in the twentieth century have the major changes examined in this book not been overtly influenced by God.

Among those who did exist, who deserves the title of the principal agent of change? The point is that no one does. Forced to select someone, I would either go for Columbus, to represent the importance of the expansion of Europe, or Galileo, to represent the victory of the scientific method over religion. But these are both personal selections and largely symbolic; really it doesn't matter. We're getting into the parlour game of opinion, and glorifying a historical character was not the reason for the exercise.

There were three purposes underlying the 'principal agent of change' section in each chapter. The first was to consider agency: how much difference can one man really make on a large scale over the course of a century? Or, rather, how little. Who in history could have *stopped* any one of the fifty changes discussed? The second purpose was to show, by way of example, that we only select as important agents of change those people who make things happen. If I had suggested Robert Malthus for the eighteenth century – the only major economist to examine the supply side of the exchange between mankind and the Earth before modern times – you might have laughed at me. He didn't *do* anything. We prefer our heroes to do things, not to stop us from doing them. This is why, when it comes to the necessity of changing our nature and giving up on the 'Go West' paradigm, our democratically elected leaders are not likely to do us any good.

Most of the principal agents of change were not even responsible for the biggest developments in their various centuries. When it comes to socio-economic change, no one is in control. No one ever has been.

The third reason for the exercise builds on this. It will not have escaped your attention that I have not considered a single woman as a principal agent of change. Had I suggested Isabella of Castile, Elizabeth I of England, Mary Wollstonecraft or Marie Curie, everyone would have seen through this as tokenism or political correctness. The influence of these women did not come anywhere near that of Columbus or Luther, Galileo or Hitler. Western society was fundamentally sexist: no woman until modern times had a chance of being the person who most profoundly affected life in the West. In highlighting this absence of truly influential women in the past, I hope to draw attention to the capacity for things to be different in the future. I wrote above: 'The challenge now is not one of expansion but self-containment: a series of problems with which the all-conquering male is ill-equipped to deal.' The emphasis on the *male* in that statement was not accidental. The character traits we commonly associate with women, which are less to do with testosterone-fuelled conquests and more to do with nurturing and protection, are much better suited to lead us into the future. If men change in their nature, then no doubt women will do too – and there is a significant danger in that: there will be no advantage for the world if women simply take on male traits. Nevertheless, if there is to be hope for mankind, we must accept that it may be better for us all if the principal agent of change in the twenty-first century is a woman.

ENVOI

Why it matters

The conclusion of this book leaves a few questions hanging. If *Earthrise* allows us to be certain that the resources of the Earth are limited, what does that mean for mankind over the next millennium? Can we determine which of these fifty historical changes will be amplified or reversed? If we are not going to enjoy a permanent state of liberal capitalism, beaming at the top of our civilisaton curves for centuries to come, what sort of world are our descendants going to inherit?

The first point I should make is that I don't believe we can consciously change our nature to suit ourselves. I may be wrong on this; perhaps we *could* all become docile, modest creatures with tiny egos and appetites, humbly tilling our own small patches of land and easily dissuaded from producing large numbers of children. In his 1985 novel *Galapagos*, Kurt Vonnegut suggested we could evolve into furry aquatic mammals with streamlined heads, smaller, simpler brains and a penchant for fish. But I doubt anything along these lines will happen. For a start, our drive to reproduce in ever-increasing numbers underpins our success as a species; historically it allowed us to rebuild our communities quickly after a famine or a plague. Also, personal ambition is part of our nature. There will always be people who will want to outperform each other and I suspect that a significant proportion of the population will continue to be attracted to them, both sexually and socially, forcing more of us to compete. Even if there was some international political agreement that moderated our behaviour, it would soon be undermined or overturned. The fact is that human beings don't like being humbled by systems, rules and limitations. We love to hear of those who break free from restriction and oppression. Our passion for freedom is intrinsic to the human spirit. I suspect therefore that we are like the Venetian Republic: doomed because we cannot bear the thought of being something other than we are.

Of all the resources that are set to shrink in availability, oil is perhaps uppermost in people's minds. It underpins all our lives – from food and transport to law and order, defence and recreation. And it will run out at some point in this current millennium, there is no doubt about that; it is just a matter of when. Proven reserves are currently about fifty times annual world consumption but that ratio is subject to considerable fluctuation. It could extend as more oilfields are found; total proven reserves were significantly greater in 2012 than they were in 2000.[1] Alternatively, it could diminish as population expansion and increasing industrialisation drain those reserves more rapidly. However, whether it takes another thirty, fifty or seventy years isn't important here. Oil supplies will cease to meet world demand at some point – and very probably in the lifetimes of our children. The same is true for natural gas, on which we currently rely to produce fertilisers. At the time of writing, proven gas supplies are roughly sixty times annual world consumption, but total consumption is increasing by 2–3 per cent every year. Shale gas has greatly extended these reserves and will probably extend them still further, but this extra energy is already being sold off cheaply. You would have thought that governments might seek to ration such a windfall, so that it lasts long enough for us to find and produce viable alternatives to fossil fuels. Aesop's fable 'The Ant and the Grasshopper' – in which the ant works hard all summer preparing for the onset of winter while the grasshopper just sings in the sun and has nothing to live on when the season turns – shows what happens to those who fail to guard against future short-ages. However, Western governments are, like grasshoppers, very present-centric: politicians sing to those who might vote for them, not to the people of the future. As mentioned in the Introduction, only dictators plan for a thousand years.

In this light, a range of possible eventualities opens up before us. At one end of the spectrum is the Sustainable Future. In this scenario, we discover how to produce all our energy and fertilisers from sustain-able sources so that society can carry on more or less as usual. At the other end of the spectrum is the Universal Crisis: a calamity of Black Death proportions, resulting from a worldwide failure to replace fossil fuels before they start to run out. My contention is that both ends of the spectrum involve society becoming more hierarchical and less liberal.

Let us begin by considering the sweeter type of outcome, the

Sustainable Future. Picture every farm with hydroelectric generators on its hillside streams, solar panels in its fields and wind turbines cresting the hills. Imagine all the houses and industrial buildings in every town glittering with photovoltaic cells on their walls and roofs, and every rural house with a biomass boiler. Vast offshore windmills harness the power of sea breezes, and with every wave, huge pistons housed in cliff tunnels pump energy into national grids. Aircraft fly on biofuel. Tractors and farm machinery use biodiesel. Electric vans carry grain and animals to urban markets from which they are transported by electric trains to their points of slaughter and processing. But even in this harmonious state there will be far greater competition for resources. In particular, there will be an uncompromising fight over land.

Take the UK as an example. Suppose that we invest significantly in solar, wind and hydro power over the next few decades so that by 2050 we can produce all our electricity from these sources.[2] This is a big supposition; nevertheless, for the sake of argument, let us assume it is possible. In fact, let us go even further and imagine that by the time oil hits a crisis point (whenever that might be), we not only meet all our electricity needs but generate so much electricity from renewable sources that we can cut our oil, gas and coal consumption in half. This would still leave the problem of replacing the remaining half of the energy we derive from fossil fuels. All the forms of biofuel currently being tried – including rapeseed, various nuts, algae, corn and sugar beet – need land. To meet just half of the UK's current road transport demand for diesel and petrol would require the exclusive use of 11.3 million hectares – almost 87 per cent of the total area of England and more than all the agricultural land available. And that does not include non-road-transport needs, such as manufacturing, plastics production, agricultural machinery and aviation fuel.[3] Nor does it account for the increase in demand as the population grows. While some might say that building several dozen more nuclear power stations is the answer, even if that were politically acceptable it would only be a temporary solution. Proven world uranium supplies are currently less than one hundred times annual world consumption, and as coal, gas and oil diminish, the likelihood is that the demand for uranium will dramatically increase, so it will not outlast oil by many decades.[4] Thus, in the long term, the Sustainable Future not only requires astronomically high levels of investment in electricity from renewable sources but also an impossibly large amount of farmland to be given over to producing biodiesel, bioethanol or some other new fuel, creating a tension between food and fuel production that is already politically explosive in some hard-pressed countries.

Population expansion and the consequent need to build more and more houses adds to this competition for land. In England, cities, towns, village centres and urban infrastructure account for 10.6 per cent of the country as a whole.[5] Woodland, coastal fringes such as dunes and estuaries, freshwater lakes, rivers, mountains, moorland and heath account for a further 15.9 per cent. The remaining 73.5 per cent is agricultural land. On the face of it, this leaves plenty of space for new housing. However, all the agricultural land at present produces only about 59 per cent of our total food requirements. Of course, some foods are imported because they don't grow here, but taking these out of the equation, we are only 72 per cent self-sufficient for the foodstuffs that are grown domestically.[6] This means that even the crops we farm in abundance, such as wheat, barley, oats, linseed and oilseed rape, are not produced in such quantities that we can afford to reduce production. We had a surplus of wheat in 2008, harvesting 10 per cent more than we needed. That was a good year. In 2012 the reverse was true: a poor crop meant we imported more than we exported.[7] We are not self-sufficient for meat either.[8] Building on agricultural land may provide roofs over people's heads but it won't help to feed them in the long run.

Some might disagree with that last statement. They would argue that it *does* help to feed them – by generating the income with which to buy other countries' surpluses. But this can only be a short-term strategy. Once a piece of land is used for housing, it ceases to be productive for food or fuel. Suppose you were to hand over a small portion of the UK's agricultural land to be developed as housing every year – an amount equivalent to current population increase. As farmland it was worth about £20,000 per hectare; now, as land for building, it is instantly worth £1,000,000 per hectare or more, depending on the locality. The national balance sheet is thus improved by at least £980,000 per hectare of development land. This extra cash feeds into the economy, supporting jobs and enhancing profits. Now suppose that we keep doing this for the rest of the century. At the current rate of population growth – 0.76 per cent per year – the population of England will almost double by 2100, reaching about 104 million.[9] Providing houses, workplaces and infrastructure for an extra 50 million people would require about 6.8 per cent of the country to be developed, depending on how much use can be made of previously developed sites.[10] This represents 9 per cent of the productive agricultural land in the UK, suggesting total domestic production would be 9 per cent less than today's levels, unless the remaining fields were farmed more intensively. But those fields would now have to feed a significantly greater number of mouths. Rather than meeting 72 per cent of our domestic food needs, our reduced agricultural land would only be able to feed 33 per cent. We would be dependent on foreign-grown food. And therein lies the problem, for two thirds of the world's nations have populations growing faster than England: the current world average is 1.2 per cent per year. All these countries are busily turning agricultural land from fields of production into houses of consumption. This is gradually eating away their ability to produce enough food for their own populations, let alone the extra needed for export. At some point the total volume of staple crops offered for sale on the international markets will reach a maximum and then start to decline rapidly. Prices will escalate worldwide and, inevitably, fewer people will be able to afford them. International aid organisations and charities will come under pressure not to spend their money feeding overseas victims of famine but to alleviate poverty at home. On top of this, the foregoing model does not take into consideration the vast amounts of land needed for electricity generation and biofuel production – and without huge tracts of land being devoted to those forms of energy, there will be no farming, no long-distance travel and no progress towards the Sustainable Future.

The above case based on England looks 86 years ahead. *Only* 86 years. We are notionally considering the rest of the millennium here, and many of us hope that there will be thousands of years' more human health and happiness after that. Hence I have no doubt that Fukuyama's 'end of history' argument – that the whole world will one day enter a permanent state of liberal capitalism – is wrong. The resources necessary to support such a vision just don't exist. Instead, capitalism will eat the heart out of liberalism. As the demand for land increases, we will have to choose between using it for food, energy or unproductive housing. We shall therefore produce some food, some biofuel, a limited amount of electricity and an ever-decreasing number of houses. But the supply of these things will be insufficient to feed, transport and house everyone to today's standards. The poorest sectors of every community will lose out, including the poorest sections of relatively wealthy countries, and thus the hierarchical character of pre-industrial society will re-emerge in the post-industrial age.

Hierarchy seems to be the default position for human society when resources are limited. As we have seen throughout this book, the desires of the wealthy are satisfied before the needs of the rest of the population; the consequent abject poverty of the poor in times of crisis results in the differences between them and the rich becoming more pronounced. Conversely, when there is a glut of resources, there is more left over for the poor and the relative wealth of the rich is lessened. In the nineteenth and twentieth centuries we ruthlessly exploited oil, coal and natural gas – thereby dramatically reducing the risks of famine and disease and lessening the relative inequalities of wealth. Now, as these irreplaceable resources start to diminish, the social structure is beginning to return to where it was before the power of fossil fuels was harnessed.

The hierarchical structure of society is not simply the result of an unsustainable economic cycle coming to an end; it is also due to the tendency of the richer sections of society to adopt exclusive patterns of behaviour – the exclusivity being in proportion to their wealth. This is why economic growth in the twentieth century could only go so far in levelling the inequalities of society: ultimately the key resources were still owned by the rich, who married other wealthy individuals, thereby retaining control of the bulk of the capital. Such patterns of exclusivity are especially significant when the population is increasing. If the population of a country were to double over the course of a century, for instance, the rich would get richer (because they tend to marry other wealthy people and retain their assets in roughly the same number of families) while the per capita wealth of the poor would diminish significantly (as a small amount of capital would have to go around twice as many people). The result is a stretching of the scale of wealth at both ends of the social spectrum: the rich become richer and the poor become both poorer and more numerous.

Marriage is not the only form of exclusive behaviour that results in this concentration of wealth. Privilege works in conjunction with meritocracy to reinforce class-, wealth- and status-related boundaries. The high achievers of one generation not only pass on their clever genes to the next, they also pay for their children to have the best education, prepare them to engage socially with other wealthy and successful people, and encourage them to pursue significant financial reward in adult life. As a result, the next generation slips smoothly into its predecessor's shoes. The 'old boy network' is another form of exclusive behaviour, as people in positions of authority choose those from similar backgrounds to fill other positions of authority. The plain fact is that, in life, 'birds of a feather flock together'. Exaggerated forms of such behaviour include corruption and political favouritism. You only have to consider Russia in the aftermath of the breakdown of the Soviet Union to see how a leader might create a hierarchy through rewards paid to his friends. Recently it was reported that 35 per cent of Russia's wealth is owned by just 110 individuals – most of them associates of Vladimir Putin. At the same time, half a million people in that country are living in slavery.[11]

These points about social hierarchy and wealth becoming more pronounced have recently been given an extra economic emphasis by Thomas Piketty in his study *Capital in the Twenty-First Century*. One of Piketty's innovations is a means to compare the unequal distribution of capital at different periods – by comparing the ratio of a nation's total capital to its income in the currency of the time. In nineteenth-century Europe this ratio was above 650 per cent; it dropped to 250 per cent in the mid twentieth century, due to the world wars, but has been rising again since 1950 and currently is about 550 per cent.[12] This means that now, as in the nineteenth century, relatively greater earning power is in the hands of those who already have substantial wealth, and by implication, less is in the hands of those who work for a living. Piketty's explanation for this tendency for capital to grow in peacetime is that the rate of return, in the form of 'profits, dividends, interest, rents and other income from capital', which he denotes as r, consistently outstrips economic growth, g.[13] His formula of $r > g$ means that the incomes of the capital-laden wealthy make them richer at a greater rate than people who work for a living. This in turn allows those with large amounts of capital to reinvest more and more of their income, whereas workers have to consume all theirs in maintaining their standard of living. Thus large amounts of wealth create ever greater amounts, in a self-perpetuating cycle. He calls the outcome 'patrimonial capitalism', because wealth is increasingly concentrated in family dynasties. Furthermore, he sees the gap between overall r and g widening in the future. As he puts it, 'if fiscal competition proceeds to its logical conclusion – which it may – the gap between r and g will return at some point in the twenty-first century to a level close to what it was in the nineteenth century'.[14]

Piketty's use of statistics has come in for some criticism from those determined to challenge his findings, but his basic formula $r > g$ is sound. Indeed, it could be said to be the hallmark of a developed economy.[15] If all the land and natural resources in a region are owned and already being exploited to the full, it is difficult for the economy to grow. If the population is expanding at the same time, there is greater competition for the use of the capital and thus a constant pressure on r to rise. Given the finite resources of the planet, it is likely that Piketty's rule $r > g$ will be the case in the West for ever more – unless the population collapses and r diminishes as a result of a lack of demand for land and other capital assets. It follows that owners of large amounts of capital will, on average, continue to get richer than the rest of us until the disparity of rich and poor reaches a saturation point – probably equivalent to the inequalities of wealth that prevailed in the nineteenth century.

The political implications of this growing economic inequality are obvious. As wealth and political power tend to go hand in hand, the rich will once again become the controllers of society. Eastern Europe already seems to be heading back in that direction: Russia, for example, is not what most people would call a liberal democracy. A similar future awaits other wealth-dominated societies. A scholarly study measuring the influence of various interest groups on political policy-making in the USA, to be published about the same time as this book, concludes that 'economic elites and organized groups representing business interests have substantial independent impacts on US government policy, while average citizens and mass-based interest groups have little or no independent influence'.[16] In short, the USA already shows signs of becoming an oligarchy. In future, this pattern will spread across the West as resource limitations start to bite. Economic hierarchy does not just mean greater disparities of capital between the very wealthy and the poor; it also implies that working people exercise very little power.

The concentration of wealth and political power in the hands of relatively few families is one way in which capitalism will undermine liberalism. The greater impoverishment of the poor themselves is another. The crucial issue here is that the small percentage of assets owned by the poorest half of the population has to be stretched even more thinly with every new person born. A flat or house in some parts of Europe is already so expensive that most young people are priced out of the market. The average price of a house in England is seven times the average salary; in some parts of the country it is much higher still. The reason is not hard to find: too many people are competing for too little land. The population density is 410 people per square kilometre – a level more than three times that of France (120 people per square kilometre), nearly three times that of China (145) and almost on a level with India (416).[17] Holland is the only European nation with a higher population density (497 people per square kilometre). In such densely settled countries as England and Holland it is not just ownership of houses that is affected. Rented property also increases in price as landlords have to charge a commensurately greater sum to reflect the value of their investment. The wealthy, who own their homes outright, are sitting pretty: they have all their income to spend on whatever they like while those renting and those newly buying are obliged to spend up to a third of their income simply on maintaining the roof over their heads. Thus the limited supply of land impoverishes those who don't own any, just as it did in the days before artificial fertilisers and cheap transport. If you were to offer all those who do not own their own home the choice of a cheap mortgage for life or the vote, many would choose the cheap mortgage. It would represent greater financial security and personal freedom. In this way it can be seen that the civilisation curve of social reform is not irreversible. The degree of political power that people enjoy, like their investments, can go down as well as up.

I suspect that when it comes to it, universal suffrage will remain a token of liberty for a long time, centuries even, maintained by wealthy Western countries as an indicator of their acceptance of certain commonly shared values. No government will want to break the pattern – at least not in an obvious way. Instead the power of the vote will be watered down, through electoral systems controlled by political parties. The threat to liberalism here is that the political parties will increasingly be dominated by the oligarchs who bankroll them; they will tailor their agendas to the requirements of that unelected political elite. As the American study mentioned above shows, ordinary people have no means of introducing government policies; they simply participate in a system that selects spokesmen connected to the unelected political elite to approve initiatives put forward by that same elite. Partly as a conse-quence, the other social reforms and improvements of the nineteenth and twentieth centuries will prove even more vulnerable. As the per capita wealth of the poor diminishes, their living conditions can be expected to deteriorate. Workers desperate for cash will take on greater risks and more dangerous trades. Families who cannot afford high rents will end up living in squalid conditions. In severe economic downturns governments will turn to extreme measures: cutting the benefits paid to the poor, as well as healthcare and other social services.

Karl Marx would weep. In his lifetime he argued vehemently against Malthus's *Essay on the principle of population* but ultimately his vision is set to be crushed by the forces that Malthus described. Although many people, including Marx, regarded Malthus as fundamentally wrong, the basis for their doing so was that Malthus failed to foresee the effects that technology would have on the world's food supply. As is now evident, technological innovation only delays Malthusian checks, it does not end them. If technological change allows us to obtain exponentially improved results from land for another two hundred years, the population will grow commensurately until even-tually the ratio of people to resources reaches a critical level; then all the self-protective, exclusive behaviour mentioned above will start rapidly to divide the rich further from the poor and starving. Even if the world population stabilised now, at just over seven billion, it would still be increasing in relation to our diminishing ability to transport food surpluses around the globe. But the world's population is far from stable: it is expected to reach 9.5 billion within the next forty years. Nor are we likely to be able to stop this: China, the one large country that has tried to restrict its population growth in the modern era, through a one-child policy introduced in 1979, has seen its numbers increase by more than 30 per cent over that time, from less than a billion to more than 1.355 billion. Therefore we can be reasonably confident that, barring a serious worldwide epidemic, the population will continue to grow, the means of production will become concen-trated in proportionally fewer hands, and the social hierarchy will become even more pronounced. It is likely that there will be revolu-tions of desperation, not ambition, in order to redistribute wealth. But revolutions in themselves do not feed people; they merely re-allocate assets. As oil grows increasingly expensive and food prices continue to rise, the poorest citizens will not be able to afford to eat properly, will not be able to travel and will not be able to rent a home, let alone fill it with heat and light. Maybe they will still have the vote. Many will not care. Democracy will be seen as an irrelevance if it fails to deliver a reasonable standard of living. For some, helping to bring enough food into a poor household will once more become a higher priority than schooling. Prostitution will again be rife, especially among the poor, leading to a dwindling of respect for women. In such circumstances a de facto slave class is likely to re-emerge across the whole of the West. The truth is that many people would rather exchange their freedom for food and shelter than see their families starve to death. The political smile of democracy, in which economic growth is confidently presented as 'normal', will increasingly turn into the anguished grimace of the disenchanted and disappointed.

This is why the *Earthrise* picture strikes me as a profound reckoning point. It shows us our finite size, the shallowness of our pockets and the unrealistic nature of our dreams of freedom, universal well-being and equality of opportunity. Before 1968 we were able to talk of going forth, increasing and multiplying without end. Gradually the awareness spread that this was very far from the case – that the limitations of the Earth allowed us to predict some things about our future with certainty. We will never have more than we have now. We will never see real economic growth return to the high levels it reached in the twentieth century. When we look to the future, we must have crisis planning in our minds, not heartfelt idealism. Utopian thinking is a thing of the past.

Obviously I find all this extremely depressing. But I'd far rather see us head towards the Sustainable Future – even if we do not quite reach it – than allow ourselves to plummet towards the other extreme, the Universal Crisis, in which solar, hydro and wind power and biofuel do not make up a significant part of the necessary energy shortfall. In such circumstances, it won't be an ordered hierarchy that gradually emerges; it will be a disordered one. The economies of marginal states will fail. Their political systems will quickly cease to function, and anarchy will break out. In the central core of the West, exports will decline. Imports of food will also diminish as trading nations' economies collapse. Prices will go up, inflation will quickly take hold and people will stop spending on non-essentials. Businesses dependent on the sale of luxury goods will fold. There will be empty shelves in stores, followed by rationing. Law and order will start to break down. International trade between Western countries will dry up. The military will be ordered on to the streets. Those who have the means to defend their personal food supplies from looters will do so – they will have no option. What then will follow, as the economy shrinks further and the armed forces themselves dissolve into the community to protect their families, is both easy and horrific to imagine. Anyone who is in any way dependent on healthcare, commercially provided assistance, benefits, and so on will be hugely vulnerable. In marked contrast, those who have a private food supply and can guarantee to feed others will have great authority – as long as they can maintain control of it. Even in rural areas it will be difficult to recover a sustainable way of life. We have lost the farming methods we employed so successfully during the Agricultural Revolution in the eighteenth and early nineteenth centuries, due to the de-skilling of the populace in general and the agricultural labour force in particular. There will be no surpluses to supply to the towns and cities. A crisis along these lines could easily see mortality rates as great as the 60 per cent or more experienced in some regions during the Black Death. The population might well revert to the level it was in the seventeenth century as it struggles to adapt to a world without cars, lorries, tractors, artificial fertilisers and motorised fishing vessels. The one thing that seems bound to survive is the ownership of property, just as it did after the Black Death and again through the crises of the seventeenth century. As a new stability emerges, with vastly fewer people, the survivors would be quick to take possession of all they could. They would start to rebuild society, retracing many of the steps highlighted in the second half of this book, probably ending up with the sort of oligarchy envisaged in the Sustainable Future. But in the immediate aftermath, those who survive will effectively be vassals like their medieval ancestors. Independent warlords will dominate a hugely dangerous neo-feudal system, their troops defending them and their fields and workers from the attacks of rivals.

Given these two extremes, is there any way of predicting whether we will find our way to an approximation of the Sustainable Future, in which the world is riven with hierarchy and poverty but is essentially stable and relatively peaceful, or whether we will have to pass through the fiery furnace of the Universal Crisis? There are factors supporting each case. One of the prime reasons to fear the calamitous outcome is the complacency of society. People who know little about history and who cannot imagine a sudden and catastrophic downturn in their fortunes refuse to admit that they or their children will have to change

Germany | France | United Kingdom | United States

—— Total primary energy consumption
—— Energy from renewable sources

Total energy consumption in relation to total consumption of
energy from renewable sources, 1965–2013 (millions of tonnes of
oil equivalent)[18]

their ideas of what constitutes a 'normal' way of life. They will
continue to demand all the privileges of late-twentieth-century society
– until it is too late.

The complacency is clearly illustrated by the above four graphs.
None of these countries is on course to meet even half its current total
energy needs through renewable sources by 2050. Only one, Germany,
currently finds more than 10 per cent of its energy requirements from
renewable sources. While today's papers and magazines often tell us

that a child born now can expect to live a shorter life than his or her parents due to the current trend in obesity, the above graphs suggest that it won't be long before the prognosis for a shorter life expectancy will be based on exactly the opposite reason: the inadequacy of the food supply.

There are two reasons why I am confident that we will avoid calamity and end up clawing our way towards the Sustainable Future. First there is the amount of time that realistic, responsible and forward-thinking people have to plan for the end of fossil fuels. This includes the very wealthy, who have the most to lose from a cataclysmic implosion of an overstretched international economy. It also includes people like me, in the middle of the social pile, and those much less privileged. Let me be clear: in saying that greater inequality is inevitable, I am not suggesting that there is no point trying to limit it. Even if the great ship of liberal democracy and social welfare is slowly sinking, and its submergence beneath the waves of inequality and hardship seems inevitable, the last thing anyone should do is start drilling holes in the hull to make the end result come about sooner. We must do what we can to keep it afloat for as long as possible. With enough time, we might be able to lower our individual needs and expectations, so that communities move in a constructive fashion towards a more self-sufficient and sustainable way of life. We should be able to improve the outlook suggested by the above four graphs. For example, if the UK were to increase its production of energy from renewable sources by the equivalent of 2.5 million tonnes of oil annually – slightly less than the increase it achieved in 2012–13 – and at the same time reduce its primary energy consumption by just 1 per cent per year – a smaller reduction than it has achieved on average over the last ten years – then it could produce enough energy from renewable sources to meet its reduced total requirements by 2059.

My second reason for optimism is that the human race is extraordinarily adaptable. We dealt with the Black Death with remarkably little social breakdown. We not only coped with the incessant wars and famines of the seventeenth century, we managed at the same time to produce some of the greatest art, architecture and literature the world has ever seen. In the great scheme of things, you'd be mad to bet against mankind coming out of the impending crisis stronger than ever before. If the correct agricultural and technical knowledge is reintroduced into society, and enough preparation takes place in terms of sustainable energy generation, planned crop planting, limited livestock production, and the reduction of unnecessary manufacturing, there is no reason why most Western countries should not support a sizeable population. With modern technologies such as hydroelectric and solar energy and the ability to make artificial fertilisers from biofuel, it is reasonable to suppose that a highly organised approach to food growing and distribution could sustain a large population without fossil fuels. But it could not provide all those people with our current standards of living. As Paul Ehrlich has pointed out in a recent address to the Royal Society:

> to support *today's* population of seven billion sustainably (i.e. with business as usual, including current technologies and standards of living) would require half an additional planet; to do so if all citizens on Earth consumed resources at the US level would take four to five more Earths. Adding the projected 2.5 billion more people by 2050 would make the human assault on civilisation's life-support systems disproportionately worse, because almost everywhere people face systems with non-linear responses, in which environmental damage increases at a rate that becomes faster with each additional person.[19]

It follows that the poor would have to be predominantly vegetarian, as peasants had to be in the thirteenth century, simply because land planted with staple cereal crops can yield more than ten times as much food as land used for livestock.[20] Nor could the population be allowed to grow freely again. If mankind can adapt to a different diet, more physical working patterns, less travelling and smaller families, there is no reason why we should not find ourselves moving in the direction of the Sustainable Future.

To end on a positive note, certain civilisation-curve benefits are likely to be with us for centuries to come. It is in the interests of both governments and individuals for everyone to be able to read. That in turn will ensure that much knowledge that is beneficial to mankind will not be lost; for example, we are unlikely ever again to be ignorant of basic medical matters, such as the circulation of blood or germ theory. People will continue to enjoy the benefits of contraception for the foreseeable future, in respect of safe sex as well as the avoidance of unwanted pregnancies. Low-energy technological devices such as phones and computers will keep us connected. It is likely that the state will continue to suppress private violence. Some negative changes might even be reversed. Throughout this book we have seen the universality of Braudel's rule – 'a dominant capitalist city always lies at the centre'. It is not hard to see how current patterns of distribution will change when fossil fuels are no longer available. Trading regions will alter massively, at the international level as well as the local. It will no longer be possible economically to fly asparagus from Peru to Moretonhampstead, for example. Thus local markets will recover their importance, as people won't want to walk more than a few miles to buy their groceries. Communities will grow stronger. Individuals will have greater reason to strengthen their relationships with their neighbours. We will probably start to reverse the process of de-skilling. Overall, resource depletion should not counteract any of the positive changes in the first eight chapters of this book, and many of the technological ones since 1800 should continue to benefit us, albeit not equally.

Whether you see the following as a positive sign or not is a matter of personal opinion. It seems to me highly likely that religion will become more prevalent in the West as the greater hierarchies take hold and the majority of people become comparatively poorer. As the Gallup survey of religion indicated, there is a strong relationship between religion and poverty across the world. This should not be taken to mean that poverty causes religious observance – the relationship between money and religion is not that simple. It is much more likely that the opposite is true: that money results in a lack of spiritual devotion. Either way, I suspect that the consolations of faith and the communities that religions create will prove important again in the future. The world's traditional religions have all catered for both lords

Per capita GDP	% stating that religion is an important part of their daily lives
< $2,000	95%
$2,000–$5,000	92%
$5,000–$12,500	82%
$12,500–$25,000	70%
>$25,000	47%

Importance of religion according to wealth, according to the international survey conducted by Gallup (2009)[21]

and servants for many centuries, seemingly being tailored to the demands of a hierarchical society. As we return to such a hierarchy, I fully expect the world's faiths to come into their own.

In conclusion, the reason why it matters that the twentieth and nineteenth centuries saw the most change is because many of the advances experienced in those two centuries were dependent on an anomalous windfall of energy and will undoubtedly be reversed at some point in the future. It is thus highly likely that society will experience an even greater set of social changes in this century or the next than it did in the twentieth. It might take a hundred years or more, but we are going to see a return to the extreme hierarchies of the pre-industrial age. Over the next thousand years, we will witness the downward curve of the common standard of living in the West and the increased power of the very wealthy. We will return to a point that, in terms of social structure, will have more in common with the world of 1800 than that of 2000. The only question is whether we will get there painfully and suddenly, or gradually.

Outside the sun is shining. As I sit here I can hear the bells of Moretonhampstead church ringing, as they have done for centuries. I can hear a motorbike, its throttle twisted hard as the rider comes out of a corner on the road from Exeter. My mind goes back to the priests who came here on foot a thousand years ago, and stood near the cross outside this house, preaching the Word of God that would eventually bind this small place into the vast network of the human race. Tomorrow the newspapers will be filled with the flotsam and jetsam of modern life – international crises, stock market reports, murder trials, sex scandals, and an aircraft lost without trace in the South China Sea. And at the end of it all, I find myself wondering what *hasn't* changed over the last thousand years, and what won't change over the next. At first those questions seem vast, and overwhelming. But then I think about them again. I picture a troubadour singing in the shadows of a hall fire. I imagine thousands of people walking beneath the overhanging eaves of narrow streets to see Shakespeare's plays. I hear the shouts of drunken farm workers in the candlelit gloom of a seventeenth-century inn as Jan Steen studies their ruddy faces, preparing to paint them. The simplicity of the answer makes me smile. What doesn't change is that we find so many things in life worthwhile – love, beauty, children, the comfort of friends, telling jokes, the joy of eating and drinking together, storytelling, wit, laughter, music, the sound of the sea, the warmth of the sun, looking at the Moon and stars, singing and dancing …

What won't change? Everything that allows us to lose ourselves in the moment.

Everything that is worth dreaming about.

Everything that is without price.

APPENDIX

Population estimates

European population figures for the early centuries before 1500 are very difficult to estimate with any accuracy. Paolo Malanima conveniently quotes several demographers' estimates for the year 1000 in his paper on medieval growth.[1] B. T. Urlanis (in 1941) estimated that Europe had 56.4 million people; J.-N. Biraben (1969) 43 million; J. C. Russell (1973) 38.5 million, C. McEvedy and R. Jones (1978) 36 million; H. Le Bras (1993) 43 million; A. Maddison (2007) 39.2 million; and Malanima himself (2009) 47 million. Leaving aside the highest and lowest of these, the mean of the remainder is 42.1 million. With regard to the population in 1500, the same demographers have the following figures: 100.4 million (Urlanis); 84 million (Biraben); 81.8 million (Russell); 81 million (McEvedy and Jones); 84 million (Le Bras); 87.7 million (Maddison) and 84.8 million (Malanima). Again, leaving off the highest and lowest, the mean of the remainder is 84.5 million. There is a close consensus on this figure for 1500. Only Urlanis, the earliest of these demographers, differs from the range 84 million +/− 3.7 million. The figure of 84 million also appeals to Massimo Livi Bacci, who newly calculated his figure from national datasets.[2]

Given the wildly varying range of figures for the period before 1500, I have revisited national estimates for the three most fully documented countries to develop a core on which to build my own estimates. Fortunately, these three countries are reasonably representative of Europe, one being northern European (England), one central (France) and one Mediterranean (Italy).

ENGLAND

The figures in Tables 1.1 and 1.2 for the period 1086–1541 are based on the annual growth statistics established from manorial data by Stephen

Date	Population	Date	Population	Date	Population
1086	1.71	1240	4.15	1400	2.08
1100	1.84	1260	4.30	1420	2.04
1120	2.07	1280	4.46	1440	1.96
1140	2.32	1300	4.35	1460	1.96
1160	2.61	1320	4.40	1480	2.08
1180	2.93	1340	4.57	1500	2.21
1200	3.37	1360	2.57	1520	2.34
1220	3.98	1380	2.44	1540	2.82

TABLE 1.1 Estimated population of England per twenty years (millions)

Broadberry, Bruce M. S. Campbell and Bas van Leeuwen of the University of Warwick in their paper 'English Medieval Population: reconciling time series and cross-sectional evidence' (2010).[3] These suggest a decline of the population in 1348–51 of 46 per cent. The studies assessed by Ole Benedictow suggest a national mortality figure for England of about 62.5 per cent for these years.[4] In trying to reconcile these different figures, we have to note that Benedictow's figure for the fall in *taxpayers* was slightly less, 50–55 per cent, and this class is nearer the population reflected in the sources that the Warwick group used. That still leaves something of a discrepancy, albeit of only 4–9 per cent. If we do apply Benedictow's 55 per cent depopulation figure for England, then this suggests the population was around 5.8 million in 1300, 4.0 million in 1200 and 2.2 million in 1100 (using back projection), which further indicates a population of 2.0 million at Domesday, which is not impossible. However, the Warwick group's figures have an integrity that cannot be replicated by taking a death rate from one sample and laying it across another. For instance, Benedictow's high-mortality samples might have seen greater growth than the Warwick group's sample before 1348. In addition, the Warwick group analysed the agrarian output of England at an estimated peak of 4.81 million (in 1348) and reckoned that it would have been difficult to feed that number, let alone a million more. At no time before 1700 had England sustained a population of more than 5.4 million. The probability is that the truth lies between the two extremes – that is, between the Warwick group model and 5.4 million – and I have chosen to use the Warwick group's figures at face value in order not to exaggerate the population of England in 1300, and thus that of Europe as a whole at that time.

The figure of 1.5 million for the year 1000 in Table 1.2 (overleaf) is simply a round figure based on the assumption that the population was growing very slowly until about 1050, and then gradually faster until it reached 0.58 per cent in the twelfth century. The estimate of 1.5 million would imply an average annual increment rate over the period 1000–86 of just over 0.15 per cent. Sources for later centuries are given in the endnotes.[5]

FRANCE

The figures in Table 1.2 for the period 1000–1400 were originally drawn from the work of J. C. Russell.[6] These high figures correspond with an estimate by Ferdinand Lot that the population of France in 1328 was in the region of 22 million, based on the hearth tax records of that year. Lot's population densities have been independently supported by later hearth tax studies conducted by Norman Pounds and Charles Roome.[7] A figure of more than 20 million corresponds with Benedictow's high mortality (50–60 per cent) in France in the years 1347–51. The National Institute of Statistics and Economic Studies (INSEE) has a web page that suggests the 2,411,149 hearths recorded in 1328 for 24,150 parishes in the kingdom of France indicate a total population of 19 million for the country, 3 million less than Lot.[8] With this in mind, the figure of 20.4 million for the country in 1300 has been presumed not to have increased much more before the Black Death, peaking at 21 million in 1340. This would be less than the maximum sustainable population of 103 per square mile that is to be noticed as a limit in England in 1700, and less than the 22.6 million (92 per square mile) of France in 1700. A population decline of 50 per cent for the years 1347–51 has been applied to this number, informed by Benedictow's estimate of the mortality. The next reasonably reliable figure for the population of France is about 19.5 million for the mid sixteenth century.[9] A population of 21 million suffering 50 per cent mortality would have required an average annual growth rate thereafter of 0.31 per cent to recover to 19.5 million by 1550. Applying this figure points to a French population in 1450 of about 14.3 million. This is very close to Pounds and Roome's estimate that the population density of France in the years around 1450 was about two thirds that of 1328. Thus it has been adopted for the figures in Table 1.2. Sources for later centuries are given in the endnotes.[10]

ITALY

Applying Benedictow's higher plague mortality figures for Italy (50–60 per cent) to the figures suggested by Federico and Malanima in their 2004 article would suggest the population of Italy was about 14.9 million prior to the plague of 1347–51.[11] At 128 per square mile, this is significantly in excess of the density of 103 per square mile noted in England for 1700 and that of 92 in France in the same year. Higher

	England	%	France	%	Italy	%	Total	%
1000	1.50	–	7.00	–	5.80	–	14.30	–
1100	1.84	23%	8.06	15%	7.00	21%	16.90	18%
1200	3.37	83%	11.96	48%	9.90	41%	25.23	49%
1300	4.35	29%	20.41	71%	12.50	26%	37.26	48%
1400	2.08	-52%	12.26	-40%	8.00	-36%	22.34	-40%
1500	2.21	6%	16.70	36%	9.00	13%	27.91	25%
1600	4.162	89%	19.60	17%	13.273	47%	37.035	33%
1700	5.211	25%	22.60	15%	13.481	2%	41.292	11%
1800	8.671	66%	28.70	27%	18.092	34%	55.463	34%
1900	30.072	247%	40.681	42%	32.966	82%	103.719	87%
2000	49.139	63%	59.268	46%	56.996	73%	165.402	59%

TABLE 1.2 Populations of England, France and Italy (millions). Please note that all the totals and percentages were calculated before rounding up or down to two or three decimal places.

population densities are possible when trade networks allow, as shown by the high densities achieved in the Netherlands and Belgium in 1700 (153 and 172 per square mile respectively), but it is difficult to see how Italy in 1300 could have sustained a population so much larger than any it achieved before 1700, despite its advanced trade, especially since few neighbouring countries would have been producing easily transportable large surpluses of food. For this reason, Benedictow's range of Italian Black Death mortality is considered to err on the high side. Nevertheless, as his Black Death mortality figures were not published at the time Federico and Malanima composed their article, it seems they did not take the possibility of a population of 13 million in 1300 into account. Therefore I have preferred the figure for Italy revised upwards in Malanima's *Pre-modern European Economy* (2009), Chapter One, which suggests 12.5 million in 1300. Sources for later centuries are given in the endnotes.[12]

EUROPE AND THE WORLD

The figures in Table 1.2 suggest a different story to the population estimates outlined at the start of this appendix. According to the figures put forward by Livi Bacci for 1550, the populations of England, France and Italy amounted to 35 per cent of the European total. Malanima's figures suggest that England, Wales, France and Italy amounted to 27.5 million of 84.85 million for 1500; deducting the Welsh population of about 300,000, this suggests that England, Italy and France represented 32 per cent of the European total. Continuing to use Malanima's estimates, the three countries amounted to 33 per cent of the European total in 1400; 34.9 per cent in 1300; 34 per cent in 1200; 35 per cent in 1100 and 34.5 per cent in 1000. This all appears very consistent: no less than 32 per cent and no more than 35 per cent. The figures in Table 1.2 for 1500, 1600 and 1700 also suggest these three countries consistently represented 33 per cent of the whole European population prior to the Agricultural Revolution. If the population of Europe may be estimated by using the three countries as a 33 per cent sample, then multiplying the population figures in Table 1.2 by 1/0.33 yields population estimates as in 'Method A' in Table 1.3. Alternatively, using the increments calculated in Table 1.2 for the three countries

Date	Malanima (2009)	% change	Method A (proportional, at 1/33 %)	Method B (back projection from 84 million in 1500)	% change
1000	47.1	—	43.3	43.1	
1100	55.6	18%	51.2	50.9	18%
1200	76.7	38%	76.4	75.9	49%
1300	93.6	22%	112.9	112.2	48%
1400	67.8	28%	67.7	67.3	−40%
1500	84.8	25%	84.5	84.0	30%

TABLE 1.3 Population estimates for Europe 1000–1500 (millions)

and projecting back from the consensus of 84 million in 1500 yields population as in 'Method B'. The figures correspond quite closely to the average of 42.1 million for the year 1000 derived above from the demographers named at the start of this appendix. They also closely correlate with Malanima's figures (which are on the high side, compared to those of others) for 1200 and 1400. However, the figure for 1300 is much higher than any demographer mentioned above has suggested.

The figures in Table 1.2 that underlie Method A and Method B are based on the most reliable data available in Europe. There is no reason to suppose that these three countries differed wildly only in 1300 from being 33 per cent of the European total. Thus it seems likely that the population of Europe rose to 112 million in 1300. It is significant that the next time the population of the three countries rose to more than 37 million, in 1700, the population of the continent as a whole amounted to 125 million. As this was before the Agricultural Revolution had taken hold, this further backs up the theory that Europe could have supported a 112 million population before the Black Death.

The reason why this has not been propounded by European demographers in the past is probably because it was not appreciated how high the Black Death mortality was. Benedictow's figures suggest a significantly greater drop in population than most pre-2004 demographers imagined. But it should not be assumed that we have simply taken his depopulation conclusions at face value; we have actually erred on the side of caution, by using significantly lower mortality figures than his findings showed. Many historians have suggested a population of more than 5 million for England in 1300, and Lot suggested 22 million for France. If we take the figure of 55 per cent as representing the depopulation of England in 1348–51 – still considerably less than Benedictow's estimate of 62.5 per cent – we can reconstruct the population of England in 1300 as 5.8 million, which inflates the three-country sample by a full million. Adding an extra million for France on the basis of Lot's research suggests the European total was nearer 120 million. The figure of 112 million for Europe in 1300 is thus a conservative one, even though it is higher than anyone has previously suggested.

I have used the Method B figures in Table 1.4 and throughout this book. For the population of Europe from 1500 on, I have used the figures in Livi Bacci's *Population History of Europe*, pp.8–9. The figure for 2000 is from a report by the United Nations, Department of Economic and Social Affairs, Population Division: 'World Population Prospects: The 2012 Revision' (2013). For the world population figures in the following table, I have used the figures of J.-N. Biraben as quoted by the US census department.[13] These have not been adjusted to account for the higher population of Europe in 1300.

Date	Europe	% change	The world	% change
1000	43	–	254	12%
1100	51	18%	301	19%
1200	76	49%	400	33%
1300	112	48%	432	8%
1400	67	–40%	374	–13%
1500	84	25%	460	26%
1600	111	38%	579	31%
1700	125	13%	679	17%
1800	195	56%	954	41%
1900	422	116%	1,633	71%
2000	729	73%	6,090	273%

TABLE 1.4 Population of Europe and the world (millions)

Notes

Unless otherwise specified, London is the place of publication.

1001–1100 *The Eleventh Century*

1   For Europe, see N. J. G. Pounds, *An Economic History of Medieval Europe* (1974), p. 99. The scarcity of coin hoards in Devon and Cornwall was drawn to my attention by Henry Fairbairn at a paper given to the London Medieval Society in April 2012.

2   C. H. Haskins, *The Renaissance of the Twelfth Century* (2nd edn, 1955, 5th imp., 1971), p. 72.

3   The Devon figure is argued more fully in Bill Hardiman and Ian Mortimer, *A Guide to the History and Fabric of St Andrew's Church, Moretonhampstead* (Friends of St Andrew's, 2012), pp. 4–5. For the Paderborn figure, see *The New Cambridge Medieval History*, vol. 3, p. 46.

4   Christopher Holdsworth, *Domesday Essays* (Exeter, 1986), p. 56; Neil S. Rushton, 'Parochialisation and Patterns of Patronage in 11th Century Sussex', *Sussex Archaeological Collections*, 137 (1999), pp. 133–52, at p. 134.

5   Rushton, 'Parochialisation', Appendix 1.

6   Quoted in Pierre Bonassie, trans. Jean Birrell, *From Slavery to Feudalism in South-western Europe* (Cambridge, 1991), p. 1.

7   The quotation from Gregory the Great is paraphrased from Frederik Pijper, 'The Christian Church and Slavery in the Middle Ages', *American Historical Review*, 14, 4 (July 1909), pp. 675–95, at p. 676. The note on St Gerald of Aurillac is from Bonassie, *Slavery to Feudalism*, p. 55.

8   John Gillingham, 'Civilising the English? The English Histories of William of Malmesbury and David Hume', *Historical Research*, 74, 183 (February 2001), pp. 17–43, esp. p. 36. I am grateful to Dr Marc Morris for bringing this to my attention.

356   HUMAN RACE

9   Plinio Prioreschi, *A History of Medicine. Vol. 5: Medieval Medicine* (Omaha, 2003), p. 171.

10   Marc Morris, *The Norman Conquest* (2012), p. 334.

11   Michael Hart, *The 100* (1st edn, 1978, 2nd edn, 1992).

1101–1200 The Twelfth Century

1   Maurice Keen, *Chivalry* (1984), p. 88.

2   John Langdon, *Horse, Oxen and Technological Innovation* (Cambridge, 1986), p. 98.

3   Geoffrey Parker, *The Global Crisis: War, Climate Change and Catastrophe in the Seventeenth Century* (2013), p. 17.

4   David Knowles and R. Neville Hadcock, *Medieval Religious Houses: England and Wales* (2nd edn, 1971), p. 494; John T. Appleby, *The Troubled Reign of King Stephen* (1969), p. 191.

5   Jacques LeGoff, trans. Arthur Goldhammer, *The Birth of Purgatory* (1986), pp. 222–3.

6   C. H. Haskins, *The Renaissance of the Twelfth Century* (2nd edn, 1955, 5th imp., 1971), pp. 38–9.

7   Ibid., p. 71.

8   Ralph Norman, 'Abelard's Legacy: Why Theology is Not Faith Seeking Understanding', *Australian eJournal of Theology*, 10 (May 2007), p. 2; M. T. Clanchy, *Abelard: A Medieval Life* (Oxford, 1999), p. 5.

9   According to Charles Homer Haskins, 'more of Arabic science in general passed into western Europe at the hands of Gerard of Cremona than in any other way'. See Haskins, *Renaissance*, p. 287.

10   Roy Porter, *The Greatest Benefit to Mankind* (1997), p. 110.

11   Plinio Prioreschi, *A History of Medicine. Vol. 5: Medieval Medicine* (Omaha, 2003), pp. 168–9.

12   Vivian Nutton, 'Medicine in Late Antiquity and the Early Middle Ages', in Lawrence I. Conrad et al. (eds), *The Western Medical Tradition 800 BC to 1600 AD* (Cambridge, 1995), pp. 71–87.

13   Stanley Rubin, *Medieval English Medicine* (Newton Abbot, 1974), p. 105.

14   Haskins, *Renaissance*, pp. 322–7.

1201–1300 *The Thirteenth Century*

1   Maurice Keen, *Chivalry* (Yale, 1984), p. 87.

2   The bishop of Winchester levied a customary poll tax of one penny on males over the age of 12 on his manor of Taunton in Somerset, and the income was recorded on the bishop's pipe roll every year from 1209. The numbers increased from £2 11s. 0d in 1209 (612 males) to £6 4s. 0d in 1311 (1,488 males), an increment of 0.85 per cent per year. See N. J. G. Pounds, *An Economic History of Medieval Europe* (1974), p. 145.

3   Samantha Letters, 'Gazetteer of Markets and Fairs in England and Wales to 1516', http://www.history.ac.uk/cmh/gaz/gazweb2.html. Downloaded 13 March 2014.

4   Quoted in Pounds, *Economic History*, p. 251.

5   Ibid., p. 100.

6   Fernand Braudel, *Civilisation and Capitalism, 15th–18th Centuries* (3 vols, 1984), iii, p. 93; Letters, 'Gazetteer of Markets and Fairs'.

7   Braudel, *Civilisation and Capitalism*, iii, p. 27.

8   Ibid., iii, p. 113.

9   Letters, 'Gazetteer of Markets and Fairs'.

10   Michael Clanchy, *From Memory to Written Record: England 1066–1307* (2nd edn, 1993).

11   Quoted in W. L. Warren, *King John* (1961), pp. 245–6.

12   William Woodville Rockhill (ed.), *The Journey of William of Rubruck to the Eastern Parts of the World 1253–55* (1900), pp. 211, 223.

1301–1400 *The Fourteenth Century*

1   Robert S. Gottfried, *The Black Death* (1983), p. 25.

2   This is adapted from the example given in Geoffrey Parker, *The Global Crisis: War, Climate Change and Catastrophe in the Seventeenth Century* (2013), pp. 19–20.

3   This is based on my revised population figures for Europe 1300 in the Appendix.

4   For the mortality of 1348–51 being in the region of 62.5% in England, see Ole J. Benedictow, *The Black Death, 1346–1353: The*

Complete History (2004). For the 45% figure, see the Appendix to this book.

5   Benedictow, *Black Death*, p. 283, quoting Marchionne di Coppo Stefani, trans. in J. Henderson, 'The Black Death in Florence: Medical and Communal Responses', in *Death in Towns* (1992), p. 145.

6   Benedictow, *Black Death*, p. 291 (mortality in Florence); Gottfried, *Black Death* (1983), p. 47 (Boccaccio).

7   Gottfried, *Black Death*, p. 49.

8   Benedictow, *Black Death*, p. 356.

9   Various writers include this story of the English ship; Benedictow, *Black Death*, p. 156, suggests the date of early July 1349.

10   Benedictow, *Black Death*, p. 383.

11   Clifford Rogers, *War Cruel and Sharp: English Strategy under Edward III, 1327–1360* (2000), pp. 40–1.

12   Sir Herbert Maxwell (ed.), *The Chronicle of Lanercost* (1913), p. 271.

13   Ian Mortimer, *The Perfect King* (2006), pp. 20–1; Rupert Taylor, *The Political Prophecy in England* (New York, 1911; rep. 1967), pp. 160–4; T. M. Smallwood, 'Prophecy of the Six Kings', *Speculum*, 60 (1985), pp. 571–92.

14   It is often said that Edward III's archers were Welsh. Jim Bradbury discusses this at length in his book *The Medieval Archer* (Woodbridge, 1985; rep. 1998), pp. 83–90, and finds the evidence for such a claim wanting. In fact the early archers credited with making significant tactical advances were English. This is borne out by the evidence for the first part of the reign of Edward III. In 1334 the king and his nobles respectively provided 481 and 771 mounted archers but at the same time the king summoned 4,000 archers from Lancashire and more than 5,000 from Yorkshire (see my biography of Edward III, *The Perfect King*, pp. 119–20). Although Bradbury concludes that the long-bow 'does not belong to any one area in particular' (*Medieval Archer*, p. 84), it is striking how much Edward III looked to the north to supply him with longbows in the early part of his reign. In later years the archers of Cheshire were renowned as the best in the country.

15 Louise Ropes Loomis, *The Council of Constance* (1961), pp. 316, 456.

16 Ibid., pp. 340–1. In 1415 England had 17 archbishops and bishops, Wales 4, Scotland 13 (who were not loyal to Henry V) and Ireland 34 (few of whom were loyal to Henry V).

17 J. R. Lumby (ed.), *Chronicon Henrici Knighton, vel Cnitthon, monachi Leycestrensis* (2 vols, 1889–95), ii, p. 94.

18 T. B. James and J. Simons (eds), *The Poems of Laurence Minot, 1333–1352* (Exeter, 1989), p. 86.

19 Chris Given-Wilson (ed.), *Parliamentary Rolls of Medieval England* (CD ROM ed., Woodbridge, 2005), Parliament of 1382.

20 Joshua Barnes, *The History of that Most Victorious Monarch, Edward III* (1688), preface.

21 Rogers, *War Cruel and Sharp*, p. 1.

1401–1500 *The Fifteenth Century*

1   For further argument on this point, see Ian Mortimer, 'What Hundred Years War?', *History Today* (October 2009), pp. 27–33.

2   C. R. Boxer, *The Portuguese Seaborne Empire 1415–1825* (1969), p. 26.

3   Accurate measurements of the Earth's diameter did exist – Eratosthenes and Posidonius had both come up with figures correct to the nearest thousand miles in the ancient world – but these works were unknown to Columbus.

4   Jean Gimpel, *The Medieval Machine* (2nd edn, 1988), p. 153. An astronomical clock powered by dripping mercury appears in a Castilian manuscript of that same decade.

5   Ian Mortimer, *The Perfect King* (2006), p. 288.

6   Gimpel, *Medieval Machine*, p. 169.

7   Lynn White Jr, *Medieval Technology and Social Change* (OUP paperback edn, 1964), pp. 125–6; Ian Mortimer, *The Fears of Henry IV* (2007), pp. 92–3.

8   White, *Medieval Technology*, p. 127.

9   The National Archives, Kew, London: DL 28/1/2 fol. 15v.

10 Lucy Toulmin Smith (ed.), *Expeditions to Prussia and the Holy Land Made by Henry Earl of Derby* (1894), p. 93.

11 Chris Woolgar, *The Senses in Medieval England* (2006), p. 139.

The cheaper two mirrors were worth 15s. 5d and 7s. 9d; the third mirror was a jewelled one worth £13 10s.

12 It is possible that Gutenberg did not 'invent' the printing press but learnt the idea. As is well known, printing was familiar centuries earlier in China than it was in the West, and the Koreans started to use movable type a few decades before Gutenberg. It is also said that Laurens Janszoon Coster of Haarlem was printing texts in Latin with movable wooden type in the early 1420s, prior to the fire that destroyed his home town. The printed woodblock was certainly in use in the West decades before Gutenberg. Asa Briggs and Peter Burke, *A Social History of the Media* (2005), p. 31.

13 Ibid., p. 33.

14 Evan T. Jones and Alwyn Ruddock, 'John Cabot and the Discovery of America', *Historical Research*, 81 (2008), pp. 224–54.

1501–1600 *The Sixteenth Century*

1 Asa Briggs and Peter Burke, *A Social History of the Media* (2005), p. 13.

2 Ibid., p. 15.

3 Roy Porter, *The Greatest Benefit to Mankind* (1997), p. 132.

4 W. B. Stephens, 'Literacy in England, Scotland and Wales 1500–1900', *History of Education Quarterly* 30, 4 (1990), pp. 545–71, at p. 555.

5 Other English examples include the bishops of Hereford and Lincoln, who actively resisted Edward II; John Stratford, archbishop of Canterbury, who tried to resist Edward III; Thomas Arundel, archbishop of Canterbury, who opposed Richard II; Richard Scrope, archbishop of York, who opposed the government of Henry IV; Cardinal Beaufort, who tried to bring about the abdication of Henry IV.

6 William P. Guthrie, *The Later Thirty Years War* (2003), p. 16; Geoffrey Parker, 'The Military Revolution 1560–1660 – a Myth?', *Journal of Modern History*, 48, 2 (1976), pp. 195–214, at p. 199.

7 The debate was started by Michael Roberts, *The Military Revolution 1560–1660* (Belfast, 1956).

8 For the introduction of the stirrup, see Lynn White Jr, *Medieval*

Technology and Social Change (OUP paperback edn, 1964), pp. 1–39, esp. p. 24.

9   In addition, the case of Japan shows that in that country at least, it was ambitious and strong government that led to the demand for firearms, not vice versa. See Stephen Morillo, 'Guns and Government: A Comparative Study of Europe and Japan', *Journal of World History*, 6, 1 (1995), pp. 75–106.

10 Parker, 'Military Revolution', p. 206.

11 Geoffrey Parker, *Global Crisis: War, Climate Change and Catastrophe in the Seventeenth Century* (2013), p. 32.

12 C. R. Boxer, *The Portuguese Seaborne Empire* (1969), p. 49.

13 110 per 100,000 people equates more or less to 165 per 100,000 adults, the Dodge City figure. See Carl I. Hammer Jr, 'Patterns of Homicide in a Medieval University Town: Fourteenth-century Oxford', *Past & Present*, 78 (1978), pp. 3–23, at pp. 11–12; Randolph Roth, 'Homicide Rates in the American West', http://cjrc.osu.edu/homicide-rates-american-west-randolph-roth. Downloaded 20 January 2014.

14 Manuel Eisner, 'Long-term Historical Trends in Violent Crime', *Crime and Justice*, 30 (2003), pp. 83–142, at p. 84.

15 This chart is based on the figures in Eisner, 'Long-term Historical Trends'. The English figure in this chart is a simple average (mean) of those for 1400 and 1600, and the Italian figure for 1700 is a simple average (mean) of the figures for 1650 and 1750.

16 Stephen Pinker, *The Better Angels of Our Nature* (2011), pp. 77–97.

17 Ibid., pp. 91–2.

18 Stephen Broadberry, Bruce Campbell, Alexander Klein, Mark Overton and Bas van Leeuwen, *British Economic Growth 1270–1870* (2011). For fuller statistics, see the GDP per capita tables in the 'Personal enrichment' section of the Conclusion, drawn from the same source.

19 B. R. Mitchell, *British Historical Statistics* (1988, paperback edn, 2011), pp. 166–9.

20 Pinker, *Better Angels*, p. 89.

21 Azar Gat, 'Is War Declining – and Why?', *Journal of Peace Research*, 50, 2 (2012), pp. 149–57, at p. 149.

22 Eisner, 'Long–term Historical Trends', p. 107, quoting Randolph Roth, 'Homicide in Early Modern England, 1549–1800: The Need

for a Quantitative Synthesis', *Crime, History and Society*, 5, 2 (2001), pp. 33–68.

23 Quoted in Henry Kamen, *The Iron Century: Social Change in Europe 1550–1660* (1971), p. 6.

1601–1700 *The Seventeenth Century*

1 Henry Kamen, *The Iron Century: Social Change in Europe 1550–1660* (1971), p. 13 (Geneva and Paris); E. A. Wrigley and R. S. Schofield, *The Population History of England 1541–1871: A Reconstruction* (1981), pp. 532–3.

2 Geoffrey Parker, *The Global Crisis: War, Climate Change and Catastrophe in the Seventeenth Century* (2013), p. 17.

3 Ibid. pp. 17, 57.

4 Cecile Augon, *Social France in the XVIIth Century* (1911), pp. 171–2, 189, quoted in the Internet Modern History Sourcebook, http://www.fordham.edu/halsall/mod/17france-soc.asp. Downloaded 22 January 2014 (200 dying by roadside); Kamen, *Iron Century*, pp. 34–5 (rotting flesh).

5 Parker, *Global Crisis*, p. 100.

6 For the height of the troops, see ibid. p. 22.

7 James Sharpe, *Instruments of Darkness: Witchcraft in Early Modern England* (1996; paperback edn, 1997), pp. 256, 257.

8 The data in this table are taken from tables 2.3–2.5 in Ian Mortimer, 'Medical Assistance to the Dying in Provincial Southern England, circa 1570–1720' (University of Exeter PhD thesis, 2 vols, 2004), pp. 92–3. The sample on which it is based is a selection of 9,689 probate accounts relating to the estates of deceased males in the diocese of Canterbury. The dates specified in the table are the central points in the date ranges 1570–1599, 1600–1629, 1630–1659 (1649), 1660–1689, and 1690–1719. Note that there are no data for 1650–1659, so the date '1645' actually represents accounts dated 1630–49. A set of charts based on this data but adjusted to reflect the status of the deceased may be found in the published version of this thesis: Ian Mortimer, *The Dying and the Doctors: The Medical Revolution in Seventeenth-Century England* (Royal Historical Society, 2009), pp. 19–20.

9   Teerapa Pirohakul and Patrick Wallis, 'Medical Revolutions? The
     Growth of Medicine in England, 1660–1800', LSE Working Papers
     no. 185 (January 2014). Available at http://www.lse.ac.uk/
     economicHistory/workingPapers/2014/WP185.pdf. Downloaded
     29 April 2014.

10   Ian Mortimer, 'Index of Medical Licentiates, Applicants, Referees
     and Examiners in the Diocese of Exeter 1568–1783', *Transactions
     of the Devonshire Association*, 136 (2004), pp. 99–134, at p. 128. Joshua
     Smith not only qualified as 'of Mortonhampstead'; he lived in
     the parish, as shown by his son being baptised here in 1666 and
     his own burial here in 1672. His son, also called Joshua, obtained
     a licence to practise surgery in 1700.

11   Quoted in Ralph Houlbrooke, *Death, Religion and the Family in
     England 1480–1750* (Oxford, 1998), pp. 18–19.

12   See Mortimer, *Dying and the Doctors*, p. 211.

13   http://www2.census.gov/prod2/statcomp/documents/CT1970
     p2-13.pdf. Downloaded 2 January 2014.

14   http://www.lib.utexas.edu/maps/historical/ward_1912/amer
     ica_north_colonization_1700.jpg. Downloaded 2 January
     2014.

15   Fernand Braudel, *Civilisation and Capitalism. Vol. 3: The Perspective
     of the World* (1979), p. 190; C. R. Boxer, *The Portuguese Seaborne
     Empire* (1969), p. 114.

16   Braudel, *Civilisation and Capitalism*, p. 522.

17   Joan Thirsk and J. P. Cooper, *Seventeenth Century Economic
     Documents* (Oxford, 1972), p. 780.

18   Jancis Robinson, *The Oxford Companion to Wine* (Oxford, 1994),
     p. 363.

19   Kamen, *Iron Century*, p. 167.

20   Faramerz Dabhoiwala, *The Origins of Sex* (2012), p. 43.

1701–1800 *The Eighteenth Century*

1   *Shakespeare's England: An Account of the Life and Manners of his
     Age* (2 vols, 1917), i, p. 202.

2   R. C. Tombs, *The Bristol Royal Mail: Post, Telegraph, and Telephone*
     (n.d.), p. 11.

3   For example, *Plymouth and Dock Telegraph and Gazette* for 4 May
     1822. An illustration of this appears in the second plate section
     of this book.

4   Fernand Braudel, *Civilisation and Capitalism. Vol. 3: The Perspective
     of the World* (1979), pp. 316–17.

5   Asa Briggs and Peter Burke, *A Social History of the Media* (2005),
     p. 81.

6   It is said that the abdication of James II was not known in the
     Orkneys for three months. J. H. Markland, 'Remarks on the
     Early Use of Carriages', *Archaeologia*, 20 (1824), p. 445.

7   *London Magazine*, 3 (July–Dec. 1784), p. 313.

8   *Gentleman's and London Magazine, or Monthly Chronologer* (1785),
     p. 86.

9   Mark Overton, http://www.ehs.org.uk/dotAsset/c7197ff4-54c5-
     4f85-afad-fb05c9a5e1e0.pdf ('caricature'); http://www.bbc.
     co.uk/history/british/empire_seapower/agricultural_revolution
     _01.shtml (derogatory assessments). Both downloaded 30
     January 2014.

10   John Mortimer, *The Whole Art of Husbandry or the Way of
     Managing and Improving Land* (2 vols, 4th edn, 1716), i, pp. 32–3,
     131, 157–60; ii, p. 177.

11   Liam Brunt, 'Mechanical Innovation in the Industrial Revolution:
     The Case of Plough Design', *Economic History Review*, New
     Series, 56 (2003), pp. 444–77.

12   Although potato farming certainly helped, it did not cover
     more than 2% of the agricultural land in 1801. See Mark
     Overton, *Agricultural Revolution in England* (Cambridge, 1996),
     p. 102.

13   E. A. Wrigley, 'The Transition to an Advanced Organic Economy:
     Half a Millennium of English Agriculture', *Economic History
     Review*, New Series, 59, 3 (August 2006), pp. 435–480 at p. 440.

14   Wrigley, 'Transition', p. 451.

15   Claude Masset, 'What Length of Life Did Our Forebears Have?',
     Population & Societies, 380 (2002), www.ined.fr/fichier/t_publi
     cation/474/publi_pdf2_pop_and_soc_english _380.pdf (down-
     loaded 27 January 2014), quoting Élise de La Rochebrochard,
     'Age at Puberty of Girls and Boys in France: Measurements
     from a Survey on Adolescent Sexuality', *Population: An English*

Selection, 12 (2000), pp. 51–80; Peter Laslett, 'Age at Menarche in Europe since the Eighteenth Century', *Journal of Interdisciplinary History*, 2, 2 (1971), pp. 221–36.

16 Ian Davidson, 'Voltaire in England', *Telegraph*, 9 April 2010.

17 http://www.constitution.org/jjr/ineq_04.htm. Downloaded 24 February 2014.

18 Faramerz Dabhoiwala, *The Origins of Sex* (2012), pp. 57–9.

19 Ibid. p. 66.

20 Ibid. pp. 103 (Locke), 108 (Hume).

21 Cyril Bryner, 'The Issue of Capital Punishment in the Reign of Elizabeth Petrovna', *Russian Review*, 49 (1990), pp. 389–416, at pp. 391 (abolition), 416 (unpopularity).

22 Between 1651 and 1690 there were 824 executions in Amsterdam; between 1761 and 1800 there were 839. Given the growing size of the city (from 180,000 to 220,000), that amounted to a decline in the execution rate by a sixth. See Petrus Cornelis Spierenburg, *The Spectacle of Suffering* (Cambridge, 1984), p. 82. At the Old Bailey in London, in the 20 years 1680–99, judges heard 6,244 cases and handed down 1,082 death penalties (17.3%). A century later (1780–99), they heard 14,971 cases and handed down 1,681 death penalties (11.2%). Figures from http://www.oldbaileyonline.org//. Downloaded 27 April 2014.

23 Murray Newton Rothbard, *Economic Thought before Adam Smith: An Austrian Perspective on the History of Economic Thought* (2 vols, 1995; 2nd edn, 2006), i, p. 346.

24 Juliet Gardiner and Neil Wenborn (eds), *The History Today Companion to British History* (1995), p. 63.

25 These were the Exeter Bank (1769), the Devonshire Bank (1770), the City Bank (1786), the General Bank (1792) and the Western Bank (1793). http://www.exetermemories.co.uk/em/banks.php. Downloaded 27 April 2014.

26 Eric Hobsbawm, *The Age of Revolution 1789–1848* (1962), p. 46.

27 A. E. Musson, *The Growth of British Industry* (1978), p. 60.

28 Gregory Clark and David Jacks, 'Coal and the Industrial Revolution 1700–1869', *European Review of Economic History*, 11 (2007), pp. 39–72, at p. 44.

29 Richard Brown, *Society and Economy in Modern Britain 1700–1850* (2002), p. 58.

30   Clark and Jacks, 'Coal and the Industrial Revolution', p. 47.

31   Eric H. Robinson, 'The Early Diffusion of Steam Power', *Journal of Economic History*, 34 (1974), pp. 91–107, at p. 97.

32   J. J. Mason, 'Sir Richard Arkwright (1732–1792), Inventor of Cotton-Spinning Machinery and Cotton Manufacturer', *ODNB*.

33   Neil McKendrick, 'Josiah Wedgwood and Factory Discipline', *Historical Journal*, 4 (1961), pp. 30–55, at p. 33.

34   The statistics in this paragraph are from Brown, *Society and Economy*, pp. 51 (cotton), 56 (pig iron), 48 (patents). The figure of 22 patents for 1700–09 has been revised to 31 in line with the official figure, http://www.ipo.gov.uk/types/patent/p-about/p-whatis/p-oldnumbers/p-oldnumbers-1617.htm. Downloaded 2 February 2014.

1801–1900 *The Nineteenth Century*

1   Robert Woods, 'Mortality in Eighteenth-Century London: A New Look at the Bills', *Local Population Studies*, 77 (2006), pp. 12–23, table 2 (1700, 1800); Geoffrey Chamberlain, 'British Maternal Mortality in the Nineteenth and Early Twentieth Centuries', *Journal of the Royal Society of Medicine*, 99 (2006), pp. 559–63, figure 1 (1900). The figure for 1900 is for England, not London specifically.

2   This chart is based on figures in Paul Bairoch and Gary Goertz, 'Factors of Urbanisation in the Nineteenth Century Developed Countries: A Descriptive and Econometric Analysis', *Urban Studies*, 23 (1986), pp. 285–305, at pp. 288, 291.

3   B. R. Mitchell, *British Historical Statistics* (1988, paperback edn, 2011), pp. 545–7.

4   The data in this chart were downloaded on 5 February 2014 from the Internet Modern History Sourcebook, http://www.fordham.edu/halsall/mod/indrevtabs1.asp. That source credits *The Fontana Economic History of Europe*, vol. 4. part 2. The statistics for the United Kingdom do not tally with those given above, taken from *British Historical Statistics*, so these have been omitted from the table.

5   These insights were gained from reading through the admissions
     registers of St Thomas Lunatic Asylum, Bowhill House, Exeter,
     which are now in the Devon Record Office, ref: 3992F.

6   C. R. Perry, 'Sir Rowland Hill', *ODNB*.

7   http://www.theiet.org/resources/library/archives/featured/fran
     cis-ronalds.cfm. Downloaded 6 February 2014.

8   Information for USA from http://www2.census.gov/prod2/stat
     comp/documents/CT1970p2-05.pdf. Downloaded 9 February
     2014. Information for UK from British Telecom's website, http://
     www.btplc.com/Thegroup/BTsHistory/Eventsintelecommun
     icationshistory.htm. Downloaded 9 February 2014.

9   Roy Porter, *The Greatest Benefit to Mankind* (1997), p. 410.

10   Ibid., p. 407.

11   Vivian Nutton, 'The Reception of Fracastoro's Theory of
     Contagion', *Osiris*, 2nd series, 6 (1990), pp. 196–234.

12   Porter, *Greatest Benefit*, p. 412.

13   These images are to be found in the *Daily Mail* publication,
     Covenants with Death.

14   This much-misquoted passage was delivered by Grey in the
     course of a debate in the Lords on 22 November 1830. See
     Hansard's Parliamentary Debates.

15   Neil Johnston, 'The History of the Parliamentary Franchise',
     House of Commons Research Paper 13/14 (1 March 2014),
     http://www.parliament.uk/briefing-papers/RP13-14.pdf. Down-
     loaded 13 February 2014.

16   Sabine Baring-Gould, *Devonshire Characters and Strange Events*
     (1908), pp. 52–69.

17   K. D. Reynolds, 'Norton [née Sheridan], Caroline Elizabeth
     Sarah', *ODNB*.

18   In 1568, Mary Cornellys of Bodmin received a licence to practise
     surgery throughout the diocese of Exeter: Ian Mortimer, 'Index
     of Medical Licentiates, Applicants, Referees and Examiners in
     the Diocese of Exeter 1568–1783', *Transactions of the Devonshire
     Association*, 136 (2004), pp. 99–134. Margaret Pelling also informs
     me that a woman called Adrian Colman and another called
     Alice Glavin obtained licences in the late sixteenth century:
     Margaret Pelling and Charles Webster, 'Medical Practitioners',
     in Charles Webster (ed.), *Health, Medicine and Morality in the*

Sixteenth Century (Cambridge, 1979), pp. 165–236, at p. 223. Also Isabel Warwike of York received a licence in 1572. By the mid seventeenth century, when the term 'doctor' became synonymous with 'physician' and formal education was seen as essential to the acknowledgement of medical expertise, women were barred from obtaining medical qualifications.

19   Deborah Simonton, *The Routledge History of Women in Europe since 1700* (2006), pp. 118–19.

20   Robert A. Houston, 'Literacy', EGO: European History Online, http://unesdoc.unesco.org/images/0000/000028/002898eb.pdf. Downloaded 14 February 2014. UNESCO, *Progress of Literacy in Various Countries: A Preliminary Statistical Study of Available Census Data since 1900* (1953); UK figures are from *Sixty-fourth Annual Report of the Registrar General* (1901), lxxxviii.

21   George Orwell, *Homage to Catalonia* (Penguin edn, 1989), p. 84.

1901–2000 *The Twentieth Century*

1   I am grateful to Nick Hasel of Woodbarn Farm, Chew Magna, for this anecdote.

2   Figures for this graph are from B. R. Mitchell, *British Historical Statistics* (1988, paperback edn, 2011), pp. 541–2.

3   Data for the period 1904–77 are from Mitchell, *British Historical Statistics*, pp. 557–8. Data for 1977–2000 are from Vehicle Licensing Statistics, http://www.dft.gov.uk/statistics/series/vehicle-licensing//. Downloaded 17 February 2014.

4   Data on cars worldwide were taken from Joyce Dargay, Dermot Gately and Martin Sommer, 'Vehicle Ownership and Income Growth, Worldwide: 1960–2030', *Energy Journal*, 28 (January 2007), pp. 143–70, at pp. 146–7.

5   Mitchell, *British Historical Statistics*, p. 561.

6   http://www.caa.co.uk/docs/80/airport_data/2000Annual/02.3_Use_of_UK_Airports%201975_2000.xls. Downloaded 18 February 2014. Note these CAA figures do not exactly tally with those in *British Historical Statistics*, quoted above, for the period 1975–80.

7   227 million in Europe, 71.1 million in USA, 4.5 million in

NOTES 369

Canada, 3.7 million in Australia and just under 1 million in New Zealand.

8   David Colman, 'Food Security in Great Britain: Past Experience and the Current View', http://www.agr.kyushu–u.ac.jp/foodsci/4_paper_Colman.pdf. Downloaded 1 July 2014.

9   The House of Lords Act 1999 permitted just 92 hereditary peers to remain as members of the House of Lords.

10   http://www.parliament.uk/documents/commons/lib/research/rp99/rp99-111.pdf. Downloaded 20 February 2014.

11   Female life expectancy at birth was as follows: Australia (82.1), Austria (81.4), Belgium (80.9), Canada (81.5), Finland (80.9), France (83.1), Germany (80.6), Greece (80.8), Iceland (81.8), Italy (82.4), Japan (84.7), New Zealand (80.9), Norway (81.4), Singapore (80.2), Spain (82.3), Sweden (82.0) and Switzerland (82.5). Male life expectancy at birth: Australia (76.6), Canada (76.0), France (75.2), Greece (75.4), Iceland (77.1), Italy (76.0), Japan (77.5), New Zealand (75.9), Norway (75.7), Singapore (80.2), Spain (75.4), Sweden (77.3) and Switzerland (76.7). Data from http://www.health.gov.au/internet/main/publishing.nsf/Content/FAEAAFF60030CC23CA257BF00020641A/$File/cmo2002_17.pdf. Downloaded 20 February 2014.

12   This 10% figure is based on the figures for the various countries in the chart: http://www.health.gov.au/internet/main/publishing.nsf/Content/FAEAAFF60030CC23CA257BF0002064 1A/$File/cmo2002_17.pdf. This indicates that the average active life ranges from 84.5% of life expectancy at birth in Russia to 92.8% in Denmark. Most countries are around the 91% figure. The UK is 91.3%.

13   Brian Moseley's online encyclopedia of Plymouth history, http://www.plymouthdata.info/Cinemas.htm (downloaded 23 February 2014), mentions the Theatre de Luxe, Union Street, opened 10 April 1909; Andrews' Picture Palace, Union Street, opened 1 August 1910; Belgrave Hall, Mutley Plain, opened 11 September 1911; Cinedrome, Ebrington Street, opened 27 November 1911; Cinema de Luxe, Union Street, licensed 23 March 1910; Cinema Picture Palace, Saint Aubyn Street, opened 21 May 1910; Empire Electric Theatre, Union Street, opened by 29 July 1910; Morice Town and District Picture Palace, William

Street, Devonport, licensed 20 October 1910; Paragon Picture Hall, Vauxhall Street, licensed December 1912; People's Popular Picture Palace, Lower Street, licensed 21 December 1910; Theatre Elite Picture Playhouse, Ebrington Street, opened 9 May 1910; Tivoli Picture Theatre, Fore Street, Devonport, opened 26 January 1911. Also the Cinedrome, Mutley Plain, opened before 6 February 1914; Electric Cinema, Fore Street, Devonport, before January 1912; Picturedrome, Cattedown, licensed before January 1912.

14   http://www.screenonline.org.uk/film/cinemas/sect3.html. Downloaded 23 February 2014.

15   Mitchell, *British Historical Statistics*, p. 569.

16   http://www.ofcom.org.uk/static/archive/oftel/publications/research/int1000.htm. Downloaded 23 February 2014.

17   http://uk.russellhobbs.com/blog/kettles-guide/the-electric-kettle-a-brief-historical-overview/. Downloaded 24 February 2014.

18   http://collections.museumoflondon.org.uk/Online/object.aspx? objectID=object-739956&rows=1&start=1. Downloaded 24 February 2014.

19   There is an interesting risk assessment from Sir John Beddington dated 18 December 2012 on this subject and the implications of a Carrington Event for the UK. It states that the risk to the UK is less than to the USA due to the much shorter power lines in this country. The National Grid estimates that 1% of transformer assets could be lost, disrupting electrical supplies for several months, and that aviation would be affected. The effect on digital communications is less well understood. http://www.parliament.uk/documents/commons-committees /defence/121220-PM-to-Chair-re-EMP.pdf. Downloaded 2 July 2014.

20   The Earl of Birkenhead, *The World in 2030 AD* (1930), p. 27.

Conclusion: Which century saw the most change?

1   Abraham Maslow, 'A theory of human motivation', *Psychological Review*, 50 (1943), pp. 370–96. http://psychclassics.yorku.ca/Maslow/motivation.htm. Downloaded 4 January 2014.

2 It perhaps should be added that, even if there was a large immigrant population, the immigration itself led to increased demands on the community's food supply, so this factor does not need to be treated separately from natural population growth.

3 Twentieth-century population growth was limited by factors other than the food supply – the choice not to have large families, for instance – and thus the relationship between population growth and increments in the food supply was broken. The achievement of being able to cater for *all* people with respect to *all* their dietary requirements was certainly a twentieth-century one. However, achievement is not the same as change. The change from meeting 90% of dietary requirements to meeting 105% of them (and therefore wasting some) is less significant in terms of need than the increase from 75% to 90%. Even if the twentieth century saw exactly the same level of change in meeting the physiological needs of society as the nineteenth, the oversupply in the twentieth would mean that some of that supply was not related to need and unnecessary, and thus a 'frippery' (to use Maslow's word). However, such is the gap that there seems to be no doubt that the nineteenth century saw more significant advances in meeting the need for food than the twentieth. Obviously this only applies to the West: circumstances elsewhere in the world were very different; the developing world, for example, changed far more in the twentieth century in this regard than the nineteenth.

4 Pitirim Sorokin, *Social and Cultural Dynamics* (4 vols, 1943).

5 Henry Kamen, *The Iron Century: Social Change in Europe 1550–1660* (1971), p. 43.

6 Sorokin, *Social and Cultural Dynamics* (1962, single vol. edn), p. 550.

7 Ole J. Benedictow, *The Black Death, 1346–1353: The Complete History* (2004), p. 251.

8 Ibid., p. 252; E. A. Wrigley and R. S. Schofield, *English Population from Family Reconstitution 1580–1837* (1997), p. 614.

9 Massimo Livi Bacci, *Population of Europe* (2000), pp. 135, 166. These tables do not give a figure for Spain in 1750, so the figure for 1800 has been used. The data for 2000 are from the list in Chapter 10, Note 11, averages of the male and female figures.

The figure for Sweden in 1950 is from *International Health: How Australia Compares*, http://www.aihw.gov.au/WorkArea/DownloadAsset.aspx?id=6442459112. Downloaded 3 March 2014.

10 An examination of the English data provided by E. A. Wrigley and R. S. Schofield in 1997 (later figures than those used in the above table) shows that life expectation at birth in 1591–1611 averaged 38.18; in 1691–1711 it was 37.98 and in 1791–1811 it was 40.19 – so the effect of the early professionalisation of medicine in the seventeenth century was negligible when combined with the greater adversities of life. The increase of life expectancy in the eighteenth century was relatively small.

11 B. R. Mitchell, *British Historical Statistics* (1988, paperback edn, 2011), pp. 166–9.

12 Angus Maddison, *The World Economy: A Millennial Perspective* (2001), p. 264

13 Stephen Broadberry, Bruce Campbell, Alexander Klein, Mark Overton and Bas van Leeuwen, *British Economic Growth 1270–1870* (2011). http://www.lse.ac.uk/economicHistory/seminars/ModernAndComparative/papers2011-12/Papers/Broadberry.pdf. Downloaded 3 March 2014.

14 R. B. Gordon, M. Bertram and T. E. Graedel, 'Metal Stocks and Sustainability', *Proceedings of the National Academy of Sciences*, 103, 5 (2006), pp. 1209–14.

15 http://www.worldcoal.org/resources/coal-statistics/coal-steel-statistics//. Downloaded 4 March 2014.

16 Various newspapers carried Professor Hawking's piece at the end of April 2010, which was a promotion for a TV series. It seems first to have appeared in the *Daily Mail* on 27 April 2010.

17 The figure of 58.6 years is derived as follows: four years speeding up, covering 1.4 light years, the same to slow down, and 21.3 years crossing the 19.2 light years in between reaching maximum speed and starting to slow down. And then the same coming home.

Envoi: Why it matters

1 1,668.9 billion barrels in 2012 compared to 1,258.1 billion barrels in 2000. Fossil fuel statistics in this section are from BP's

Statistical Review of World Energy (2013) unless otherwise stated, http://www.bp.com/content/dam/bp/excel/Statistical-Review/statistical_review_of_world_energy_2013_workbook.xlsx. Downloaded 7 March 2014.

2 Currently 14% of the electricity in this country is produced by sustainable methods, equating to 5.2% of our total energy consumption. See *Renewable Energy in 2013*, pp. 1–2, https://www.gov.uk/government/uploads/system/uploads/attachment_data/file/323429/Renewable_energy_in_2013.pdf. Downloaded 28 June 2014. Obviously, to increase the sustainable energy supply from 14% to 100% will require not just another 86% of today's needs but a further 100% of the needs of the additional population living in the country by 2050, which could easily be another 12 million people.

3 One hectare of rapeseed produces 1.1 tonnes of biodiesel per year, which equates to 1,311 litres. One acre of sugar beet produces 4.4 tonnes of bioethanol, 5,553 litres (figures from http://www.biomassenergycentre.org.uk, downloaded 23 March 2014). UK road fuel consumption is currently 68 million litres of diesel and 56 million litres of petrol *per day* (http://www.ukpia.com/industry_information/industry-overview.aspx, downloaded 22 March 2014). So, all other things being equal, we need almost 18.9 million hectares of land for all our diesel needs and almost 3.7 million hectares for our petrol. The total is 22.6 million hectares: halving this gives the figure of 11.3 million hectares in the text.

4 Current world annual consumption is about 68,000 tonnes of uranium per year. Known recoverable resources amount to 5.327 million tons in 2011. http://www.world-nuclear.org/info/Nuclear-Fuel-Cycle/Uranium-Resources/Supply-of-Uranium/. Downloaded 28 June 2014.

5 Woodland accounts for 9% (a small proportion compared to our European neighbours); coastal fringes such as dunes and estuaries make up just under 1%; freshwater lakes and rivers represent just over 1%; and 4.8% is mountain, moorland and heath, most of which is in a national park or similarly protected landscape. See UN National Ecosystem Assessment (2012), Chapter 10.

6   Statistics from http://www.agr.kyushu-u.ac.jp/foodsci/4_
     paper_Colman.pdf. Downloaded 16 June 2014.

7   Robyn Vinter, 'UK Becomes Net Importer of Wheat', *Farmers
     Weekly* (10 October 2012).

8   We only produced 82% of the beef we needed in 2008, 52%
     of the pork, 88% of the mutton and lamb and 92% of the
     chicken. Figures are from Table 7.4 of the UN National
     Ecosystem Assessment (2012).

9   The incremental figure 0.76% is the annual increment between
     the English census population in 2001 (49,138,831) and 2011
     (53,012,456). The figures released by the Office for National
     Statistics in June 2014, after these calculations were made, indi-
     cated that the population of England is still growing at 0.7%
     per year.

10   At present about 75% of new housing is on previously devel-
     oped land. But there is a limit to how much land of this sort
     is available. This sum presumes that for every decade that
     passes, two thirds as much development can take place on
     brownfield sites as in the previous decade; so, after 2020, only
     50% of new housing is on previously developed sites, 33% after
     2030, 22% after 2040, and so on, and only 10% after 2060 – but
     that it stays constant at that level thereafter. The actual current
     rate of building is difficult to determine. Looking purely at the
     domestic housing stock, according to the Office for National
     Statistics, in 2007–8 there were 223,000 new homes built in
     England, 1.1% of the total housing stock of 22,288,000. In
     2012–13 only 125,000 houses were built: 0.54% of the total
     housing stock of 23,236,000. The average of those two figures
     is slightly higher than our assumption of 0.76%; however, the
     current government and all other parties want to see the
     number of houses built dramatically increase, as illustrated in
     the 2014 Queen's Speech. We should expect house building
     therefore to be significantly in excess of the figures quoted.

11   The figure of 110 comes from Credit Suisse, *Global Wealth Report*
     (2013), as reported widely in the UK press. As for slavery,
     although officially abolished, it exists in many countries. The
     top ten offenders are India, China, Pakistan, Nigeria, Ethiopia,
     Russia, Thailand, Congo, Myanmar and Bangladesh. The estimate

was just under 30 million people in slavery in 2013, India dominating with almost 14 million. See http://www.ungift.org/doc/
knowledgehub/resource-centre/2013/GlobalSlaveryIndex_2013_
Download_WEB1.pdf. Downloaded 23 March 2014.

12 Thomas Piketty, trans. Arthur Goldhammer, *Capitalism in the Twenty-First Century* (2014), p. 165.

13 Ibid., p. 25.

14 Ibid., p. 356.

15 To illustrate this, imagine the opposite: a completely undevel-
oped economy, in a sparsely inhabited region. If a man has
surplus seed and wants to farm an extra few acres for his family,
then he can just claim them from the wasteland. There is no
payment for the use of the new land and thus r does not have
a value. But g does: his work results in economic growth.
Therefore while an economy is in its initial stages of development, it is inevitable that the return on capital is less than
growth, $r < g$. However, after all the land is claimed, r will start
to be positive, as any new farmer looking for land will have to
pay rent to the landowner. Initially, while there are no other
people competing for the land, it is likely that r will remain
low, but as the population increases and the demand for land
grows, the price farmers have to pay for it will increase. It is
still possible for r to be less than g at this point, for the discovery
of a glut of new resources such as oil might permit some
unprecedented levels of economic output to be achieved, but
ultimately these will prove unsustainable. Either the resources
will dry up or they will increase the demand for the land and
act as an inflationary force on r, as people compete more
intensely for the land and pay higher rents. After a while, when
all the land in the region is claimed and fully exploited, and
non-renewable resources start running out, it will become
harder and harder to maintain positive economic growth, so g
will diminish and $r > g$ will become the norm.

16 Martin Gilens and Benjamin I. Page, 'Testing Theories of
American Politics: Elites, Interest Groups and Average Citizens',
Perspectives on Politics (forthcoming, 2014).

17 Figures for countries other than England are from the World
Bank statistics: http://data.worldbank.org/indicator/EN.POP.

DNST. Downloaded 12 July 2014. The 2013 English population estimate of 53.5 million is from http://www.ons.gov.uk/ons/rel/pop-estimate/population-estimates-for-uk-england-and-wales-scotland-and-northern-ireland/mid-2011-and-mid-2012/index.html. Downloaded 12 July 2014. The area of England was taken as 130,400 square kilometres.

18   Statistics from BP's *Statistical Review of World Energy* (2014).

19   Paul R. Ehrlich and Anne H. Ehrlich, 'Can a Collapse of Global Civilisation be Avoided?', *Proceedings of the Royal Society B*, 280: 20122845. http://dx.doi.org/10.1098/rspb.2012.2845. Downloaded 24 March 2014.

20   Geoffrey Parker, *The Global Crisis: War, Climate Change and Catastrophe in the Seventeenth Century* (2013), p. 19.

21   http://www.gallup.com/poll/142727/religiosity-highest-world-poorest-nations.aspx. Downloaded 8 March 2014.

Appendix: Population estimates

1   Paolo Malanima, 'Energy and Population in Europe: The Medieval Growth' (2010), pp. 3–4. http://www.paolomalanima.it/default_file/Papers/MEDIEVAL_GROWTH.pdf. Downloaded 12 February 2014.

2   Massimo Livi Bacci, *Population of Europe* (2000), pp. 8–10.

3   Draft version downloaded from http://www.lse.ac.uk/economicHistory/pdf/Broadberry/Medievalpopulation.pdf, 15 January 2014.

4   Ole Benedictow, *The Black Death 1346–1353: The Complete History* (2004), p. 383.

5   The figures in Table 1.2 for the period 1541–1871 have been taken from the revised figures for the country in E. A. Wrigley and R. S. Schofield, *English Population History from Family Reconstitution 1580–1837* (1997), p. 614. The figures for 1900 and 2000 are actually from the censuses of 1901 and 2001: Office for National Statistics, *Census 2001: First Results on Population for England and Wales* (2002), p. 5.

6   They are actually taken from David E. Davis, 'Regulation of Human Population in Northern France and Adjacent Lands in

the Middle Ages', *Human Ecology* 14 (1986), pp. 245–67, at p. 252. The figures for 1798 boundaries have been amplified by a factor of 1.3 to represent the whole of France and make it compatible with later calculations.

7   Norman Pounds and Charles C. Roome, 'Population Density in Fifteenth Century France and the Low Countries', *Annals of the Association of American Geographers*, 61 (1971), pp. 116–30.

8   'Le recensement de la population dans l'Histoire', http://www.insee.fr/fr/ppp/sommaire/imethso1c.pdf. Downloaded 3 February 2014.

9   Livi Bacci, *Population of Europe*, pp. 8–10.

10   The figures for 1600 and 1700 are from ibid., p. 8; those for 1800 and 1900 from Jacques Dupaquier, *Histoire de la population Française* (4 vols, Paris, 1988), and those for 2000 from the French 2001 census.

11   Giovanni Federico and Paolo Malanima, 'Progress, Decline, Growth: Product and Productivity in Italian Agriculture 1000–2000', *Economic History Review*, 57 (2004), pp. 437–64.

12   The figures for 1500–1800 have been taken from ibid., p. 446. Those for 1900 and 2000 are from the census data published by ISTAT.

13   https://www.census.gov/population/international/data/worldpop/table_history.php. Downloaded 3 February 2013.

Picture Credits

View of Moretonhampstead, Devon (*author's collection*).

Gatehouse of Exeter Castle (*author's collection*).

Speyer Cathedral, Germany (*author's collection*).

Mural in Chaldon Church, Surrey (*author's collection*).

Arab physician performing a bleeding, *c.*1240 (*Bridgeman Art Library*).

Window depicting a wine merchant in Chartres Cathedral, France (*Bridgeman Art Library*).

Hereford world map (*Bridgeman Art Library*).

Cadaver effigy in Exeter Cathedral (*author's collection*).

Golden rose of Pope John XXII (*copyright Brian Shelly*).

Image of cannon from Walter de Milemete's treatise on kingship (*Bridgeman Art Library*).

Portrait of a Man in a Turban by Jan van Eyck (*Bridgeman Art Library*).

Printing press, from a book printed in 1498 (*Bridgeman Art Library*).

Clock in the chapel at Cotehele House, Cornwall (*author's collection*).

Portrait of Columbus by Sebastiano del Piombo (*Bridgeman Art Library*).

Map of the world from Abraham Ortelius's *Theatrum Orbis Terrarum*, 1570 (*Bridgeman Art Library*).

Wheel-lock hunting pistol dating from 1578 (*Bridgeman Art Library*).

Iris from Leonhart Fuchs's *De Historia Stirpium*, 1542 (*Bridgeman Art Library*).

Bamberg witch house (*Staatsbibliothek Bamberg, shelf-mark V B 211m*).

Johannes Hevelius's telescope (*public domain*).

Isaac Newton's telescope (*Bridgeman Art Library*).

London opera rehearsal, painted by Marco Ricci, 1708 (*Bridgeman Art Library*)

Thomas Newcomen's steam engine, 1718 (*Getty Images*).

The Tennis Court Oath, after Jacques-Louis David (*Bridgeman Art Library*).

Power looms, painting by Thomas Allom, 1834 (*Bridgeman Art Library*).

Advertisement for the Plymouth to London stagecoach, *Plymouth and Dock Telegraph and Chronicle*, 4 May 1822 (*author's collection*).

The Boulevard du Temple, Paris, photographed by Louis Daguerre, 1838 (*public domain*).

SS *Great Britain* in Cumberland Basin, photographed by William Fox Talbot, 1844 (*public domain*).

The Wright Brothers' *Flyer*, airborne on 17 December 1902 (*Library of Congress*).

Autochrome photograph of French soldier, June 1917, by Paul Castelnau (*Ministère de la Culture – Médiathèque du Patrimoine, Dist. RMN-Grand Palais/Paul Castelnau*).

Dr Nagai in Nagasaki after the nuclear bombing, August 1945 (*Bridgeman Art Library*).

Park Row Building, New York (*Library of Congress*).

Petronas Towers, Kuala Lumpur (*Bridgeman Art Library*).

Earthrise: the Earth photographed from Apollo 8, 24 December 1968 (*NASA*).

Index

www.vintage-books.co.uk